Praise for *The Bio-Integrated Far*

D1296823

"Nature knows no waste, and cutting-edge farmers should start today to incorporate the bio-integration principles that Shawn Jadrnicek describes in detail in this book. Shawn shares his in-the-field experience using easy-to-understand formulas and charts to encourage the reader to develop a plan and translate project descriptions into hardworking results. I have seen Shawn's transformational power in person at Clemson University, and his visions are addictive, teaching you not just to think outside the box by harmonizing natural systems but to act outside the box to create inexpensive and highly functional growing environments that are much more profitable and efficient than traditional growing systems. Even at our mushroom farm, we are incorporating Shawn's design approach with aquaponics, black soldier fly composting, and passive heating. Farmers need all the financial help they can get, and *The Bio-Integrated Farm* will help readers prevent costly experimental failures. Every farm's needs and layouts are unique, and this book allows the reader to customize and hybridize systems that harness the power of nature to transform ordinary farms into models worthy of world-class learning centers for permaculture."

— TRADD COTTER, author of *Organic Mushroom Farming and Mycoremediation*

"*The Bio-Integrated Farm* is an invaluable resource for market farmers, homesteaders, and serious gardeners who are interested in improving their relationship with the land. Shawn Jadrnicek's creative use of materials, animals, and space, which he clearly and thoroughly explains in this book, will inspire and teach you how to improve the efficiency and resiliency of your farm or garden. I can't wait to implement some of Shawn's designs at my site. While reading through this book, I was reminded of the ancient gardening techniques mentioned in F.H. King's *Farmers of Forty Centuries*. Shawn has integrated these simple and effective technologies into his modern-day working farm; his book will allow readers to access some of that ancient wisdom, too."

— JEROME OSENTOWSKI, author of *The Forest Garden Greenhouse*

"*The Bio-Integrated Farm* provides practical solutions for farmers and homesteaders facing the dual challenges of sustainably feeding an expanding global population and building resilience into their systems in response to climate disruption. Shawn Jadrnicek's designs for greenhouses, irrigation, composting systems, and more are based on the most enduring answer: let's look to nature. From experience, Jadrnicek knows how to create organic, resilient, and highly productive systems based on creative use of water, solar energy, and other free forces of nature."

— COURTNEY WHITE, author of *Two Percent Solutions for the Planet*

"What is unique about permaculture, at its heart, is integration of elements like fish ponds, greenhouses, chickens, and crops. *The Bio-Integrated Farm* highlights real-world farm management experiences, including particularly innovative practices such as raising freshwater shrimp in greenhouse tanks that also provide thermal mass. The authors also respect and acknowledge the ancient roots of many of these ideas."

— ERIC TOENSMEIER, author of *The Carbon Farming Solution*

"Harvesting, multiplying, capturing, conveying, sloping, fertilizing, irrigating, heating, cooling . . . 'at least seven functions' is the genius of Shawn Jadrnicek's stacked systems in *The Bio-Integrated Farm*. Shawn turns almost everything we think we know about farming on its head, beginning with the notion that the odds are stacked against the small farmer and permaculturist. Instead, he stacks the odds in our favor by requiring every component of a farm to serve at least seven functions. These stacked systems create biological and mechanical efficiencies that maximize production and ecological diversity. While innovators abound in the permaculture world, Shawn is unparalleled in the practicality and detail of his innovative designs.

As a connoisseur of college farms, I've visited several dozen such operations across the United States, and I have yet to see another academic farm demonstrate the level of careful design and innovation found on the Clemson University Student Organic Farm, one of several permaculture masterworks Shawn has helped create. As much an engineer as he is an ecologist, Shawn teases out nature's secret systems with a covey of collaborators — students, prawns, soldier flies, chickens, and thermophilic bacteria to name but a few — and he shares all his best designs in this paradigm-altering guide. Be prepared to have your vision of a farm's limits shattered. Fear not, Shawn shows you how to reassemble those broken pieces into a '7-plus mosaic' that can take any homestead or farm to the next level."
— PHILIP ACKERMAN-LEIST, professor, Green Mountain College;
author of *Rebuilding the Foodshed*

"As climate change drives increasingly challenging weather variability and extremes, ecosystem-based strategies such as those presented in *The Bio-Integrated Farm* will be crucial to successful adaptation efforts. In this practical, clearly written, and beautifully designed book, Shawn Jadrnicek shares his unique ability to conceptualize, design, and manage water for whole-system benefits at multiple scales. A must read for anyone interested in design and management of water systems for resilient homesteads and farms."
— LAURA LENGNICK, author of *Resilient Agriculture*

"Shawn Jadrnicek has spent the last decade getting his hands dirty and taking risks, experimenting with how to create systems that actually work. *The Bio-Integrated Farm* covers areas often neglected in the current permaculture literature. Shawn's systems-based designs show permaculture's relevance beyond typical gardening scenarios. He offers a load of detailed practical advice based on personal experience, demonstrating how to make connections that result in greater yields and ease. His zeal for making the most out of the resources on hand has inspired to me to find ways to further integrate my own permaculture homestead!"
— RAIN TENAQIYA, author of *West Coast Food Forestry*

THE
BIO-INTEGRATED FARM

THE
BIO-INTEGRATED FARM

A Revolutionary
Permaculture-Based System
Using Greenhouses, Ponds,
Compost Piles, Aquaponics,
Chickens, and More

SHAWN JADRNICEK
with Stephanie Jadrnicek

Chelsea Green Publishing
White River Junction, Vermont

Project Manager: Alexander Bullett
Acquisitions Editor: Makenna Goodman
Developmental Editor: Fern Marshall Bradley
Copy Editor: Eileen M. Clawson
Proofreader: Eric Raetz
Indexer: Linda Hallinger
Designer: Melissa Jacobson

Printed in the United States of America.
First printing February, 2016.
10 9 8 7 6 5 4 3 2   16 17 18 19 20

Our Commitment to Green Publishing
Chelsea Green sees publishing as a tool for cultural change and ecological stewardship. We strive to align our book manufacturing practices with our editorial mission and to reduce the impact of our business enterprise in the environment. We print our books and catalogs on chlorine-free recycled paper, using vegetable-based inks whenever possible. This book may cost slightly more because it was printed on paper that contains recycled fiber, and we hope you'll agree that it's worth it. Chelsea Green is a member of the Green Press Initiative (www.greenpressinitiative.org), a nonprofit coalition of publishers, manufacturers, and authors working to protect the world's endangered forests and conserve natural resources. *The Bio-Integrated Farm* was printed on paper supplied by QuadGraphics that contains at least 10 percent postconsumer recycled fiber.

Library of Congress Cataloging-in-Publication Data
Names: Jadrnicek, Shawn, 1976– author.
Title: The bio-integrated farm : a revolutionary permaculture-based system using greenhouses, ponds,
    compost piles, aquaponics, chickens, and more / Shawn Jadrnicek with Stephanie Jadrnicek.
Description: White River Junction, Vermont : Chelsea Green Publishing, 2016. | Includes bibliographical references and index.
Identifiers: LCCN 2015036595 | ISBN 9781603585880 (pbk.) | ISBN 9781603585897 (ebook)
Subjects: LCSH: Permaculture. | Integrated agricultural systems.
Classification: LCC S494.5.P47 J34 2016 | DDC 631.5/8 — dc23
LC record available at http://lccn.loc.gov/2015036595

Chelsea Green Publishing
85 North Main Street, Suite 120
White River Junction, VT 05001
(802) 295-6300
www.chelseagreen.com

To my mother, Lori Kase, and father, Kimo Jadrnicek,
who spent an extensive amount of time
teaching me all the important things in life.
My father's constant reminder that
"everything you need is all around you"
inspired every design in this book.

CONTENTS

Introduction

Putting Nature to Work

Imagine engaging the free forces of nature to benefit you and your environment. Water, wind, sunlight, convection, gravity, and decomposition — all of these energy sources can serve your needs.

By making the right choices when we interact with nature, we create sustenance in our landscapes with little effort or work on our part. If we choose to let rainwater flow aimlessly from our rooftops and gutters, it will erode centuries of soil from under our feet. Or we can choose to harness the potential of rainwater. When we harvest rainwater in a carefully planned way, we can use it to heat and cool greenhouses and homes, to spread fertility through the landscape, to grow fish, to flush toilets, to clean chicken coops, and to irrigate plants and mushroom logs. Water has the power to destroy life or support life. The difference is in the design.

Likewise, sunlight can simply bounce off the landscape in winter, its energy wasted as we burn fossil fuels to heat our homes and greenhouses. However, through design we can capture sunlight and use it to provide free heat in winter and move heat through buildings. The powers of nature can be wasted or used to our benefit through intelligent design.

Design can transform waste into a resource. Wood chips and food waste burden our urban and suburban waste collection systems. But through careful design we can combine the waste into active compost piles to capture heat for homes and greenhouses, while producing a high-quality source of fertility to feed the soil of farms and gardens. We have the opportunity to capture and use the forces of nature to benefit from every component in our landscape, rather than continuously attempting to mitigate the environmental damages resulting from poor (or no) design.

The benefits of design are not limited to the landscape. Properly designed roofs and homes protect and sustain their inhabitants. Sometimes the first step in practicing good design is recognizing bad design. For example, an inverted gable roof — resembling a butterfly taking flight — funnels rain, cold winds, and intense sunlight directly into the building it should protect. Similarly, the front door of a traditionally designed home faces the road, regardless of whether the road lies to the east, west, north, or south. But if the house lies on a north-south axis, it inevitably overheats throughout the summer and remains dark and cold all winter. Instead, it makes sense to orient a house to face the winter sun, so that sunlight can enter and heat the house in winter and the roof will block intense summer rays. By observing one simple rule, "function comes first," we end the futile attempt to swim upstream and instead float downriver, navigating any obstacles with agility and ease.

If you are interested in spending less time maintaining your landscape and more time enjoying it, then this book is for you. If you are concerned about decreasing your dependence on outside resources, then this book is for you. If you are curious about how to inexpensively extend your growing season, then this book is for you. The process of achieving all three of these goals is simple — put nature to work through intelligent design. That's the basis of bio-integration.

In the Beginning

I remember the moment I first realized a plant could save me money and hours of backbreaking labor. Someone had planted comfrey under grapevines at my home. Surprisingly, the vines closest to the comfrey plants grew faster and produced more fruit. The comfrey had prevented the growth of weeds and may have also fertilized the vines by pulling nutrients up from deeper in the soil. A helpful neighbor, Rain Tenaqiya, author of *West Coast Food Forestry*, later informed me that planting comfrey under fruit trees is a common practice in permaculture.

Immediately intrigued, I began experimenting with permaculture patterns at our farm in the Santa Cruz mountains of California. My wife Stephanie and I had recently been blessed by the birth of our daughter Sage. With little time and resources, I found that permaculture offered me a way to spend time with our growing family and run a nursery and edible flower operation. I integrated our nursery operation under our fruit trees to capture water and nutrients, and I reduced inputs by cover cropping and employing no-till farming techniques. As a result, I saved an enormous amount of time, money, and energy.

After seven years in this Mediterranean-like climate, we moved to the East Coast — a new canvas for practicing permaculture. I taught Master Gardener classes as an Extension agent for Clemson University, and in my off time I established the Urban Permaculture Institute of the Southeast. In the mild subtropical coastal climate, I turned ½ acre of poor sandy soil into a fertile ecosystem. Located in the heart of town, the institute's no-maintenance, low-input system supplemented our diets through aquaculture, chickens, bees, and food forestry while decreasing our energy consumption using passive solar techniques and greywater systems.

Another life-changing permaculture project occurred at the Ashevillage Institute in the temperate climate of Ashville, North Carolina. At this location I directed the design and built the landscape, developing systems to harvest rainwater into ponds from steep slopes on urban property. This was my first experience integrating ponds with greenhouses. Combined with raising chickens, this system became a sustainable example of food production in an urban area with difficult terrain.

Finally, the culmination of my experience has taken place at the Clemson University Student Organic Farm (SOF) and my current residence in the foothills of the Blue Ridge Mountains. These locations lie at the cooler northern edge of a humid subtropical climate that extends west to the middle of Texas and north along the coast to New Jersey. Working with students at the Clemson farm, I've created permaculture patterns — governing everything from raising transplants and field design to freshwater prawn production and composting. These patterns have simplified the operation of our 125-share subscription farm while reducing our reliance on outside resources. At my 2-acre homestead, in less time than it takes to mow, I'm building a you-pick fruit farm using permaculture patterns. The only labor my landscape requires is harvesting, and the only outside input I buy is a small amount of chicken feed. By carefully employing gravity, rainwater harvesting ponds, and poultry systems, I maximize the power of nature to do the work for me.

A Unifying Principle

Creating new and innovative permaculture patterns is my passion. Permaculture patterns are replicable models easily repeated in diverse design settings. What exactly is permaculture? Permaculture is a

More about Permaculture

The term "permaculture" is a contraction of the words "permanent and agriculture" or "permanent and culture." Permaculture encompasses ethics, principles, and techniques rooted in ecological designs, as well as the practices of ancient cultures and modern innovators. Permaculture founder Bill Mollison believes people can create systems that function as effortlessly as natural ecosystems, providing food, energy, shelter, and other needs in a sustainable way.

Permaculture ethics call for care of the earth, care of people, and setting limits on population and consumption.[1] Many permaculture teachers also focus on a set of principles or truths to define the foundation of permaculture. Though sometimes differing in details, all the principles and truths aspire to the final goal of harmony. Here are the principles of the founders of permaculture.

Bill Mollison's Principles

1. Work with nature, rather than against the natural elements, forces, pressures, processes, agencies, and evolutions, so that we assist rather than impede natural development.
2. The problem is the solution; everything works both ways. It is only how we see things that makes them advantageous or not. A corollary of this principle is that everything is a positive resource; it is just up to us to work out how we may use it as such.
3. Make the least change for the greatest possible effect.
4. The yield of a system is theoretically unlimited. The only limit on the number of uses of a resource possible within a system is in the limit of the information and the imagination of the designer.
5. Everything gardens, or has an effect on its environment.

David Holmgren's Principles[2]

1. Observe and interact.
2. Catch and store energy.
3. Obtain a yield.
4. Apply self-regulation, and accept feedback.
5. Use and value renewable resources and services.
6. Produce no waste.
7. Design from patterns to details.
8. Integrate rather than segregate.
9. Use small and slow solutions.
10. Use and value diversity.
11. Use edges, and value the marginal.
12. Creatively use and respond to change.

design system based on ethics and principles created by Australian naturalist Bill Mollison and his informal student David Holmgren. Bill Mollison presented five guiding principles in his book *Permaculture: A Designers' Manual*. David Holmgren uses twelve guiding principles, which he described in his book *Permaculture: Principles and Pathways Beyond Sustainability*. Ben Falk, a designer living in Vermont, created seventy-two guiding principles in his 2013 book, *The Resilient Farm and Homestead*.

In my own work I continuously run into an underlying rule or directive that, if done properly, accomplishes most of the other permaculture principles. I believe it's a unifying principle that underlies the heart of permaculture and all good ecological designs. In the permaculture community it's known as *stacking functions*.

What is this principle of stacking functions? Mollison defines permaculture design as "a system of assembling conceptual, material, and strategic components in a pattern which functions to benefit life in all its forms. It seeks to provide a sustainable and secure place for living things on this earth."[3]

Mollison also states the prime directive of function and the source for the stacking functions principle: "Every component of a design should function in many ways. Every essential function should be supported by many components."[4]

When we look at the comfrey plant example, the possible functions are many:

1. Prevents potential weeds from invading
2. Accumulates nutrients fertilizing adjacent trees
3. Serves as feed for animals
4. Serves as a natural medicine
5. Serves as a garnish (flowers)

One could also argue that a comfrey plant harvests and filters water while producing oxygen and supporting wildlife. But it is difficult to quantify energy savings or place a monetary value on these functions; therefore I did not include them in the functional analysis. In addition, aesthetics is never considered a function. Stacking functions is about adding more beneficial functions to design components. Every function added conserves energy and resources. More functions equal more energy and resources saved.

One famous multifunctional permaculture pattern is Mollison's original, and much replicated, herb spiral pattern, which he created in 1978 as a kitchen-door design. Containing all the basic culinary herbs, the ascending spiral garden sits 6 feet wide at its base and rises 3 feet above ground level. The form creates dry sites for Mediterranean herbs such as thyme, sage, and rosemary and shaded areas for herbs such as mint, parsley, chives, and coriander. Mollison's herb spiral performs the following functions:

1. It creates dry areas on top of the mound.
2. It creates wet areas toward the bottom of the mound.
3. It creates shaded cool areas on the north side.
4. It creates sunny warm areas on the south side.
5. It increases growing area in a small space.
6. It provides culinary and medicinal herbs and food for the kitchen.

Building Up to Bio-Integration

After years of experiencing Mollison's permaculture systems from the inside out, I've come to the conclusion that something very special happens when a component within the design exceeds seven functions. Once the magic odd number of seven is breached, the design takes on a life of its own. For a component to perform seven functions it must be so connected with the surrounding environment that it takes on a new autonomous, lifelike quality. I refer to this quality as bio-integration, to represent the new life born into the design once seven functions are breached.

I define a component as one part of a larger design. For example, a greenhouse would be one component of a farm. Another could be a pond or a chicken coop. Each component is a simple pattern within itself but should also connect with other components in a larger pattern. Sometimes, however, the construction of one component depends on the construction of another, such as the greenhouse built on a sloped platform with an exterior pond. I consider this pattern to be a single component, since digging the pond builds the sloped platform.

For example, we could bio-integrate Bill Mollison's herb spiral. Since the herb spiral requires bringing in soil to make the 3-foot-high mound, we could harvest the soil from an area adjacent to the herb spiral to form a pond. The herb spiral harvests rainwater as the water spirals down the mound and creates a wet area at the bottom. If we built a small pond in this wet area on the south side of the spiral, the new bio-integrated functional analysis would add the following five functions to the previous six to create a total of eleven functions:

7. The elevated spiral planting bed harvests rainwater for a pond.
8. The pond on the south side of the herb spiral reflects light onto the herb spiral during winter.
9. Excess soil from digging the pond provides soil for the raised mound.
10. The pond attracts toads, frogs, and other predators of plant pests.
11. The pond moderates the microclimate, absorbing heat during the day and releasing heat at night.

Since digging the pond provided the soil to raise the herb spiral, the components are now considered a single component. For this design to truly work, it requires water from an additional source to keep the pond full and combat evaporation. Therefore, we would need to add one more functional connection, a catchment surface, to increase the total number of functions to twelve:

12. The pond harvests water from the roof of the adjacent house.

I often compare bio-integration to an individual's success. One individual may achieve success; however, it is through connections with others that he or she becomes successful. Likewise, each component becomes bio-integrated by performing its own seven functions, but it performs those seven functions through its connection to other components. In essence, bio-integration forms cohesion between components by connecting the dots.

After I created the greenhouse systems at the SOF, Dr. Geoffrey Zehnder, director of the Sustainable Agriculture Program, said the greenhouses were so different from anything else he had seen, he thought the technique should have its own name. We coined the term *bio-integration*. I've seen some farms use the term, and it's also used by the medical industry (with a different definition). Throughout this book I hope to establish firmly the definition of *bio-integration* in relation to farms, landscapes, and homes.

When I sat down to begin writing this book, I realized there's a profound difference between working with nature and working with the written word. Sometimes it's much easier to *do* something than to *explain how* to do it. That's where Stephanie's writing skills came into play. She is not only an award-winning journalist, but she has lived alongside me for nearly twenty years watching my projects evolve. So it was only natural for us to tackle this endeavor as a team. The book is written primarily from a first-person perspective, and the "I" refers to me. The ideas and concepts are my creation; however, the expression of those ideas and concepts was a collaborative effort between Stephanie and me.

Each chapter in this book explores an advanced, innovative, bio-integrated pattern. Each of the patterns exceeds seven functional connections, and some exceed twenty functional connections. Chapter 1 begins with an ancient pattern, but most of the other patterns are my original designs. The book is divided into two parts: The first part focuses on component patterns of larger designs, such as a multifunctional small pond or a multifunctional composting system—but many chapters overlap, since some patterns share components. To understand the big picture and how components fit together, I refer readers to the landscape plan of the SOF on page 297, my home in Anderson on page 348, the Urban Permaculture Institute of the Southeast on page 336, and the Ashevillage Institute on page 349.

The second part of the book guides the reader through case studies composed of some new design components and new combinations of components mentioned earlier in the book. It's not necessary to read the book in consecutive order, but readers may need to cross-refer from one chapter to another. Cross-references from chapter to chapter throughout the book will help you find information more easily.

You don't need to have prior knowledge of permaculture to read and understand this book.

I've included in-depth information about how to design and build the systems; however, I haven't written step-by-step instructions for each project. You'll need some previous experience with building projects, managing a greenhouse, and growing crops to get the most out of this book. Many of the chapters start with the basics, then move into the more advanced engineering required for complex integrations. If you don't understand a topic, simply skip to the next chapter and refer back to the engineering as needed when you're designing and building projects.

My hope is that you'll be inspired to replicate these patterns, much like the renowned permaculture herb spiral pattern. I also hope these bio-integrated patterns inspire new and improved patterns that will advance the language of permaculture design. Finally, I hope everyone who reads this book can experience the joy and ease of living and working in a landscape where all the components are bio-integrated.

PART ONE

BIO-INTEGRATED FARM SYSTEMS

CHAPTER 1

The Chinampas
An Ancient Example of Bio-Integration

Sometimes the oldest patterns are the best. The story of bio-integration starts over two thousand years ago on the edges of the Valley of Mexico. The pre-Aztec indigenous people, ancient yet advanced, recognized the productivity of swamps and ponds. Swamps and ponds produce more biomass — living matter — than any other type of ecosystem, including forests, grasslands, and farm fields. These wet areas have three times the productivity of agricultural land and are almost twice as productive as a forest.[1] Using a system called chinampas, these indigenous cultures learned to farm the special edge created where water meets land.

By dredging a series of canals, the Aztecs drained swamp areas, then layered soil and vegetation to create raised beds between the canals. The beds extended up to 3 feet above the waterline, close enough to receive moisture from the canals but raised sufficiently so the plants would not become waterlogged.[2] The rich soil was high in organic matter from the addition of mud and plants added to the beds. Chinampa productivity relied heavily on the high organic matter content of the soil.[3]

Nutrients from the land washed into the canals, promoting dense growth of aquatic plants and fish. Prior to the planting season, the indigenous farmers would gather the aquatic plants and add them to the raised bed areas with layers of soil dredged from the bottom of the canals.[4] The process retained the fertility of the soil and allowed the Aztecs to grow several crops per year — with no need for a fallow period. Canals also served as avenues for shipping goods and produce, as well as reserves of irrigation water during dry times.

With chinampas the Aztecs developed a highly productive system of agriculture. Everything necessary for successful plant growth — water, nutrients, and soil — was found within footsteps of the site, close enough to gather by hand. Even the most modern organic farming systems usually

Figure 1.1. Chinampas of Xochimilco, Mexico, ca. 1905, show canals integrated with homes and fields. Photo courtesy of ETH-Bibliothek

Figure 1.2. In a present-day scene, a farmer scoops up mud from the chinampa canal bottom to spread on farm fields for fertilizer or to make soil blocks for transplants. Photo courtesy of Dr. Philip Crossley

Figure 1.3. The fertile beds among the canals produce excellent yields of lettuce and many other crops. Photo courtesy of Dr. Philip Crossley

bring in nutrients from off the farm. The Aztecs designed for ease and efficiency with these multifunctional canals.

By the early part of the sixteenth century, the Aztec system of chinampas had reached a masterful level of sophistication. Aztec farmers diverted springs to feed the canals and built long dikes to prevent saltwater intrusion.[5] It's estimated the Aztecs fed between 140,000 and 190,000 people in the Lake Chalco–Xochimilco chinampa zone containing 23,475 acres.[6]

Another interesting innovation was the production of soil blocks for transplants using fertile mud. Farmers piled mud from the bottom of canals onto adjacent chinampa land. Once the mud had dried sufficiently, they divided it into blocks using a large knifelike tool, making sure to leave air space between each block. Farmers then planted seeds into the mud. As the roots developed, the air spaces separating the blocks stopped the roots from growing. Thus, fewer roots had to be cut or pulled apart at transplanting time, which reduced transplant shock. With seedbeds close to canals, seedlings were easily watered by hand with nutrient-rich canal water.

Research shows the mud used to make the soil blocks contained up to 40 percent organic matter, with a mineral content derived from volcanic ash and diatomite, making the blocks excellent for seedling establishment.[7]

The Aztecs' two-thousand-year-old planting innovation paved the way for the modern practice of making soil blocks for transplants. Eliot Coleman popularized this technique in his book *The New Organic Grower*. Soil blocking uses a special tool for compressing soil into blocks, and it saves purchasing and storing plastic pots and reduces transplant shock.

Chinampas also provided a microclimate, protecting the plants from cold nights. The canals and moist soil absorbed heat during the day and released the heat at night, thus decreasing the severity and duration of cold. Research testing some of the few remaining chinampas found air temperature above the land and adjacent to the canals up to 6°F warmer than a similar area without the canals.[8] The temperature difference may not seem significant, but it actually provides more protection than an unheated greenhouse or row cover over plants on cold nights. The chinampa farmers had discovered a way to extend the growing season long before greenhouses were a reality. They successfully combined water and land into a landscape more productive than either could provide individually. In modern times we either destroy and drain swamps or relegate them to nature. But these early indigenous people recognized the potential productivity of swamps, then modified them and put them to good use. The challenge lies in adapting the chinampa system to a dry location.

As you delve into the upcoming chapters, you'll find echoes of the chinampas in a web of bio-integrated designs: Capturing reflected sunlight, integrating water into dry landscapes, and growing fish and vegetables in symbiotic systems — all of these functions can be stacked on a solid foundation of chinampas.

Figure 1.4. Seedling blocks made from canal mud and manure for transplanting into chinampas fields. Photo courtesy of Dr. Philip Crossley

SUMMING UP THE FUNCTIONS

1. Canals reflect sunlight into adjacent plant beds, increasing solar energy.
2. Canals moderate temperature extremes and extend growing season by absorbing heat during day, then releasing heat at night.
3. Aquatic plants produce organic matter and decomposing waste in canals.
4. Accumulated fish and plant waste in canals provide nutrients for plants.
5. Canals transport people, goods, muck, and soil.
6. Mud layered with plant matter creates raised beds to reclaim wetlands.
7. Canals irrigate plants and store water for later use.
8. Soil blocks formed from mud provide fertile soil for transplants and seedlings.
9. Chinampas produce vegetables and fish.

CHAPTER 2

A Pool of Resources

The Bio-Integrated Pond

Everyone's looking for a sunny place — a spot to retreat to for a breath of fresh air or to feel the warmth of the sunshine. Vitamin D is not only good for the body, it's also good for the soul. No one knows this better than Christopher Alexander, author of *A Pattern Language* and one of my favorite designers.

Alexander recognized that people are drawn to sunny places — courtyards, gardens, or patios — located on the south side of buildings. On the sunny south side, buildings bask in sunlight during the winter, while roof overhangs provide shady retreats during the heat of summer. The south side of a building creates the framework for the perfect human outdoor microclimate: warm in winter and cool in summer.

In contrast the north side of a building is cool, damp, and shaded year-round; it's an area better suited for storage. It's a mistake to locate design elements intended to attract people on the north side of a building. For example, a recent addition to the public buildings in our community includes a beautifully designed courtyard, with benches for sitting, grills for cooking, and tables for eating. Unfortunately, the courtyard was positioned on the north side of the building and thus sits in cold, dreary shade throughout most of the year. Needless

to say, the courtyard generally lies vacant. When the grills are used, the cooks wait in the warmth inside the building, peering through windows to keep a close watch on the food. They rush outside to flip the food, then quickly return to the warm building. The benches in the courtyard also lack use because no one wants to sit in a cold, dark spot. In comparison, the sunny south side of the same building creates the perfect warm microclimate, enjoyable even on a cold, sunny winter day. But there's not a single bench to offer a spot to sit and enjoy the warmth of the sun.

Christopher Alexander developed the design pattern called "A Sunny Place" to address this common problem in architecture. He suggests making the south side of a building a special place that draws people in, like an outdoor room. He recommends providing a bench for people to sit and enjoy the warmth and insists on blocking cold winds to create a warmer microclimate. Alexander also recommends giving people something to do in this special sunny place, such as "a swing, a potting table for plants, a special view, [or] a brick step to sit upon and look into a pool." The bio-integrated pond builds upon Alexander's pattern with a rainwater-harvesting pond in a "sunny place."

Stacking Functions

The pond collects the rainwater to grow the fish to feed the chickens to feed the farmer to build the pond . . . sounds like a nursery rhyme, but it's actually the basis of bio-integration: stacking functions.

Much like multitasking, stacking functions allows us to accomplish several goals at once by interrelating tasks. Our sunny place contains a pond, often built aboveground, that brings water within hands' reach from a bordering bench. The additional height of the raised pond lets the free force of gravity increase the water pressure for use. Besides storing water for irrigation, ponds also provide habitat for fish. But just like any other ecosystem, there's a balance in the beauty. Whether it's a big fish in a small pond or a small fish in a big pond — that fish still needs a healthy habitat to survive.

"What about mosquitoes?" is the most common question I hear regarding small ponds. Without fish a pond becomes a breeding ground for mosquitoes. With fish in the water, ponds become traps for mosquitoes. I've actually noticed a reduction in mosquito populations when small ponds stocked with fish are added to a neighborhood. The fish promptly eat any mosquito that lands on the pond, thereby reducing the population of mosquitoes in the surrounding community.

Minnows make great mosquito hunters. And after those minnows get their fill they become tasty treats for chickens. Kids of all ages (even the adult kind) enjoy catching minnows with an umbrella net that hovers over our pond at the Clemson Student Organic Farm (SOF), then tossing the slippery little morsels to an eager flock of chickens.

But stacking functions doesn't stop there. When properly placed, the pond reflects winter sunlight

Figure 2.1. This pond at the Clemson Student Organic Farm provides a host of activities to draw people to the "sunny place." Filled with flowering aquatic plants, the water is screened from view, becoming visible as you approach the bench.

into adjacent buildings, providing extra heat and light during the dimmest time of year. In summer, cool winds are captured and funneled across the pond, providing a haven on a hot day.

Imagine a self-sustaining garden — capable of watering itself, fertilizing itself, and even moderating its own climate. That's a reality attained by cultivating edible plants in and around ponds. A mixture of edible, useful, and beautiful plants surround our raised pond at the farm, taking advantage of fourteen different microclimates and nourishing every visitor drawn to this sunny place.

Ponds provide the perfect location for an edible landscape. Supplying food all year long, this efficient ecosystem survives with little to no external input. Plants rooted within the water nurture a symbiotic relationship with the pond, which offers the plants water and nutrition, while the plants pump oxygen into the water and absorb excess nutrients to maintain a pristine water quality. The bio-integrated pond creates such a powerful sunny place you may find it hard to leave it behind when it's time to go inside.

Using a Pond to Reflect Sunlight

We've all fallen victim to the surreptitious sunlight that reflects off the surface of a body of water. Playing at the beach or riding in a boat, all too quickly we end up looking like lobsters. The same thing happens in winter when sunlight reflects off snow. Water, in its diverse forms and phases, creates a perfect yet complicated reflective surface. For example, calm water reflects more sunlight than disturbed water because waves or moving water scatter light in all directions. Ice, on the other hand, reflects more sunlight than liquid water.

A pond located on the south side of a building reflects winter sunlight into the building. During

Elements of the System

Small ornamental ponds — sometimes called "water gardens" — can be very appealing, but they're usually filled and replenished from a municipal water line or domestic well, and they perform few useful functions. By contrast, the bio-integrated pond is designed to capture rainwater from roofs and roads. A strategically located pond improves microclimates for adjacent buildings. The techniques I've developed for bio-integrated ponds are most applicable to areas east of the Mississippi and in the Pacific Northwest because of their plentiful rainfall year-round. However, the techniques can be applied in any climate. This chapter shows you how to:

▶ Orient ponds to heat buildings in winter and cool them in summer
▶ Use small ponds to grow minnows for chicken feed and bait
▶ Design ponds to create up to fourteen different microclimates
▶ Grow edible and ornamental perennial food crops in and around a pond
▶ Elevate ponds to maximize gravity flow for better rainwater harvesting

winter the sun's path is lower in the sky, which creates a broad angle of reflection off a pond surface that can send light through windows of the building. Besides creating a pretty, wavy pattern on the ceiling, the reflected sunlight conveys extra light and heat into the room. During the summer the sun

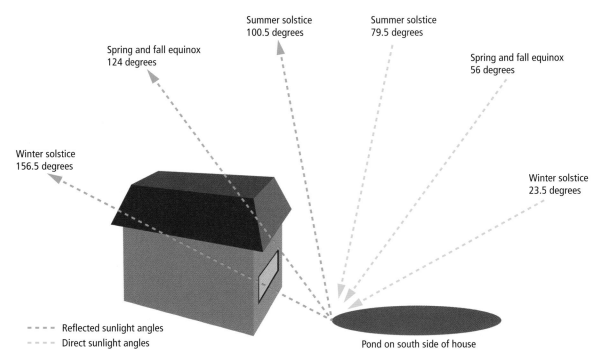

Summer solstice
100.5 degrees

Summer solstice
79.5 degrees

Spring and fall equinox
124 degrees

Spring and fall equinox
56 degrees

Winter solstice
156.5 degrees

Winter solstice
23.5 degrees

- - - - Reflected sunlight angles
- - - - Direct sunlight angles

Pond on south side of house

Figure 2.2. Reflected and direct sun angles at different times of the year at 34 degrees north latitude.

is at a higher elevation in the sky, which changes the angle of reflection and prevents the reflected light from shining through the windows of the building.

A properly placed pond is a perfect reflecting surface because it naturally adjusts in sync with our heating and lighting needs. In the morning and evening, when temperatures are cooler, the pond reflects more light through the windows to warm the room. Similarly, in the winter, when we need additional light and heat, the sun sits lower in the sky throughout the day, so more light is reflected off the pond and into the house. The reflective pond works with Mother Nature to provide more light when we need it most (see table 2.1).

The sun has fascinated humans since ancient times. The national patron of the Inca was Inti, or the Incan sun god. Each province in the Incan territory established a temple to Inti and set aside one-third of its crops and herds to pacify him. In ancient Egypt the sun god Ra was known as the creator. Although ancient civilizations worshiped the sun and its daily path, modern humans have all but forgotten the importance of the sun as the source of life for our planet. We just don't pay attention to the sun. Most students I teach are surprised to learn that the elevation of the sun above the horizon changes so drastically from season to season. I have to sympathize, since I didn't realize this myself until I was in my midtwenties.

Regardless of the time of the year, people like to retreat to the tropics for vacations because of its constant warm climate, which is due to the fact that in areas near the equator, the sun's path barely shifts. Nearly all year long, the sun's pathway arches through the sky like a rainbow linking the western horizon to the eastern horizon.

In all areas other than the tropics, the sun travels at a higher elevation during summer and also rises

Table 2.1. Solar Reflection from Water Based on Elevation of Sun from Horizon

Elevation of Sun (degrees)	Average Reflectivity
2	0.9
3	0.8
4	0.7
5	0.6
7	0.5
8	0.4
12	0.3
16	0.2
24	0.1
35	0.05

Source: Adapted from Dan Pangburn, Wikipedia: Albedo

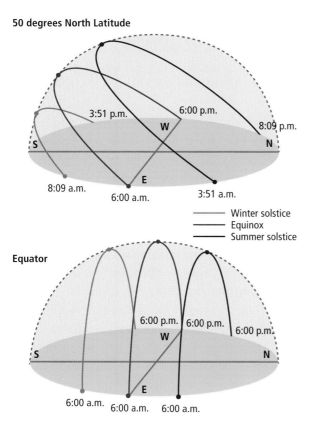

Figure 2.3. The closer to the North Pole we travel, the lower the sun's pathway is during the winter. Thus a winter in New York is drastically cooler than a winter in Florida.

and sets closer to the true eastern and western horizon. In winter the sun rises in the southeast, travels at a lower elevation across the sky, and sets in the southwest, as shown in figure 2.3.

The website of the University of Oregon Solar Radiation Monitoring Laboratory includes a great feature for printing sun charts that allow you to see how the path of the sun at a particular location on the planet changes throughout the year.[1] Simply enter the zip code or longitude and latitude for the site, and the chart is generated for printing.

Finding the Right Spot

To determine the proper placement of a pond to provide reflected winter light for a building, we must first consider the angle of incidence — loosely defined as the angle between a ray of light diving into the pond and an imaginary line standing perpendicular to the surface of the pond. Based on the law of reflection, the angle of sunlight reflecting off the pond is equal to the angle of elevation of the sun.

DISTANCE CALCULATIONS

When I constructed a reflective pond at my home, I first determined the angle of elevation from the horizon of the sun at midday when the temperature begins to rise. In my climate in Anderson, South Carolina, the weather starts warming around the spring equinox on March 20. The angle of the sun's path from the horizon at the spring equinox is 56 degrees (see figure 2.2). This means that sunlight will reflect off a pond at the same angle at this time. Next, I measured the distance from the ground to the top of the windows on the south side of my house, which equaled 8 feet. Then I used a simple algebraic formula to figure out the distance to locate the pond from the house so that reflected light would enter the windows only during the cold time of year from late September through late March (see figure 2.5).

Figure 2.4. This pond (complete with minnow net) is a mere 5.4 feet away from the building so that light will reflect through the windows only after the fall equinox and before the spring equinox. The building is in Anderson, South Carolina, at a latitude of 34.5 degrees.

Even simple algebra is not my strong suit, so I requested assistance from my sixteen-year-old daughter. We followed the formula:

$$\tan c = a \div b$$

where

c = angle of incidence;
a = distance to top of window; and
b = distance between pond and house wall

The actual calculations looked like this:

$$\tan 56 \text{ degrees} = 8 \div b$$
$$1.48 = 8 \div b$$
$$1.48b = 8$$
$$1.48b \div 1.48 = 8 \div 1.48$$
$$b = 5.4 \text{ feet}$$

Once I crunched all the numbers, I determined that 5.4 feet is the ideal distance to locate the pond from my windows to allow reflected light into the building only during the cold time of year (i.e., from about September 20 through March 20 for my climate). The edge of the pond closest to the building is the point of focus in these calculations, and the width of the pond determines the width of the reflective area. As the sun continues to move lower, the full width of the reflective area will send light into the window, like a sundial marking the time of year. For those of you not inclined to remember high school math, online calculators are available to determine the pond distance from the building. Simply search for "right triangle angle and side calculator."

Since 5.4 feet is a pretty precise number, I commonly hear the question, "How precise do you have

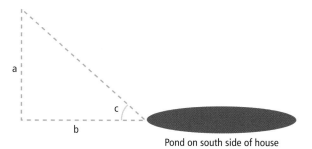

Pond on south side of house

Figure 2.5. In this schematic, (a) is the distance to the top of the window of the building to be warmed with reflected light, (b) is the distance between the pond and the house wall, and (c) is the angle of incidence of the sun. Since (a) and (c) are known quantities, we can use an algebraic formula to solve for (b). The equation used to solve this is: $\text{Tan}(c) = a \div b$.

Stick with the South Side

Proper placement of a reflective pond is essential to cutting energy costs, and that applies not only to the distance from a building but also the directional position relative to the building. What are the results from poor pond placement? A pond located on the east, west, or north side of a building won't reflect winter sunlight into the building. Instead, a pond located on the east side will reflect *summer* sunlight in the morning, and a pond located on the west side will reflect *summer* sunlight in the afternoon. Unfortunately, this unwanted summer sunlight will raise indoor temperatures, resulting in more expensive cooling bills. A pond positioned on the north side of a building remains shaded for most of the year. Since sunlight drives the food chain that makes a pond productive, a shaded pond will hold little life, and it won't reflect sunlight into the building at any time of year.

to be?" The short answer is, "as precise as possible," because every foot makes a big difference. For example, if I located a pond 8 feet from my house at my latitude, sunlight and solar energy wouldn't start entering the windows for another entire month. Good design is based upon careful observation and planning prior to implementation. In other words — don't put the cart in front of the horse.

SUN ANGLES DURING THE YEAR

While sun charts and online calculators are available to determine specific sun angles for various times of year, it's easy to figure out the highest and lowest sun angles during midday summer and winter and the equinox using the site's latitude:

Step 1. Check an atlas and determine the latitude of the location of interest. For example, my latitude in South Carolina is approximately 34 degrees.

Step 2. Subtract the latitude of the location from 90. This gives the angle of the sun during the spring and fall equinox. In my example, I subtract 34 from 90 to equal 56 degrees, or the elevation from the horizon of the sun at midday during the spring and fall equinox.

Step 3. Compensate for the tilt of the earth during the winter and summer equinox. Subtract 23.5 from the spring and fall equinox value to calculate the winter solstice value and add 23.5 to calculate the summer solstice value. In my example, I subtract 23.5 from 56, which equals 32.5 degrees. That's the elevation from the horizon of the sun during midday on the winter solstice at my location. Adding 23.5 to 56 totals 79.5 degrees or the elevation from the horizon of the sun during midday on the summer solstice. More complicated math is necessary to determine the sun elevation angle for a specific day of the year other than the solstice. I would recommend using an online sun elevation angle calculator or a sun finder app on a smart phone or tablet or looking at the sun chart for your site to nail down a specific date if needed.

Obstructions That Block Light

Trees and other obstructions can ruin the productivity of a solar reflecting pond if they block sunlight from reaching the pond during the winter. Even deciduous trees potentially reduce the amount of light reaching a pond surface by 50 percent. Furthermore, ponds are leaf traps. Leaves from nearby trees will find their way into the pond, decreasing the water quality as they decompose. If many trees surround a pond, it will quickly turn into black water, which is not necessarily a problem, except

for aesthetics. A fountain can mitigate water quality and potential odors by aerating the black water.

To determine if a tree, building, or other obstruction will block sunlight during the winter, print out a sun chart from the University of Oregon Solar Radiation Monitoring Laboratory, as shown in figure 2.6. Standing in the planned location of the pond, use a sun angle finder to site potential obstructions during the time of year when you need the sunlight. Draw the obstructions on the sun path chart to depict when they will block the sun. The Build It Solar website has helpful resources such as a printable sun angle finder and azimuth gauge along with easy-to-follow instructions.[2]

Alternatively, use a sun finder app for a smart phone or tablet to site obstructions. Unfortunately,

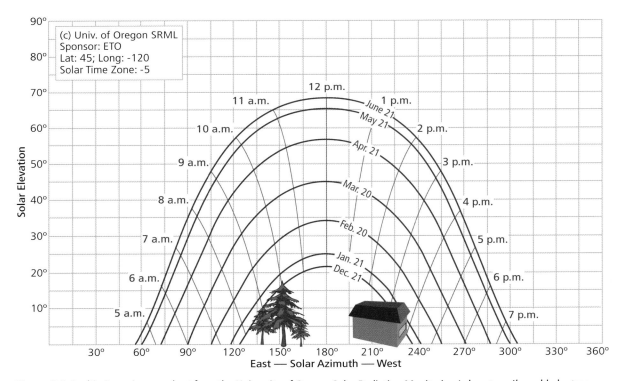

Figure 2.6. In this Cartesian sun chart from the University of Oregon Solar Radiation Monitoring Laboratory, I've added a tree and building in the landscape to show how obstructions can block the sun during winter. The location analyzed with this sun chart would receive five hours of direct sunlight during the coldest time of year, between about 10 a.m. and 3 p.m., because of the shadows created by the tree and building. Image courtesy of University of Oregon Solar Radiation Monitoring Laboratory

the apps only give a snapshot in time and don't generate a chart. To compare sun angles at different times of the year in relation to obstructions, use the University of Oregon Sun Chart Program.

Unfortunately, your research may reveal that an obstruction will block sunlight from reaching your pond site on the south side of your building. If the obstruction is a tree, it may be worth cutting down the tree or removing branches, especially if this would allow sunlight to directly enter windows on the south side of the building and passively heat the building. While this idea may seem contradictory from an environmental standpoint, the passive solar energy generated by the sunlight greatly contributes to your overall energy conservation. Removing the sun-blocking tree and planting new ones in another location on the north side of a building may be a more sustainable solution in the long term.

However, if it's a building obstructing the light, there's nothing you can do. I recommend carefully examining a site using a sun chart before purchasing property or placing a building to avoid shading that prevents access to free solar energy.

Using a Pond for Cooling

Evaporation is one of nature's most powerful tools. Similar to sweat drying on a human body, evaporation — which is simply the change of water from a liquid to a gas — will tend to cool rising temperatures of any surface. It's nature way of restoring balance to unstable conditions. When we design bio-integrated systems, we mimic Mother Nature by applying these same principles. Ponds cool the surrounding air through evaporation, so if positioned properly, they counterbalance the wrath of summer's heat. Carefully considering a pond's position next to a building and the vegetation next to the pond allows ponds to heat buildings during winter and cool buildings during summer.

Winter winds generally blow from the north and bring cooler air. Thus, placing a pond on the north side of a building worsens the effects of winter. Since ponds cool air through evaporation, a pond located on the north side of a building, exposed to wind, creates even colder winter winds. These ponds intensify ice and snow accumulations next to the building, generating maintenance issues and increasing heating costs.

Cooling winds are beneficial during summer. A good design uses ponds to cool captured summer winds but blocks winter winds from blowing over the ponds. The first step in pond design is to determine the derivation of summer and winter winds. When I first began dabbling in design, the Internet didn't exist. Determining wind patterns for a specific location required real work. I'd set up a series of flags around the site and note wind directions during different times of year and at various times of day. The website Weather Underground (wunderground.com) has changed the game of weather observation. Hosting a sophisticated series of personal weather stations across the country, the website allows users to easily access information with a few clicks of the mouse.[3]

For my location, winter winds derive predominately from the north. Since my pond is on the south side of my house, the house blocks these cooler winds and limits unnecessary evaporation. However, winter winds occasionally come from the southwest; my summer winds are from the southwest, too. This poses a design problem for my site. I welcome the summer winds from the southwest but not the winter winds. Luckily, the solution was simple: I diverted the southwest winds into the pond microclimate by planting trees and shrubs to the east of my pond. This promotes evaporation and cools the area. I was careful to select deciduous trees and shrubs, which allow the winter winds to pass through — thus, no worries about winter

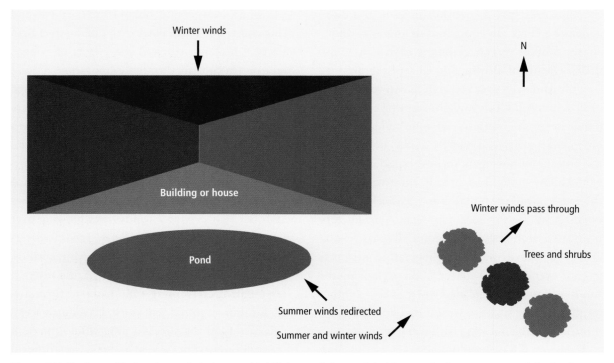

Figure 2.7. The house blocks winter winds that would chill the pond. A row of deciduous or herbaceous trees and shrubs direct cooling wind toward the pond and house area in summer but allow winter winds to pass through.

cooling from evaporation, as shown in figure 2.7. Alternately, in other designs I have exaggerated the effect by recommending herbaceous shrubs that grow quickly. For example, elderberry can be cut back to the ground in winter, and by late summer it stands 8 feet tall.

Pond Construction Overview

Once you've determined the perfect location for your pond, you're ready to start preparing the site, laying out the pond, and excavating. Let's start with a summary of the overall process of pond construction for a natural pond with no pumps or waterfalls. Following this summary, I describe some of the stages of construction in detail.

If you're planning on putting a pump or waterfall in your pond, refer to chapter 5 for more information.

Here's a step-by-step overview of pond construction.

Step 1. Call the local utility companies or the organization in your state responsible for locating underground utilities. In South Carolina a nonprofit company called Palmetto Utility Protection Service coordinates with all the utility companies to locate underground utilities before digging projects begin, free of charge. Most states have a similar service. Don't skip this step — it's saved my life several times. The process usually takes at least three days, so plan ahead and allow time for locating all utility lines.

Step 2. Determine the size, depth, and shape of your pond, and decide whether you will construct the pond rim at ground level or raise it to form an elevated pond. I recommend making the pond as large as the watershed (roof area) will allow because a larger pond is deeper and more stable than a small pond. The size of the pond ultimately determines its depth because of the angle of the slope required to prevent erosion. Two feet is the typical maximum depth for a small pond.

Step 3. Mark the location for the pond with a hose, rope, or marking paint. Place grading stakes around the pond in at least four locations. Mark the final water level on the grading stakes using a water level or transit level (don't use a line level). If you plan to raise the pond above the existing grade, mark the stakes a foot or so above ground level. If you plan to excavate the pond below ground level, simply mark the stakes closer to ground level. Tie a string between those marks during the final stages to visualize the final water level of the pond; this will help guide your soil distribution between stakes.

Step 4. Remove and stockpile all topsoil from the pond area and store it next to the pond site. If you plan to elevate the pond, excavate topsoil from the area where you will construct the berms, too.

Step 5. Excavate soil to pond depth, build berms in areas where the marks on grading stakes are above the existing grade, and shape slopes on all sides of the pond. If you're going to use a pond liner, excavate the additional soil necessary to cover the liner. For example, if you plan to cover the liner with 6 inches of soil, excavate an additional 6 inches beyond the desired depth of the pond.

Step 6. If you plan to use a drainpipe in the pond, dig the trench for the drainpipe and install it.

Step 7. If your pond will have a liner, prepare for installing it by digging a trench 12 inches deep along the top edge of the pond to secure the upper edge of the pond liner. Place excavated soil from the trench along the outer edge of the trench. If you're not installing a liner, skip ahead to Step 10.

Step 8. Continue preparation by removing all debris and rocks the size of a pea and larger from the floor and sides of the pond. If you're using a protective membrane such as carpet or geotextile fabric, place it on top of the soil and position the liner on top of the membrane. Large liners and liners made of EPDM (a type of synthetic rubber) may need several people to position in place. Leave the liner somewhat loose to prevent overstretching when it's filled with soil and water.

Step 9. Bury the edge of the liner inside the trench. To prevent water from wicking out the sides of the pond through the soil covering, leave ½ inch of liner protruding above ground level even after the topsoil is in place. Place stockpiled subsoil or topsoil over the entire liner. Place topsoil on the outer edges of berms to support plant growth around ponds.

Step 10. Seed and mulch all bare soil areas. If you plan to fill the pond immediately with water from a well or your local municipality, you don't need to mulch and seed the pond floor and sides. However, if you're going to rely on rain to fill the pond over time, seed and mulch the inside of the pond to protect the soil from erosion.

Step 11. After the pond has filled, inoculate the pond with water from an older pond if you want to promote growth of natural foods in the water. Plan to stock the pond with fish as soon as possible to prevent mosquito infestations.

Pond Parameters: Size, Shape, and Profile

Now that we have a good grasp of the overview of pond construction, let's zoom in to take a look at the specifics. The first step is determining the size and shape of your pond, and whether it will lie flush with its surroundings or be elevated above ground level.

Determining Pond Size

Ponds lose an enormous amount of water to evaporation. On a hot summer day, a loss of ¼ inch of water depth to evaporation is not uncommon. While ¼ inch doesn't sound like much, a square pond measuring 25 feet wide and long could lose 390 gallons of water every day, or 2,700 gallons of water a week. Using city and well water to replace the lost water is not only expensive but also squanders a natural resource. On the other hand, roofs and roads shed copious amounts of water that's usually wasted as it runs off the property. By using ponds to capture runoff from roofs and roads, we secure and store a free natural resource. It's important, though, to match the size of the pond to the amount of water available for harvest from a roof or other feeder source.

Different regions of the country have diverse climates producing varying amounts of rain. Regions with limited rainfall require larger watersheds to maintain ponds between rain events, whereas regions with abundant rainfall require smaller watersheds. Generally, areas east of the Mississippi River and in the Pacific Northwest receive abundant amounts of regular rainfall and require only small watersheds to keep ponds full all year. However, areas in the Mediterranean climate of California receive only seasonal rain.

Figure 2.8. Rainwater flows off the roof of the house, through the downspout, and into the pond. Chapter 6 shows additional techniques for harvesting rainwater.

Winter rains fill up ponds, but then the ponds dry up during summer drought. Distinctive plant communities inhabit these vernal ponds. Southwest of the Mississippi River, sparse rainfall and increased evaporation limit the feasibility of small watershed ponds. Often vernal, ponds in this region also require larger watersheds to remain full.

The Natural Resources Conservation Service (NRCS) developed a map to estimate the amount of drainage area necessary to maintain ponds based on rainfall and evaporation in specific regions of the United States. The map determines the drainage area for every foot of depth in the pond. The NRCS map is based on runoff from soil, but we are capturing runoff from a roof. When runoff flows across the ground some of the water infiltrates into the soil, resulting in a loss of runoff. Therefore, the NRCS map is an overestimate because we will not lose water from infiltration. Still, I've found that the map serves well for estimating the total drainage area needed to keep small ponds full. For example, my ponds in the upstate region of South Carolina are estimated to need 1.5 times the drainage area for every foot of pond depth. I wanted to use rainwater harvested from my roof as my "watershed" in place of a natural drainage area. Thus, I measured the square footage of the house and used that figure as the drainage area to calculate the maximum size of the pond I could build. I applied the following formula:

Max area of pond = drainage area (sq ft. of roof)
÷ (depth of pond × drainage area required
per foot of pond depth)

For example, the roof of our house extends over the garage, porch, and house, with a total area of 2,564 square feet. I want the pond to be 2 feet deep, and according to the NRCS drainage map, I need 1.5 times the surface area of the pond per foot of pond depth in drainage to keep the pond full. Note: Maximum pond depth will be based on slope angles and pond size, but a depth of 2 feet is a typical average depth for a pond between 15 and 25 feet wide.

Max area of pond = 2,564 ÷ (2 × 1.5)
Max area of pond = 855 square feet

Therefore, based on the drainage map provided by the NRCS for our climate, the roof area of our house, and the proposed depth of the pond, the maximum size of the pond would be 855 square feet, or approximately 29 feet by 29 feet.

Choosing a Pond Shape

Before I started building ponds, my vision of the ideal pond stemmed from storybooks. Much like everyone else, I pictured a circular pond. It wasn't until I started engineering my dreams into reality that I realized "it's hip to be square."

Circular ponds, though fanciful and fairy-tale-ish, are not always functional. For example, pond liners are manufactured in square or rectangular shapes, so fitting them into a circular pond is much like forcing a square peg into a round hole — it's possible to make it fit by trimming off the excess, but it's not practical. To install a pond liner in a circular pond, the liner corners must be cut. So at the end of the day, there's a pile of large pieces of liner leftover that usually end up in the garbage.

Another issue arises from the folds of the liner. Have you ever tried wrapping a ball with cloth or paper? Crease after crease forms, because a circular shape lacks edges. In a pond liner those creases can create havoc. Folds are difficult to bury and tend to push their way to the surface. Exposure may lead to puncture or sun damage, which defeats the purpose of using a liner.

Square and rectangular ponds present a cheaper and less risky alternative. Sure, a few folds may occur at the corners — but overall it results in much less loss and hassle. The liner stays put and prevents damages from punctures or the sun. And over time, as vegetation grows in and around the pond, the perceived edge softens, and the shape comes to resemble that fairy-tale image we've all grown to love.

Elevating Ponds

Gravity is inevitable. We can work either with it or against it, but we cannot change it. Within the scope of bio-integration, we can incorporate gravity to create more efficient and sustainable designs, such as an elevated pond.

A functional pond fits into the landscape like a missing piece in a puzzle. It fills and releases water easily without much maintenance. In flat terrain, raising a pond above the existing grade allows gravity to drain the water directly into an adjacent basin or landscape. Conversely, a recessed pond requires a pump to move the water. Without an expensive large pump, the water will discharge slowly, which limits the ability to use the water for flood irrigation. A raised pond also brings the water within hands' reach from a nearby bench. Besides the aesthetic appeal, this effect mimics the practicality of a raised garden bed — offering easier access without straining the back.

Figure 2.9. The berm for this turkey nest pond still in construction contains 11 cubic yards of soil. The drainpipe will allow us to drain the pond water into the basin on the far side.

Unfortunately, building a raised pond involves moving a lot of soil. The first raised pond I built was a turkey nest pond, which is a pond that sits atop the soil like an aboveground pool. It's surrounded by a raised berm and collects water draining from nearby rooftops. Building the berm requires transferring soil from off-site or the surrounding area. I used soil from excavating nearby recessed ponds to build the berms. I was shocked by how many trips I made with the front-end loader full of soil and thankful the other ponds were close by. I realized the benefit of planning ahead and calculating the amount of soil needed to prevent future surprises.

Excavating a Pond

I'm not going to describe all the details of excavating a pond. Suffice to say, if you don't have any excavation experience, it's probably best to hire someone for this task.

If you're building an elevated pond, you will need to use grading stakes to guide the work. I also like to use a water level or transit level (not a line level) to mark the final water level on the grading stakes. Also, when building an elevated pond, it's smart to do some calculations in advance to figure out how much soil you'll need to build a berm of the desired size.

If you plan to use a pond liner, remember to excavate extra soil, as described in the section on "Pond Liners" on page 34.

Building a Water Level

A water level is a cheap, accurate device for determining level or grades between two points spaced far apart. To build a water level like the one shown in figure 2.10, I start with two tomato stakes. I then use staples to secure tailor's measuring tapes to both stakes, making sure the tapes are aligned at the bottom of each stake. Next, I fasten a 50-foot-long

section of clear, ⅝-inch-diameter vinyl tubing to the stakes using cable ties. The tubing is then filled with water until only a few feet of tube is left unfilled at each end. Since water naturally levels itself, the water inside the tubing shows whether the landscape is level or sloped.

You'll need two people to use the water level, one to hold each stake. Or you can add a spike to the bottom of one stake to make it freestanding. Start by removing all the air bubbles from the tubing. Leave the ends of the tubing open to allow a free flow of air in and out. Position the stakes of the water level next to two grading stakes of the pond, and mark a single grading stake to indicate the preferred water level in the pond. Next, position the bottom of the water level stake at the mark on the grading stake of the pond, then move the opposite

Figure 2.10. This simple homemade water level works well for measuring level and grades between two distant points.

water level stake up or down until the water level in the tube gives the same reading on both stakes. Then transfer a marking onto the opposite grading stake. With every measurement, compare the difference between the water level on each stake to determine level or grade.

Figuring Out Slope Angles

One important aspect of building a pond is setting the slope of the pond walls. The angle of a slope determines the stability of soil and its ability to withstand weathering. Soil that is too steeply sloped will collapse or slump, like a fragile cliff falling into the ocean. Every soil type has its innate stable maximum slope angle capable of withstanding the natural elements without eroding. For most soils this slope is very gradual. Vegetation and protective coverings of rock and plastic help stabilize steeper slopes.

The slope, tilt, or inclination of the soil can be measured in three ways:

1. As a ratio of run to rise. For example, 3 to 1 (3:1) would mean that for every three horizontal units the height of the pond wall would rise one vertical unit.
2. As an angle. For example, a 45-degree slope.
3. As a percentage or "grade" determined by calculating rise/run × 100. For example, a 1 percent grade is equal to a rise of 1 foot for every 100 horizontal feet of run.

Slope ratios of 3:1 are recommended to maintain a stable structure to the soil. This means that for every foot of depth the pond needs to be 6 feet wide to achieve a 3:1 slope angle on both sides. To reach a maximum depth of 2 feet at the center, a pond would need to be 12 feet wide. If the sides of a pond are built more steeply than 3:1, the soil will slough off or collapse at the edges over time. A gradual slope is especially important when a pond will include a liner covered with soil. If the slope angle is too large, soil on top of the liner will simply slide down the liner to the bottom of the pond, negating the protective qualities the soil should provide. The slope ratio of 3:1 provides the natural angle that will keep soil stable — so don't fight it.

Berm Calculations

When you're building a berm, be sure to measure and mark the required bottom width to achieve the proper slope ratio. For example, if a 1-foot-tall berm should have a top width of 1 foot and side slopes of 3:1, the base of the berm should be 7 feet wide. As you build up a berm, use heavy machinery or a tamper to compact the soil every 3 to 6 inches. Plan on berms shrinking 10 percent over time. For example, a 1-foot-tall berm will shrink to 10.8 inches.

The berm takes on the shape of a trapezoid or a triangle with a flat top. By first determining the area of the trapezoid, you can then determine the amount of soil needed to build the berm.

To compute the area of a trapezoid with given slope angles and volume of a berm, use the following formulas:

$$\text{Area of trapezoid} = [w + (z \times h)]h$$
$$\text{Volume of berm} = (\text{area of trapezoid} \times l)$$

In these formulas:

h = height of berm; w = top width of berm; z = side slopes; and l = length of berm

Example: I want to build a berm to raise a pond by 1 foot above the surrounding grade. The top width of the berm will be 1 foot, and the slope ratio of the berm is 3:1. The shape of the pond is a square with each side measuring 20 feet, for a total length of berm measuring 80 feet.

Area of trapezoid = $[1 + (3 \times 1)]1$
Area of trapezoid = 4 square feet
Volume of berm = (4×80)
Volume of berm = 320 cubic feet
320 cubic feet ÷ 27 cubic feet per yard
= 11.8 cubic yards

One dump truck of soil holds approximately 7 cubic yards. So if I want to build a berm 1 foot high and 80 feet long, I would need nearly two full dump truck loads.

Another, easier alternative to hauling in soil to build berms is to build a ring pond. With this more conventional method, the soil excavated from digging the pond is used to build the berm. The result is that part of the pond lies below ground level and part sits above. One disadvantage of a ring pond is that draining the pond may necessitate a pump instead of simply using the assistance of gravity, depending on the existing topography.

Managing Drainage and Overflow

Ponds are water banks. We store up water until we have enough to disburse. We might want to empty the pond to irrigate our landscape or so we can easily harvest fish, plants, and nutrients. A good pond design includes a drainage plan that also accommodates all the "what ifs." For example, what if relentless rain falls for days? A rainwater-harvesting pond will not only fill up, it will probably flood if too much rain falls within a short period of time — potentially damaging the landscape.

I install watershed ponds with two types of overflow systems. The overflow systems control the direction and placement of the excess water and prevent water from washing out berms and embankments. (An undersized pond in relation to roof catchment area will favor water overflow to areas outside the pond.) For principal overflow I install a pipe at the bottom of large ponds to remove excess water. A drainpipe installed through the side or bottom of a pond can quickly and easily release large volumes of water. Sometimes the principal overflow can't handle all of the water, though, so I fit an additional secondary overflow to direct excess water through a broad low point at one edge of the pond. Secondary overflow systems guide water away from fragile berms, embankments, and buildings, directing it safely to another area.

Small rainwater harvesting ponds designed to catch water from roofs and small watersheds don't need complex overflow systems. However, it's still critical to manage the small amount of overflow that may occur.

It's easy to construct a secondary overflow system. To slow the flow of water, create an area on the rim of the pond that's 4 to 6 inches lower than the rest of the rim. The overflow area should span a breadth of 4 feet and should direct water into a retention basin, rain garden, or swale. I usually construct secondary overflow systems after initial pond construction is complete. I remove 6 inches of the protective soil laid over the pond liner in the area needed for the secondary overflow. This usually exposes the pond liner material. I cover the exposed pond liner with a piece of scrap pond liner material as extra protection; then I spread a thin layer of pea gravel or pavers to hide the liner material. When the water level of the pond rises high enough during a heavy rain, water will spill over the secondary overflow area into the basin or swale, preventing a general overflow and flood. I always locate the basin 15 feet from buildings to catch excess water and prevent it from saturating the foundation.

It's also possible to construct multiple secondary overflow systems to feed multiple rain gardens or

Figure 2.11. The retention basin in the foreground accepts overflow from the larger turkey nest pond at the Clemson University outdoor farm market. We think about sustainability for small details, too: the stepping-stones in the retention basin are made from recycled concrete!

swales. This way the overflow rotates between different rain gardens, preventing oversaturation by staggering the overflow. To accomplish this, install all the secondary overflows at the same height, 6 inches lower than the rest of the pond. Next, install what's called a flashboard riser on each secondary overflow. The flashboard riser is basically a dam made from a 2 × 4 wedged between bricks. By damming up all but one secondary overflow, water is forced out of the unblocked secondary overflow. Moving the dam between secondary overflows determines where the water goes.

Drainpipes and secondary overflow systems usually empty into two separate locations. I sometimes install an adjustable drainpipe, and this allows me to choose whether to send overflow through the pipe or through the secondary system; it also allows me to control the amount of water each overflow location receives.

The adjustable overflow works by using a riser—a pipe with holes in the sides near the top end. This pipe extends upward to the surface of the water. The riser connects (using two threaded elbows) to a drainpipe that penetrates the bottom of the pond. The threaded elbows are joined using a nonhardening liquid thread sealant; this allows for the riser to be moved up or down by turning the threads on the elbows. If the threads are engaged to a tight position before the riser is added, the riser will stick in any position at which it is placed. This allows the water level in the pond to be controlled simply by changing the placement of the top of the riser (see figure 2.12).

Installing an Adjustable Drainpipe

Installing a drainpipe isn't difficult, but it requires careful attention to detail, especially if you're installing the pipe through a pond liner. Part of the pipe installation process happens before the liner is installed, and the rest happens afterward.

Drainpipes for large ponds are usually fitted with valves and a mechanism to filter out debris. I've built much simpler and cheaper systems for smaller ponds by crafting an adjustable overflow system out of male- and female-threaded elbows and a riser. The threaded elbows allow me to push the riser under the water, and gravity takes care of the rest. Here's how to install an adjustable drainpipe:

Step 1. Determine if a drainpipe is feasible for your pond. Drainpipes need an outlet area lower than the bottom of the pond. If an outlet lower than the pond is not available, only a portion of the water will drain out of the pond. In flat areas a lower outlet may not be feasible unless the pond is elevated.

Step 2. Decide on the size of pipe. If the purpose of the drainpipe is to divert excess stormwater, a pipe with a diameter of 4 inches should suffice — unless the pond harvests water from an area larger than ¼ acre. If the drainpipe will be used solely for draining the pond, then a pipe with a diameter of 2 inches is sufficient.

Step 3. Locate underground utilities before digging.

Step 4. Dig a trench through the pond wall from the lower outlet to the pond bottom or side next to the deepest area in the pond. The trench should be sloped, ideally at a 2 percent grade, toward the lower outlet. Insert the drainpipe in the trench. Alternatively, install the pipe and bury it during berm construction to avoid the need to dig a trench.

Step 5. At this stage, you'll need to install the pond liner, as described in the "Pond Liners" section on page 34, and use the proper type of fitting to seal the liner around the drain pipe.

Step 6. Once the pipe has been installed safely through the pond liner or clay membrane, install one male- and one female-threaded elbow onto the end of the pipe that's inside the pond. Be sure to apply liquid thread sealant to the threads on the elbows and tighten before gluing the elbows to the drainpipe. Once the elbows are installed on the drainpipe, you may not be able to turn the fittings again to tighten. Check levelness as you work, as shown in figure 2.12.

Figure 2.12. When gluing threaded elbows to a pipe to make an adjustable drainpipe in a pond, check with a level to ensure that the riser will sit plumb in the elbow opening, ensuring water spills evenly over the top of the riser.

Step 7. Cut a length of pipe equal to the depth of the pond to serve as the riser. Install an endcap on the pipe without using glue to allow for easy removal if quick drainage is needed. Alternatively, a sewage cleanout with threaded fitting installed at

Figure 2.13. A black pipe boot surrounds the drainpipe of this newly constructed, unfilled pond. The remainder of the pond liner is covered with soil, which has been topped with straw to prevent erosion. The upright portion of the pipe is an adjustable riser that can be positioned either higher or lower than the level of the pond's secondary overflow.

the end of the riser is easier to remove than an endcap, since the endcap may stick. Drill a series of holes ¼ inch in diameter, or make a series of cuts using a skill saw over a 4-inch section of the pipe just below the endcap. This design allows water to enter the pipe but prevents fish and debris from getting sucked into the pipe.

Step 8. Using a water or transit level, determine the difference in height between the top of the threaded elbows and the level of the emergency overflow system on the pond. Then cut off the uncapped end of the riser pipe to a length such that when the riser is installed, the holes near the endcap will sit at the same level as or just a bit higher than the level of the pond's secondary overflow. This ensures that excess water will drain out the secondary overflow unless the riser pipe is lowered. However, if you don't want water to drain out through the secondary overflow unless you have an extreme rain event, position the riser so the holes will sit lower than the secondary overflow. This will cause water to always drain out through the adjustable drainpipe, but if a heavy rain fills the pond faster than it can drain away through the pipe, the excess

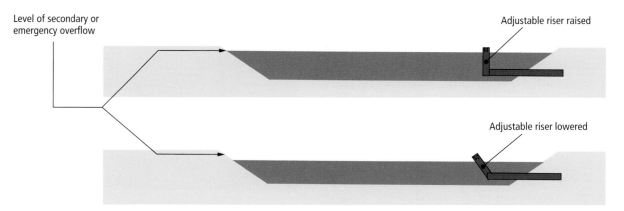

Figure 2.14. When the adjustable riser on the principal overflow pipe is raised, excess water will drain out of the pond through the secondary overflow. When the riser is lowered, excess water will enter the holes in the riser and drain through the principal overflow.

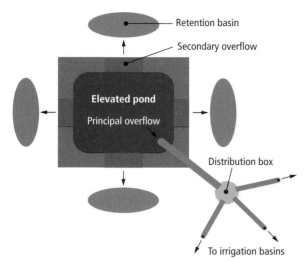

Figure 2.15. An elevated pond can have multiple overflow options. Raising the outlet of an adjustable principal overflow forces water into secondary overflows, which can be controlled by wooden flashboard risers. The principal overflow also drains into a distribution box that can split the flow into multiple irrigation areas.

Figure 2.16. This distribution box is made from a plastic pot. Once water from the pond drainpipe enters the distribution box, three other pipes channel the water to three different destinations. Caps or sewage cleanout fittings can be placed on the ends of the pipes to control where the water goes.

will escape the pond through the secondary overflow, as shown in figure 2.14.

Step 9. Install the riser pipe onto the threaded elbows.

To operate the riser once the pond has water in it, you'll need to walk out into the water or use a stick to push the riser pipe sideways and down under the water surface. Gravity will then force water through the holes in the riser and down and out through the drainpipe. If you want to drain all the water out of a pond, make sure that the top of the riser is fully submerged to the lowest point of the pond floor.

Additionally, it's possible to connect a principal overflow drainpipe to a transfer box where water can be diverted into one of three pipes, as shown in figures 2.15 and 2.16. For example, by adjusting the drainpipe in one of my small rainwater harvesting ponds, I can direct nearly all the excess water into the drainpipe and the transfer box. Depending on which pipe I open, the water flows to one of three different destinations: an irrigation ditch, a diversion channel, or a swale. Or I can send water through the secondary overflow into a basin. I make the choice depending on my irrigation needs at the time.

Pumps and Siphons

I like to use drainpipes to remove water from a pond because of their versatility and because they take advantage of gravity to move the water. They're a free, low-maintenance system. There are two other ways to remove water from a pond — pumps and siphons — and I've used them as well.

Using an electric or gas-powered pump is the most costly and time-consuming method of draining a pond, but it is straightforward. Two types of pumps are available for removing water: submersible and external. Both types require the presence of water inside the pump to keep mechanisms lubricated and cooled while the pump is operating. Without water inside the pump, damage happens quickly. Submersible pumps are less likely to run dry because they are operated underwater. Installation is a breeze — simply place the pump in the pond, connect it to a pipe, and turn it on. However, submersible pumps are usually smaller than external pumps, and they move less water and require more energy for the amount of water moved.

External pumps can run on gas or electricity. To prime the pump I fill it with water. This prevents air from damaging the pump and allows suction to occur. Overall, I find that pumps are expensive and require installation, maintenance, and eventually replacement.

Simple and cheap, siphons are a tried-and-true method for moving liquid from a high location to a low location. To create a siphon to drain a pond, simply install a valve in one end of a pipe or hose, fill the pipe with water, and lay it on the ground with the open end extending over the side of the pond and into the water. Place the valve end consistently downhill, avoiding any dips (which can cause sediment to collect in the siphon pipe). Once the valve is opened, the force of gravity pushes or siphons water out of the pond. The volume and rate of water siphoned out of the pond depends on the pipe size and how far the valve is located below the pond. The larger the pipe and the lower the release point, the faster the siphon will remove water from the pond. Usually small, flexible pipes are used to siphon water, limiting the volume to a slow trickle. With flood irrigation techniques, larger volumes and rates of water removal are necessary to move water through the flood irrigation systems. If only a small height difference is available from the pond to the drain, a small pipe works better. (See chapter 6 for more

information on siphons. See chapters 3 and 10 for more information on flood irrigation.)

Pond Liners

Small rainwater harvesting ponds are like large water tanks; they store a precious resource. Lining a pond with a synthetic pond liner protects this precious resource against loss from leakage and seepage. Unlike tanks, pond liners are cheap to ship since they are folded or rolled into a tight package. Tanks take up space, which means higher shipping costs. The savings are significant: Lined ponds store water for five cents per gallon, compared to tanks that store water for one dollar per gallon (considering the cost for materials only, not installation). Pond liners are a small investment with a dependable dividend.

I have also used clay soil. Clay prevents seepage by creating an impermeable layer if the soil contains enough clay; generally, the soil should consist of about 20 percent clay for a good seal. However, some seepage may still occur in clay-lined ponds. Since small ponds don't hold a lot of water, any seepage is unacceptable. A synthetic liner is like a cheap insurance policy. And a liner topped with a thick layer of soil provides the best available coverage for the long run.

Covering pond liners with soil offers several advantages. The soil protects the pond liner from UV degradation and punctures. Without a soil covering, UV light will degrade the liner within ten to twenty years, depending on the material. A covered liner will last indefinitely (unless it gets punctured), so adding an extra layer of soil is well worth the work. It's not a good idea to cover a pond liner with soil, however, if the water will recirculate through a waterfall or hydroponic system. The fast movement of water may keep soil in suspension, constantly clouding the water.

Figure 2.17. Part of this ring pond under construction is below grade, and part is above grade. The raised berm is built from some of the excavated soil, and more soil is stockpiled on the outer edge of the berm.

Figure 2.18. Next a Permalon pond liner is installed. At the stage shown here, the edge of the liner is buried in a trench to hold it in place and some soil has been added to the middle of the pond to help weigh the liner down.

Figure 2.19. Here's the same pond covered with soil, mulch, and seeds and ready to receive harvested rainwater from the house below (see chapter 6).

Pond liner manufacturers recommend covering pond liners with 12 inches of soil. Be sure to factor this in while you're digging the pond. You'll need to excavate 12 additional inches of soil and plan for wider berms to accommodate the width of the soil that covers the liner. I usually compromise by covering pond liners with only 6 inches of soil. Although the shallower covering is not ideal, it does supply some protection.

Covering pond liners with soil also provides a medium for plant growth. A host of edible and ornamental plants thrive in the soil atop a pond liner, giving the pond a natural appearance. A layer of soil on top of the liner also buffers water pH and nutrients to help improve water quality.

I recommend two different types of pond liners. For ponds that will have soil covering the liner, I purchase a product called Permalon directly from the manufacturer, Reef Industries. Permalon liners are lightweight and easy to put into place. For extra security I sandwich the liner between layers of carpet or geotextile fabric. For ponds in which I don't want to cover the pond liner with soil, I use a heavier liner product called EPDM, which withstands UV sunlight and wear.

A bit of advice: remove all rocks and debris the size of a pea and larger from the floor and sides of a pond before laying out the liner. It may seem tedious, but just as in the famous fairytale by Hans Christian Andersen, the liner is as sensitive as a princess.

If your pond has a drainpipe that passes through the pond wall, you'll need to install a fitting around the pipe that is especially designed for the type of pond liner you're using. The fitting allows the pipe to pass through the liner safely without creating leaks. Different pond liners require different techniques for installing a fitting, so check with the liner manufacturer, and be sure to have the proper fitting and instructions on hand. Permalon pond liners require a "pipe boot" and specialty tape to seal the opening around the pipe. EPDM liners require a plastic fitting with a rubber seal. If using clay to seal the pond, purchase or build a baffle plate to prevent water from seeping out along the length of the pipe.

Once the liner and pipe fittings are in place, it's time to apply soil over the top of the liner. Ideally, if machines were used to dig the pond, the same machine covers the liner with soil. I work from the outside toward the middle of the pond. This way the soil on the outer edge protects the liner as I work my way toward the inside. A thick layer of soil even protects the liner from the weight of heavy machinery, allowing the bucket of the tractor to enter the middle. If the liner doesn't have a layer of carpet or geotextile fabric over the top to protect it, be sure to sift any rocks or debris out of the soil (see chapter 8 for sifting techniques).

Filling the Pond

Sometimes we don't have the luxury of waiting for a newly built pond to fill up with water from natural sources. If you fill the pond with water from a local municipality, consider the composition of the water. Most municipalities use chlorine to sanitize the water. It's very important to remove chlorine from the pond prior to stocking it with fish. Two types of chlorine are used to treat water, chlorine gas and chloramine. If your municipality uses chlorine gas, simply let the chlorine evaporate for a few days before stocking. However, if chloramine has been used to treat the water, the pond will either have to sit for several weeks before stocking or you can use a dechloramination chemical to remove the chloramine.

When a pond that has a liner covered with soil fills up for the first time, the water may look murky or brown and dirty, especially if clay is present in the soil. After a week the clay settles out of the water, and the water clears. Mulching and planting the area around the edge of the pond helps keep the brown at bay by preventing incoming runoff from washing clay back into the pond. If the water hasn't cleared after a week, I apply 2 grams of gypsum for every 5 gallons of water in the pond to help settle the clay. If the clay doesn't settle after several days, I repeat the application.

To figure out how much gypsum to add, you need to know the total volume of water in the pond. You can use an online calculator or the following formula to figure volume:

$$V = [LW + (L - 2SD)(W - 2SD) + 4 (L - SD)(W - SD)] \times (D \div 6)$$

where

V = volume; L = length; W = width; S = side slope (example: 3:1 would be entered as 3); and D = depth

Pond Covers

While the old saying "a watched pot never boils" is not completely accurate, it does offer a bit of truth. The water eventually boils; it just requires a longer amount of time. But if a lid is placed on the pot — which inevitably prevents the watching — the water will come to a boil faster.

A pond is much like a pot of water, and though we do not want our pond to boil, a warm pond

Figure 2.20. A pond cover prevents evaporation and warms the water.

Figure 2.21. It's possible to use evergreen vegetation to screen a pond cover from view. Viewed here from a different vantage point, the pond cover is barely visible.

Pond Safety

Ponds inevitably attract the curiosity of children. Build ponds with shallow slopes to allow an easy exit. You may need to add a net or ladder on the edge of ponds to allow easy entry and exit. Floating pool covers also add a level of danger since children may think it's a solid surface. Fencing or cactus around a pond provides the most protection.

creates microclimates helpful in extending the growing season. Covering a pond may detract from its aesthetic appeal, but it serves beneficial functions when necessary.

Surrounded on all sides by impermeable material, a covered pond essentially becomes a giant tank. The cover halts evaporation. Without the loss from evaporation, a smaller watershed can preserve a larger reserve of water. Additionally, pond covers can raise the temperature by 10°F and create drastically different microclimates. I've used the warm water to flood-irrigate plants in basins in an effort to protect the plants from freezing. However, I measured air temperatures above the basin and above adjacent land and found no distinguishing difference. We do use the warm water to top off our greenhouse aquaponic systems to replace water drained or evaporated, thus saving on heating demand.

I use long-lasting polyethylene "bubble wrap" swimming pool covers. The tiny air pockets inside the plastic provide a layer of insulation to keep the water warmer. Clear covers allow sunlight to penetrate and heat water while stimulating algae growth for fish feed. While the cover does restrict oxygen exchange at the water-air interface, plants and algae in the pond still pump oxygen into the water during photosynthesis. Alternatively, if I desire clear water and I don't need fish and plants, I use a light-blocking cover to prevent algae growth. A light-blocking cover will also kill any vegetation growing in soil on top of the liner, which allows a quick, easy weeding if the pond growth gets out of hand.

In purely functional ponds maximized for irrigation, the reduction in evaporation could make covers useful year-round. In other ponds, seasonal use extends the growing season for fish and plants. While pond covers aren't pretty, their practicality makes them a valuable tool.

Fish for Small Ponds

Building a pond is like building a house. Once the backbreaking construction labor is done, it's time to decorate. But instead of selecting shades, flooring, and furniture, we're developing a sustainable ecosystem of plants and animals. Stocking the pond with fish is first on our to-do list. If we don't stock the pond within four or five days, we may have a major mosquito fiasco on our hands.

Growing a large amount of fish in a small pond is feasible only if the water is properly aerated and filtered. Small populations of fish can flourish in small ponds without elaborate filtration systems, and they will control mosquitoes and aquatic weeds while keeping the water clear (if algae-eating fish are stocked). If fish production is favored over aesthetics, unaerated algae-rich ponds can grow high densities of minnows, which serve as a sustainable feed for chickens and fertilizer for the garden.

Mosquitofish (*Gambusia* sp.) minnows are the prime candidates for stocking small ponds. Reaching a maximum size of less than 3 inches, the minnows thrive in small, isolated ponds barely larger than a puddle. Native to warmer areas of North America, the fish have now been introduced all over the world. They have even been described as the most widespread freshwater fish on the planet.

The worldwide introduction of mosquitofish has been a blessing for some countries and a curse for others. In South America and Russia the minnows have been used to eradicate malaria, a mosquito-vectored disease. In Sochi, Russia, a marshy city once plagued by malaria, a monument in the central town square commemorates the mosquitofish. Because of the introduction of the mosquitofish to Sochi, the city has been malaria-free since the 1950s. However, introduction of the minnows in Australia has threatened native fish and frog populations. Check with your local authority before releasing mosquitofish into your pond to ensure they're acceptable in your community.

Mosquitofish thrive in shallow ponds free of larger predatory fish. Practically indestructible,

Figure 2.22. Widely adapted and tough, mosquitofish are great minnows for small or large ponds.

they tolerate poor water quality and extreme water temperatures. The minnows subsist on natural growth in the pond and never need feeding or care, unless you want to increase the population size.

"What about algae and pond weeds?" is the second most common question I hear regarding ponds. While most small pond owners employ an arsenal of chemicals, filters, and pumps to keep their ponds looking clean, stocking a few plant-eating fish works just as well. Sterile grass carp, koi, goldfish, and tilapia can all control rampant algae blooms so ponds remain crystal clear.

A single sterile grass carp can keep most vegetation at bay in a small pond. Carp are hardy fish, capable of overwintering in any pond in the United States that doesn't freeze solid. Check with your local authority to ensure sterile grass

Predator Watching

One sunny afternoon I was sitting out by our pond in the backyard with my then eight-year-old daughter. We were reviewing her science homework, studying the basics of the animal kingdom. (As a biology major, I shine in this particular subject. Now that she's bringing home precalculus homework, I've had to move backstage.)

Just as we were discussing birds and fish, my daughter directed my attention toward the treetops. "Papa, look!" As I turned my head, I heard the shriek of two hawks combating each other in midair over the prey below. The loser flew off to find another fishing spot, but the winner swooped down to grab its prize. Though I lost a tilapia that day, my daughter watched the animal kingdom in action—a lesson she's never forgotten.

Small ponds attract all kinds of wildlife. I've looked out my window to see a great blue heron standing by my pond like a statue, hopefully hunting its next meal. I could predict precipitation better than the local meteorologist by listening for the chorus of frogs drawn to my pond for a romantic rendezvous. My daughter has watched the edge of our pond month after month as gelatin-like eggs hatched into tadpoles, grew teeny-tiny legs, and eventually hopped out of my pond to explore their new world. Once, we had so many little frogs hopping across our driveway that each step required conscientious care.

Predators are part of life. The old saying "build it and they will come" applies to ponds

Figure 2.23. Great blue heron. Photo courtesy of Scott Davis

as well. Birds, raccoons, turtles, snakes, frogs, and even domestic animals such as cats will gravitate toward a pond to feed on fish and frogs. Although fencing and nets will prevent some predators from accessing the pond, these devices usually diminish the pond's natural beauty. Instead, I place water plants along the edges of the pond so minnows and small fish find shelter in the shallows. Also, open-ended pipes, rocks, and hollow logs submerged in the deepest part of the pond provide refuge.

The easiest way to combat predators is to accept them as a part of the ecosystem. Seeing a bird swoop down and catch a fish might be worth the price of the fish, especially if it was free. Growing cheap, self-sustaining fish such as mosquitofish makes the loss of a few even more bearable. I say sit back and enjoy the show.

carp are accepted in your community, and never stock nonsterile carp. A nonaerated pond can hold about one 6-inch fish for every 5 square feet of pond surface area.

Tilapia are also great allies in weed control for small ponds. Native to tropical Africa, tilapia consume a diverse array of aquatic vegetation, including microscopic plants such as plankton. A single tilapia will keep a small pond clean. Multiple tilapia will reproduce, providing a bunch of small fish to use as a feed or fertilizer. Unfortunately, tilapia are tropical. Once water temperatures dip below 48 degrees, the fish will die. The fish slow down as temperatures drop, making them easy to scoop out with a net before they meet their chilly demise. Tilapia can be purchased yearly or overwintered in a warm greenhouse pond or indoor aquarium. For non-aerated ponds stock at a rate of one fish for every 5 square feet of pond surface area.

Koi and goldfish are popular for their beautiful bright colors. Related to carp, koi and goldfish will also eat aquatic plants and help keep a pond clean. Because of the expense of these fish and the risk of losing them to predators, I choose not to grow them. Any fish that's bright orange and white is simply a floating bull's-eye for every predator in the neighborhood.

Maximizing Minnow Production

Growing minnows is a great way to put a small pond to work. Chickens squabble over these little delicacies like wild beasts at a watering hole. If you don't have a flock of chickens, you can also use minnows as bait for fishing or as a fertilizer for your garden, or you can sell them for mosquito control. Minnows feed on microscopic zooplankton, algae, and a host of other aquatic organisms, turning unviable plant life into a highly valuable resource. If you want to get serious about raising minnows,

Figure 2.24. A fertile green pond grows more minnows, but if I can't see my hand when my arm's submerged up to my elbow, I know the pond has too much algae.

you have to forget about preconceptions with pond aesthetics. Maximum production of fish occurs in ponds with a dense algae bloom. My aquaculture mentor called it "pea soup green."

Algae are the green plants that coat the bottoms of ponds and submerged rocks. A unicellular form is called phytoplankton. Long, thin strands of filamentous algae give the pond a green color, like new grass covering a freshly mown field. Algae pump oxygen into ponds during daytime photosynthesis, then the fish "breathe" the oxygen through their gills. At nighttime the algae removes oxygen from the water through respiration, so too much algae may cause problems with low oxygen in the morning. I like to refer to it as an algae hangover — sometimes too much isn't a good thing.

Zooplankton and minnows feed on algae the way cows graze a field of grass. You can keep a pond looking clean and clear by stocking tilapia or sterile grass carp to eat vegetation, but for maximum production of minnows or any other fish, green is the color of choice.

To achieve an algae bloom in the pond, fertilize when water temperatures reach 65°F. Ideally, use a liquid fertilizer high in phosphorous to stimulate the algae bloom. Fortunately, we are blessed with a free shot of liquid fertilizer several times a day — our urine. Urine is sterile (unless you have a bladder infection) and perfectly safe to use as a fertilizer concentrate (if you're not taking toxic pharmaceuticals). Unless you live out in the country, or have minimal inhibitions, peeing straight into your pond may not be a viable option. I know several people who collect their urine in the privacy of their bathroom and pour it into the pond just as they would a store-bought liquid fertilizer. Simply adding several shots of urine to your pond triggers a chain of events that begins with an algae bloom and leads to an abundance of minnows to feed to

the chickens. Alternatively, a shovelful of chicken manure or a handful of commercial fertilizer will do the trick. Also, when fish are fed, the extra fish manure generated by the addition of feed will fertilize the pond and stimulate an algae bloom.

However, if your goal isn't production and you desire a clean, clear pond, avoid feeding the fish, keep nutrients out of your pond, and definitely don't pee in your pond. A variety of aquatic plants placed in and around the pond will help absorb excess nutrients, keeping the pond algae-free.

Harvesting Minnows

Our farm is a field trip destination. Teachers love our hands-on approach and return year after year

Figure 2.25. This fulcrum makes it easy to raise a drop net to harvest minnows, and it entertains kids for hours.

with their students. Inevitably, the kids gravitate toward a turkey nest pond situated on the south side of our farm building. Sitting aboveground with a stonework ledge, the pond is like a stage drawing the children's curiosity. It doesn't take them long to discover the drop net.

The drop net attaches to a 10-foot metal galvanized pole. Balanced on a fulcrum atop a 4 × 4 wooden post, the pole moves like a lever, dropping the net in and out of the pond. I can't take credit for the design; ancient Egyptians developed the fulcrum technique to draw water from the Nile River. However, I use it to harvest minnows.

Minnows prefer to congregate in shallow, warm, sunny areas of ponds protected from wind. Designing ideal minnow microclimates next to pathways and access points makes harvesting easy. I simply leave the drop net submerged in the minnow microclimate. If I feed the minnows, I place the feed over the drop net so they grow accustomed to feeding in that area. When I want to harvest minnows I pull down on the fulcrum to raise the net and catch the fish while they're feeding or lounging in the warm microclimate. Someday we'd like to set up a gumball machine with fish feed. The

Figure 2.26. A typical catch of minnows using the drop net.

kids will buy the feed to catch the minnows, then feed the minnows to the chickens.

Another way to catch minnows is by using a dip net. A simpler design, this net resembles skimmers used to remove debris from swimming pools, but the dip net is stronger. I lower the dip net into the water a few feet from shore and pull the net back toward the edge of the pond, scooping up minnows as they try to swim away.

I remember reading an agricultural magazine that described aquatic chickens capable of wading out into shallow water and submerging their heads to hunt schools of minnows. Another story recalls chickens raised with ducks that would dive for minnows. The fabled aquatic chickens hunting for minnows inspired me to develop yet another technique for minnow harvesting. I usually start the process at a time when it hasn't rained for several weeks and I need water from the pond for irrigation. Using my movable electric chicken fence, I completely surround the pond with fencing and release the chickens into the fenced area. Next, I drain all the water from the pond to irrigate the landscape. As the water level drops, the chickens start pecking at all the aquatic snails and organisms stranded on the exposed pond floor. Once the water level recedes further, the minnows start thrashing about. The chickens notice the commotion, and a feeding frenzy ensues that is reminiscent of grizzlies catching salmon swimming upstream. With time the chickens improve their minnow-catching skills and wade into deeper water.

A slightly more humane approach is to place the drop net at the low point in the pond. When the water is drained the minnows can be harvested as they congregate over the net, which saves them from suffering a slow death in the shallows.

Whatever harvesting method I use, I find that a few baby minnows usually survive the harvesting process, and thus restocking isn't necessary.

Figure 2.27. I'm preparing to make a grand scoop with the dip net to catch minnows. Photo by Stephanie Jadrnicek

However, saving a few minnows in a bucket or bringing some from an adjacent pond will speed up repopulation.

Harvesting mosquitofish is a seasonal adventure. During summer the population booms, then declines drastically through winter. Once pond temperatures warm, the fish become active, and breeding begins. Allow the fish to have at least one brood in spring before harvesting begins. The females have up to nine broods per year, starting in midspring and continuing through midautumn. Each brood contains between five and one hundred offspring born live. Summer is peak harvest season for minnows. All harvesting should cease by late summer to let the population rebuild before winter. Mosquitofish go dormant during winter and

seem to disappear because they sit on the bottom of the pond.

A note of caution: Never give animals access to your entire pond area long enough to denude the area of vegetation. If chickens are allowed to completely remove the vegetation along the edges, water quality will quickly decline, and erosion will ensue. Limit animal activity in pond areas from a few days to a few weeks, depending on stocking densities.

Growing Tadpoles for Chicken Feed

In nature the tables turn quickly. A strong, vicious predator may have once been vulnerable young prey. So goes the cycle of life.

Mature frogs eat small fish, but tadpoles (baby frogs) are tasty treats for chickens. Large

Figure 2.28. Chickens temporarily fenced around a small pond eat minnows after the pond has been drained.

Figure 2.29. The drop net holds a typical harvest of tadpoles from a small pond.

populations of minnows tend to keep tadpoles away as they eat frog and toad eggs, nip at the tails of tadpoles, and compete for food resources. After a pond has been drained of water and chickens have eliminated minnows, the stage is set for tadpole production. Once the pond is refilled with water, tadpoles will predominate while the minnow population rebuilds.

Similarly to minnows, tadpoles flourish on a diet of algae in nutrient-rich ponds. I harvest tadpoles using the same techniques I employ for harvesting minnows, by using a drop net or by draining the pond. The drop net is useful for a partial harvest, which ensures that some tadpoles will remain in the pond to grow to adult stage and populate the landscape, ridding the area of plant pests and reproducing to supply the next round of tadpoles.

Tadpoles represent a huge food source for chickens. However, some may contain toxic alkaloids in their skin, so I advise caution. The toxins are species specific — toad tadpoles are more likely to contain toxins than frog tadpoles. Identifying particular species of tadpoles is difficult; however, observing to see what type of adult amphibians live in and around your pond may present a clue to the tadpoles' origins and offer you enough assurance to feel safe using the tadpoles as chicken feed. Alternatively, boiling the tadpoles will remove the toxins — and my hens happen to love baby frog stew.

Selecting Edible Plants for Ponds

Reflected light creates special microclimates around a pond. Plants rooted by a pond enjoy the benefits of this temperature-buffering effect — warmer in the winter, cooler in the summer. I have witnessed an additional four weeks tacked on to a growing season from a pond's microclimate. Conversely, I have successfully grown hops and other cool-climate plants in the heat of a southern summer by moderating microclimates with ponds. In essence, the pond creates a miniature

Mediterranean environment in which water helps moderate temperatures and nearly all plants can thrive. I focus on edible ornamental plants to maximize the functionality of small ponds. All of the plants discussed below are perennials except watercress, a self-seeding annual with pretty yellow flowers and nutritious leaves.

To maximize the potential of your pond, I recommend experimenting with the following microclimates:

- Free-floating plants on water surface
- Submersed plants growing completely underwater
- Plants rooted to bottom of pond but floating
- Plants growing from roots fixed in sediments on bottom of pond with tops of plants emerging above water level
- Marginal plants in areas fluctuating between submerged and dry
- Plants in wet areas next to ponds

You'll also want to experiment with the microclimates that exist a few feet from the edge of the pond. These areas are warmer in winter and cooler in summer than the surrounding landscape. And the sloping sides of berms create special microclimates, too:

- South-facing slopes are warmer in winter and drier year-round.
- North-facing slopes are cool in winter and stay moist. Areas on the north side of a pond catch reflected winter sunlight useful for ripening late-fruiting plants.
- East-facing slopes dry quicker in the morning. This effect is useful for disease suppression in fruits.
- West-facing slopes are hot and dry during the summer.

Plants for Wet Areas

The wet areas on the edges of ponds create a self-watering nexus for low-care edible plants.

Groundnut (*Apios americana*) is native to North America and is found as far north as Canada and as far south as Florida. The 6-foot vine usually grows along streams, tolerating the fluctuating wet and dry conditions. Groundnuts have edible tubers that are high in protein and a taste that resembles a cross between potatoes and peanuts. A staple of Native Americans and early colonists, these little tubers possess a bitter latex, and they require cooking before consumption. The plant's pretty maroon flowers give way to edible bean pods that are cooked and eaten when small. Considered a low-maintenance plant, groundnut is also a nitrogen fixer.

Groundnuts flourish along the edges of ponds and north slopes. A good spot to plant groundnuts is in the soil covering the pond liner but above the water level. The liner keeps the tubers near the surface where they're easy to harvest. If soil isn't used to cover the liner, the plants themselves form a protective flap along the upper edge of the pond. Since the flap of roots and soil cannot adhere to the pond liner, simply lift up the flap and pull out the visible tubers hidden under the detritus.

Jerusalem artichoke (*Helianthus tuberosus*), a relative of sunflowers, grows to a stout 6 feet or more. The plant peaks with a showy display of small sunflowers, then quickly declines with age. A groundnut vine, if planted alongside the Jerusalem artichoke, will climb the sunflower-like stalk as if it were a trellis, concealing a decomposing mess in a jungle of leafy green. The tasty tubers of the Jerusalem artichoke are eaten raw, cooked, or pickled. I recommend caution — they may cause flatulence.

Wild ginger (*Asarum* sp.) is a group of native and foreign plants growing in the shaded understory

Figure 2.30. During late summer, the crimson-colored groundnut flowers bloom and then bear edible pods.

Figure 2.31. A patch of Jerusalem artichokes growing in front of a persimmon tree.

The Three Brothers

Most growers have heard of the Native American "three sisters" garden. According to Iroquois legend, corn, beans, and squash planted together sustain a synergistic relationship. The corn provides the stalk for the beans to climb, the beans fix nitrogen, and the large squash leaves shade out weeds. Native Americans also practiced a lesser-known tradition — the perennial "three brothers" garden. Jerusalem artichoke replaced the corn, groundnut (*Apios americana*) replaced the beans, and wild ginger (*Asarum* sp.) replaced the squash.

All three plants prefer the moist habitat around ponds and furnish food for years to come with little care. However, "little care" means they are nearly impossible to eradicate. Once they find a good spot on the backside of a wild pond, the three brothers can become quite unruly.

Figure 2.32. Spiderwort.

of forests. The wild ginger covers the ground in the "three brothers" trio (see sidebar), preventing other weeds from taking over, especially during the winter. Though the roots resemble ginger in flavor and Native Americans used them, modern science indicates they contain toxic compounds. The showy leaves keep the landscape green when the other two brothers are dormant.

Spiderwort (*Tradescantia virginiana*) is a common garden plant and native of the East Coast of North America that thrives in the moist area around ponds. This low-growing grasslike plant has edible showy purple blossoms that beautify any fresh salad. I've also added the leafy greens to stir-fry dishes, cooked slightly to tenderize.

Hot tuna (*Houttuynia cordata*), an aggressive groundcover, will quickly take over wet areas. The leaves have a strong flavor reminiscent of citrus and fish. Also a useful medicinal herb, it acts as a lung tonic and more. Make sure buildings, roads, or pathways enclose the pond area to contain this prolific plant. Otherwise, it will spread indefinitely.

Horseradish (*Armoracia rusticana*) grows freely in sunny, wet areas. This 3-foot-high perennial produces roots used to make horseradish sauces and preserves. If not harvested yearly, this aggressive plant will spread.

Watercress (*Nasturtium officinale*), a hardy European perennial, self-seeds freely around the edge of ponds. The pungent mustard-flavored leaves add heat and nutrients to salads and stir-fries.

Figure 2.33. A prolific ground cover, hot tuna grows in and around these yucca plants.

Figure 2.34. Horseradish (foreground) and Jerusalem artichoke make great companions. The horseradish grows in early spring before the chokes have a chance to take over.

Daylily (*Hemerocallis* sp.) is a perennial plant grown widely for its showy flowers. Also edible, these flowers add spectacular decor to salads and can be stuffed with dried fruit or sweetened cottage cheese for an eye-catching dessert.

Emergent Plants

Emergent plants take root in soil in the shallow water at the edge of a pond and can tolerate fluctuating water levels.

Katniss (*Sagittaria latifolia*) is infamous because the heroine of *The Hunger Games* trilogy bears its name. Native to North and South America, katniss prefers shallow, wet areas but also tolerates the fluctuating water level on the edge of a rainwater-harvesting pond. Reminiscent of stomping grapes to make wine, harvesting of katniss tubers is done with bare feet. Simply sink your toes into the mud

Figure 2.35. Katniss.

Figure 2.36. Duckweed.

along the edge of the pond and wiggle your feet around; the tiny tubers will soon float to the surface. When roasted, they have a taste and texture similar to potatoes and chestnuts — but be sure to remove their bitter skins.

Pickerelweed (*Pontederia cordata*) is native to North and South America. This shallow-water plant helps clean the pond. The young tender leaves resemble spinach in flavor, and the purple flowers make a striking presentation (see Figure 2.38).

Floating Plants

Duckweeds (subfamily Lemnoideae) are small floating plants that coat the surface of still ponds, resembling a lawn of freshly mown grass. A friend of mine, thinking it was solid ground, once fell into a pond coated with duckweed. Duckweed flourishes in shady, fertile ponds and serves as a great feed for chickens. I either scoop the plant with a net or simply give the chickens access to a small edge of the pond where they'll readily forage

for duckweed. Chinampa farmers used duckweed to fertilize garden beds adjacent to canals. Tilapia also eat duckweed and will slowly eliminate it over a growing season. Ponds undersized in relation to the water catchment area will have a difficult time growing duckweed, as it's purged out of the pond with every rain.

Plants for South and West Slopes

Prickly pear (*Opuntia* sp.) comprises a group of about two hundred cacti native to the Americas. These plants thrive on the sunny, dry, south- and west-facing slopes around ponds. The warm microclimate pushes the hardiness boundary for some of the choice edible varieties of prickly pear, such as *Opuntia ellisiana* (usually hardy only to zone 9) and allows them to thrive in zones 8 and 7. I grow some prickly pear for the spineless edible pads and some for the watermelon-strawberry-flavored fruit. Since prickly pears are evergreen, they provide crucial winter interest to the garden.

Figure 2.37. A prickly pear cactus thrives on the well-drained and warm south slope of a raised pond. An uncommon companion, the watercress with yellow flowers sits adjacent to the prickly pear on the wet edge of the pond.

Harvesting the fruit is a great excuse to build a fire in the outdoor fire pit. Burn off the spines, and place the fruit in 1-gallon ziplock bags for freezing. Unfortunately, the rock-hard seeds in the fruit are entangled with the pulpy goodness. To navigate this problem, simply add the whole fruit to smoothies. Once blended, the seeds fall to the bottom and the pulpy juice pours off the top.

Adam's needle (*Yucca filamentosa*) looks tough but is actually quite gentle. Since the soft leaves bend on contact, the spines are no threat. Another useful evergreen, it has stunning white flower spikes that reach 8 feet tall, bearing clusters of edible flower petals. Reminiscent of artichokes, the petals are used in many Hispanic dishes.

Figure 2.38. Another unlikely set of companions, white-blossomed Adam's needle sits on the well-drained edge of a pond with hot tuna covering the ground beneath it and purple pickerelweed thriving in the pond shallows.

Plants for East and North Sides

East-facing slopes dry quickly in the morning, reducing disease-causing moisture. Blackberries and raspberries find a nice home on east slopes next to ponds. Water acts as a natural barrier to the spread of brambles, keeping the sprawling plants contained.

Reflected light shines up onto the north edge of a pond during winter. Ideally, a home or building catches the reflected winter light. Alternatively, evergreen fruiting trees and shrubs that are marginally hardy for the climate can be grown in this area. I've planted pineapple guava (*Acca sellowiana*) and hardy citrus to take advantage of this special microclimate.

Evergreen pineapple guavas (which are not true guavas) are hardy to about 12°F. They've withstood dips to 9°F on the northern side of a pond at the Clemson farm. The reflected light also helps ripen the fruit in October. Similarly, a whole host of hardy citrus would also suit the special microclimate on the northern side of ponds.

I've found that deciduous fruit trees perform poorly when planted on the northern sides of ponds. The warm microclimate and reflected light may cause the trees to leaf out early, making them susceptible to late winter freezes. We planted ten Asian persimmons around the farm in various locations. The Asian persimmon is marginally hardy in our area and is known to leaf out early, which makes it susceptible to subsequent cold spells. Sure enough, the Asian persimmon on the north side of the pond was the only tree to perish during a late cold spell. I suspect the warm microclimate caused the tree to leaf out earlier than the rest of the trees we planted.

In contrast, an evergreen Meyer lemon tree I planted at my grandmother's house bounced back from temperatures as low as 19°F. Normally, this tree is hardy to only 20°F. But sandwiched between a brick wall and the north side of the pond, the lemon tree survived the bitter cold nestled in a modified microclimate.

Floating Transplant Trays

The buffered microclimate of a pond presents an ideal setting for starting seedlings. Seed trays, floating on rafts on the surface of the water, wick moisture from the pond and eliminate the need for irrigation. I use Winstrip brand seed trays, which have air slits in the sides to promote aeration.[4] Alternatively, you can use floating speedling trays presented in chapter 12. The floating speedling trays have issues with roots hanging in the water, but it's not a problem if plant-eating fish prune them for you.

I fill the trays with a lightweight potting soil mix containing 50 to 75 percent perlite—50 percent more perlite than I use in my standard seed-starting mix (described in chapter 8 under Compost Potting Soil Mixes). This not only lightens the load for the floating trays but also provides additional aeration, since the plants are sitting in a pool of water. Next, I plant my seeds and fully saturate the soil by placing the trays in the water at the edge of the pond.

I build floating rafts out of ¾-inch-thick extruded polystyrene and lay the saturated trays atop the rafts. By placing rocks on the raft, I can cause the raft to sink a bit, until the bottoms of the trays are submerged by about ⅛ to 1/16 inch into the water. I adjust the weight of the rafts as needed. Pond plants usually prevent the floating rafts from ending up beached along the edge; if not, upright stakes protruding out of the water will keep the rafts in deeper water. Caution: Hammering stakes into the pond bottom would puncture the liner. Instead, I place the stakes in pots filled with soil, then place the pots into the pond. Alternatively, you can anchor it to a concrete cinderblock on the bottom of the pond.

Figure 2.39. Transplant trays float on Styrofoam rafts, wicking moisture up from the pond in the perfect microclimate for seed starting.

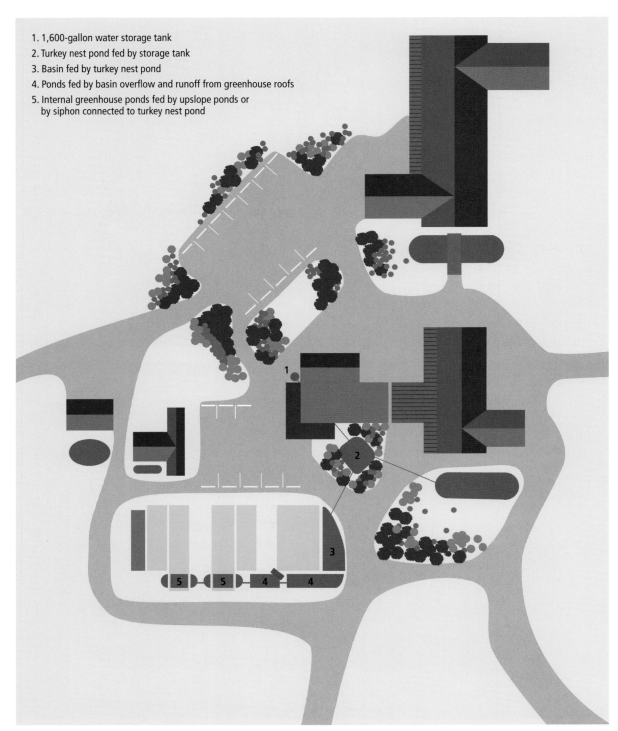

1. 1,600-gallon water storage tank
2. Turkey nest pond fed by storage tank
3. Basin fed by turkey nest pond
4. Ponds fed by basin overflow and runoff from greenhouse roofs
5. Internal greenhouse ponds fed by upslope ponds or
   by siphon connected to turkey nest pond

Figure 2.40. Water flows through the Student Organic Farm from building roofs through tanks, ponds, and basins, slowly making its way through the landscape.

Beyond the Pond

The turkey nest pond at the SOF is just one component of a larger water supply system, as shown in figure 2.40. Rainwater from a building adjacent to the pond first fills a 1,600-gallon tank, which is located at a higher elevation than the turkey nest pond. When the tank fills, the overflow enters the turkey nest pond. The water in the tank also responds to a float valve to keep the pond filled between rain events. Having the extra water reserve uphill extends the amount of time the pond stays full using natural rainfall. A rule of thumb applicable in most situations is that it's always a good idea to have backup.

When the pond fills and overflows, the overflow can be diverted downhill to a basin (I explain what a basin is in the next chapter). Overflow from the basin feeds into ponds located at the south end of a series of greenhouses (described in chapter 4). Rainwater runoff from the greenhouses also fills these ponds. Another component of the system is two ponds located inside greenhouses (described in chapter 5). These ponds can be filled from basins adjacent to the greenhouses or directly via a siphon from the turkey nest pond. Thus, the pond itself — with its minnows and surrounding plantings — is a bio-integrated system that is also part of a much larger bio-integrated system, all connected by water, slope, and gravity.

SUMMING UP THE FUNCTIONS

1. Pond provides sunny place on the south side of a building.
2. Overflow from cistern flows into pond.
3. Water from cistern fills pond when it's shallow, using gravity and float valve.
4. Pond produces minnows to use as fertilizer, feed, and bait.
5. Pond serves as holding area for tilapia and prawns before sale, transport, or use.
6. Solar reflection from pond creates fourteen different microclimates, allowing cultivation of wide range of edible plants.
7. Diversity of microclimates produces large variety of plants for propagating nursery stock.
8. Pond reflects light into adjacent building during winter and provides passive solar heating.
9. Pond cools air in summer through process of evaporative cooling.
10. Pond stores water for mushroom-log soaking and nursery plants (see chapter 3).
11. Fish waste in pond water provides fertilizer for nursery plants (see chapter 3).
12. Predators of plant pests find habitat at pond.
13. Pond supplies solar-heated water to greenhouse nursery ponds.

All About the Basin

The Bio-Integrated Basin

Flooding shallow basins planted with crops is one of the oldest forms of irrigation. Dug with a simple spade, basins provide an easy and inexpensive way to use stored water.

A basin is a shallow depression in the ground, no more than 6 inches deep. Though it's level like a pond, it serves a different purpose — a pond holds water, but a basin allows water to infiltrate

Figure 3.1. This small basin adjacent to a greenhouse is lined with landscape fabric. Nursery plants perched on the basin floor soak up water. Seedlings on the benches are hardening off. The shade cloth draped over the bench protects mushroom logs underneath.

slowly, because a basin has no impermeable clay or manufactured liner. We design basins so water will evenly spread across their surface, then soak into the soil. Because of this, a basin is the perfect place to irrigate plants or mushroom logs.

Bunds are to a basin what berms are to a pond: small earthen embankments that contain water. In flat areas bunds completely surround a basin. But on sloping land the bund is only necessary on the downhill side, and the uphill side is excavated below grade.

As I described in chapter 2, at the Student Organic Farm (SOF) at Clemson, a turkey nest pond captures harvested rainwater and discharges the excess water into a basin next to the greenhouse. We keep potted nursery plants in the basin, and when the pond floods the basin, our nursery plants are watered. Since there are fish in our pond, the plants also wick up a rich dose of nutrients (from fish manure) along with the water.

Ancient Basins and Modern Basins

One of the first irrigation techniques in recorded history originated in ancient Egypt on the floodplains of the Nile River. During the monsoon season the Nile produced a predictable annual flood-and-retreat cycle in the headwaters of the Nile around Lake Victoria. Beginning in July and continuing through August, floods spread into the lower reaches of the river.

Basin irrigation was based on raised earthen berms surrounding leveled fields adjacent to the Nile River. When floods arrived, the raised berms restrained the floodwater and allowed it to soak into the soil of the basin. A gridwork of canals fed and drained the water from one basin to another. The Egyptians extended the entire flood cycle by draining water from upper basins into lower basins before allowing it to return to the river. As the water retreated, they planted crops in the saturated fields that would grow through the mild winter. The roots of the plants would follow the retreating floodwaters deep into the soil, which kept them alive until April and May harvests.

Nutrients from the highlands of Ethiopia were carried downriver with the floodwaters, and those nutrients settled out in the basins to fertilize the crops. The irrigation system was simple enough to control on the local level, free of the vagaries of war and politics. With basin irrigation the Egyptians improved upon the natural hydrologic and nutrient cycles of the Nile, allowing their civilization to become the breadbasket of the Roman Empire. Providing proof of the antiquity of this technique, a relief on the head of a five-thousand-year-old mace (a heavy club) depicts the Egyptian king Scorpion with a hoe cutting a grid pattern into the landscape.

The modern bio-integrated basins I've designed build upon the irrigation pattern created by the ancient Egyptians. The rooftops of buildings serve as the headwaters of the watershed, much like the Ethiopian highlands form the headwaters of the Nile. Instead of large lakes such as Lake Victoria receiving water to feed into the Nile, small

Elements of the System

Learn how to use an ancient technique, with modern modifications, to irrigate your plants and mushroom logs. With the turn of a knob, your stored rainwater floods into bio-integrated basins and distributes itself evenly over the surface of the soil. Easy and inexpensive, these basins provide an alternative to pricey irrigation systems. This chapter shows you how to:

▶ Save time and money by using rainwater stored in ponds to irrigate and fertilize potted nursery plants and inground plants in basins
▶ Create multidepth basins to handle diverse irrigation needs
▶ Produce shiitake mushrooms in a labor-efficient way by flood-irrigating inoculated logs in a shaded area in a basin
▶ Size basins based on available water storage

bio-integrated ponds store the falling water caught by a rainwater-harvesting system. Water from the ponds can be drained into small basins, mimicking the monsoons that cause the flooding and draining cycle of the Nile.

The Egyptians grew crops directly in the soil of their basins, but in the bio-integrated basins I've designed, the flooded basin is used to irrigate mushroom logs and containerized plants. Benches set up in the basins provide a shaded shelter for the mushroom logs and also serve as a surface for propagating and hardening off plants. Additionally, solar-heated water in basins adjacent to greenhouses helps modify the microclimate, promoting warmer temperatures and season extension.

I first experimented with constructing basins when I was in my early twenties. Nestled in the Santa Cruz Mountains of California with my new budding family, I ran a small grassroots nursery. I cultivated over a hundred different varieties of culinary and ornamental salvias—from the more common pineapple sage and *Salvia leucantha* to the rarer *Salvia africana* and *Salvia gesneriiflora*.

The nursery plants grew in the shade of fruit trees my father had planted long ago—peaches, plums, apricots, and pineapple guavas. Looking back now, it seems only natural that I started my first nursery and fostered my first irrigation ideas beneath the protective canopy of his orchard.

Because there was no rainfall from late spring through autumn in that region, I had to become innovative with irrigation techniques. I decided to irrigate from the bottom up, which meant the plants' leaves stayed dry and disease resistant.

In hopes of conserving precious California water, I placed semipermeable fabric on the sloping ground around the fruit trees. Impervious to weeds but porous enough to allow some water to seep through, the fabric was perfect for lining a series of basins with which to irrigate my nursery stock. I positioned the potted salvia plants on the fabric, in the dappled light of the fruit trees. Then I turned on the spigot, and as the water filled the fabricated basins, the salvias wicked up the water through their roots. The remaining water overflowed to the next basin downslope to irrigate another batch of salvias.

I hadn't started experimenting with rainwater harvesting back then (and arid summers don't present the best conditions for harvesting rain). Instead, I used well water, conveyed through irrigation tubing. Although the climate was

challenging, the terrain was accommodating. The slight slope of the hill passed the water from basin to basin, employing the physics of gravity to do most of the work.

I've built many basins since then and discovered the benefits of rainwater-harvesting ponds as the water source for basins. My California system was labor saving, but by incorporating rainwater-harvesting ponds, I now save myself the cost of paying for municipal or well water. I locate my ponds uphill from the basins, yet as close as possible to limit the expense of running pipe.

This chapter focuses on homemade earthen basin construction using passive techniques to fill and drain the basin. However, prefabricated nursery basins designed to flood-irrigate potted plants inside greenhouses are commercially available. The floodwater drains into a sump, then a pump returns the water to an upper reservoir to prevent waste. You can find instructions and parts for these systems online.

Connecting Basins to Greenhouses

The basins at my California nursery served a limited purpose of irrigating plants. But when basins and greenhouses go hand in hand, exciting possibilities for multiple functions open up.

A basin filled with solar-heated pond water modifies the microclimate surrounding a greenhouse. Like giant solar panels, small ponds collect the sun's energy during the day in the form of heat. At the SOF, when I drain the ponds into the basins in the evening, the pools of water release the sun's energy as heat, creating a warm buffer around the greenhouse overnight. Ideally, a small amount of water remains in the basin the following morning to reflect sunlight into the greenhouse.

Basins also serve as the perfect transition spot between the greenhouse and the field. Tables placed in basins become extensions of greenhouses. When the temperature inside a greenhouse rises too high for plant propagation, the basin table becomes the logical location to move plants to, since it's right outside the greenhouse. I also start seedlings in the greenhouse, then transfer the young plants to the basin bench to "harden off" before transplanting them into the field. The area underneath basin tables also provides the perfect shady spot for protecting logs inoculated with edible mushrooms.

Basin Basics

Building my first basin was like preparing a recipe for the first time. I had the right ingredients, and I'd read the instructions, but the results were not perfect. Experience was essential to refine the process. Every basin requires a different design. Size and shape depend upon the size of the pond serving the basin, the recharge rate of the pond, the soil type of the basin, the slope of the land, and your aesthetic tolerance to low pond levels, as summarized in table 3.1. In general, I start with a basin the same size as the pond, then add or subtract proportions depending on the environmental constraints.

Factoring in Water Table and Soil Type

I once thought soil type was the main environmental factor to consider in determining basin size, but I've since realized that water table is equally important.

The coastal plains of South Carolina are called the Lowcountry for a reason: The land lies only

Heat Exchange Experiments

I've experimented with insulated covers to capture more heat in pond water (see figure 2.20). In trials at SOF we examined climate modification using water from a small pond before and after solar heating with a clear pool cover. The solar pool cover placed over the pond increased the pond water temperature by approximately 10°F. I hoped that flooding this warm water into a basin on a cold night would modify the temperature for seedlings growing in flats inside the basin. I measured this by placing temperature sensors on top of tables inside the basin and on top of seedling flats just above the water surface. I also placed additional temperature sensors outside the basin at the same height above the ground in areas without floodwaters. To expand the experiment I placed the temperature sensors over and under the row covers.

The results unfortunately indicated the warmed water flooded into the basin didn't protect plants in the basin from cold temperatures. Wind may have contributed to the lack of cold protection, blowing any heat released from the water away from plants. More research is necessary to determine the effectiveness of warmed water for cold protection.

a few feet above sea level. When I dug a basin in Jasper County, South Carolina, the heavy clay soil in combination with a high water table created an instant pond. But I wasn't building a pond for water storage; I was constructing a basin for water infiltration. So the effect was the exact opposite I was shooting for. I've learned it's best to consider both factors when designing basins.

In areas with a high water table, digging a basin even 1 foot deep will create a shallow pond or wetland. I recommend digging a hole and conducting a simple percolation test to help identify a problematic situation. Simply dig a test hole 1 foot deep, then fill the hole with water. When the soil is saturated, it should drain at a rate of 0.3 inches per hour; if it doesn't drain at this rate, then the area may not be suitable for an irrigation basin. Also keep in mind that in some areas water tables fluctuate with the seasons.

In addition to percolation, soil type will determine basin performance. The three basic soil types are sand, clay, and loam. Soil types are like people: Each type of soil has its strengths and weaknesses, so it's best to know how to work well with all three.

Sandy soils absorb water quickly. In a sandy soil area, water will sink into the ground before it has a chance to spread throughout a basin. For example, by the time a wave laps onto the beach and recedes back into the ocean, the sandy shore has already begun to dry. In basins sandy soil promotes good drainage; however, the trick is in retaining the water long enough to irrigate the plants. When I'm creating a basin in sandy soil, I use a large-diameter pipe to fill the basin, and I limit basin size. The large pipe swiftly floods the basin with a greater volume of water, and the basin fills quickly because of its small size. It's like filling a baby pool with a fire hose instead of filling an inground pool with a water hose. The goal is retention in sandy soil basins, because the water will drain naturally.

Table 3.1. Basin Size Based on Environmental Constraints

Basin Size	Small	Medium	Large
Sandy Soil	X		
Clay Soil	X	X	X
Loam Soil	X	X	X
Steep Slopes	X	X	
Flat Areas	X	X	X
Dry Climate	X		
Wet Climate	X	X	X
High Tolerance to Low Water Levels	X	X	X
Low Tolerance to Low Water Levels	X		

Note: This table is based on the assumption that a medium-size basin has a surface area equal to that of the supply pond. A small basin would have only half the surface area of its supply, and a large basin would have twice the surface area of the pond.

Basins aren't just for pots and mushroom logs. Long ago, before the advent of the plastic pot or the development of mushroom cultivation, the Egyptians discovered the best use for basins — grow the plants right in the ground. If you're planning to grow plants directly in a basin that has sandy soil, there's no need to choose varieties tolerant of wet areas. Most vegetables and plants will grow successfully in well-drained sandy soil.

Heavy clay soils hold water longer. I adapt to heavy soils by growing leafy vegetables such as chard and basil that are more genetically tolerant of moist areas. A tried-and-true technique for dealing with clay soil is raising the plant above the soil level. Either I pot the plant and bury the pot halfway into the soil or I place the plant into a raised mound.

The roots of the raised plant can still wick water when the basin is flooded, but they won't be left sitting in water too long, drowning from lack of oxygen, as the basin slowly drains. (Some vegetables may die from a lack of oxygen — a condition called "scalding" — if their roots sit in pooled water for several days.) So if you're not growing rice in your basin, I advise raising the plants for better drainage in heavy soils.

If soil types were fairy tales, loamy soil would be Goldilocks's favorite. While clay holds water for too long and sand dries out too fast, loam drains just right. Ideal for basin irrigation, loam allows water to spread swiftly but also prevents ponding that damages plants' roots. The water-holding capacity of loamy soil conserves water, but I recommend avoiding loamy soils that have a tendency to crust or crack. Although it's fantastic to find the perfect soil for basins, it's advantageous to learn how to adapt to any.

If your soil doesn't percolate fast enough, but you're determined to have a basin, there is a last resort. Excavate another 18 inches deep at the bottom of the basin and install an underdrain of perforated pipe surrounded by three inches of gravel. Replace the removed soil with a sandy loam to cover the underdrain and make sure the pipe slopes downhill to an open outlet. Remember that tree roots can easily clog the pipe perforations and might render the drain useless with time.

The Slope Factor

When I lived in the Lowcountry of South Carolina, where the land is flat as far as the eye can see, I thought the lack of slopes would be a disadvantage.

Figure 3.2. Water floods from an uphill pond into this narrow irrigation basin on a steep hill. Grains or vegetables may be grown directly in the irrigated soil of the basin.

Table 3.2. Basin Width Based on Slope Angle

% Slope	Maximum Basin Width (feet)
1	50–100
2	16–50
3	16–32
4	10–26

Source: C. J. Brouwer. *Irrigation Water Management: Training Manual.* Rome: Food and Agriculture Organization of the United Nations, 1985.

However, I soon realized that flat land is a blessing in disguise, because basins need a level bottom. So flat land makes constructing basins easy, because very little soil needs to be moved to create a level basin. There's no restraint on size, either, so basins can be broad, capable of retaining a large quantity of water.

Steeply sloped areas limit the shape and size of basins. On a slope I design small, narrow basins that follow the contour of the land to reduce the amount of soil I have to move to level the bottom of the basin. The best method to use in steeply sloped areas is either terracing or forming steps in the land, with the length of the basin on contour. The steepness of the slope will ultimately determine the maximum width of the basin, as shown in table 3.2.

How Much Water?

The size of the supply pond dictates the amount of water available for basin irrigation. A 1-inch-deep pool of water in a basin should supply a reasonable amount of irrigation for plants grown in the soil of the basin or submerged pots. Since some of the water that enters a basin will be absorbed into the soil before a pool forms, basin irrigation has an average efficiency of 60 percent. Therefore, to fill the basin with 1 inch of water requires 1.67 inches of water or more. The actual amount of water will vary depending on soil type. Complex engineering solutions are available to determine the amount of water and flow needed, based upon the advance rate of the water, to fill a large basin. Consult an engineer for these types of projects. However, for small basins supplied by small rainwater-harvesting ponds or tanks, making a guesstimate has always worked for me.

Creative Shaping

I had a few basins under my belt by the time I built the one next to the greenhouse at the SOF, but this basin brought its own challenge. With loamy soil at the site and a turkey nest supply pond backed up by a tank already in existence, size wasn't the issue. I knew I could build a basin with a surface area as large as that of the supply pond because the soil would retain the water and there would be no lack of water supply.

The problem was shape. The basin had to fit into a compact area sandwiched between the greenhouse, the road, and the pond. After considering all the options, I decided to form the basin in the shape of a half-moon to maximize the space available.

Here's an example of how to figure out how much water a pond can supply to a basin: If a basin has the same amount of surface area as the pond and the pond has an average water depth of 18 inches, the pond can irrigate the basin nine times with 2 inches of water if all the water in the pond is available. However, in a more realistic scenario, a basin usually needs irrigation after a dry spell, during which water in the pond has evaporated. For example, the average depth in the pond might be only 12 inches because of the drought, and the tolerable limit for low water in the pond is 8 inches (to keep fish alive), leaving only 4 inches of water available for irrigation, or two irrigation cycles. When plants are established and growing in the soil of a basin, a single flood cycle should provide enough water for one week of plant growth, and a properly sized basin and pond can supply enough water to get through a few weeks of drought.

When a pond is drained to supply irrigation for a basin and must be refilled by rainfall, a seasonal ebb and flow between wet and dry occurs. The effect is exaggerated in dry climates and when excessive water is used for irrigation. Planting plenty of marginal plants along the edge of a pond helps fill the aesthetic void as water retreats. In addition, timing minnow harvests with droughts finds a use for a drained pond, making the loss of water more bearable. If a loss of water in the pond would not be tolerable, design a smaller basin, so less water will be needed for irrigation.

Constructing Bunds for Basins

Basin construction is similar to pond construction; however, a level bottom is critical. Building basins requires marking the basin dimensions, setting out grading stakes, forming bunds, then leveling the bottom of the basin.

The desired depth of the water level in a basin determines the height of the bund. Bunds should sit at least 6 inches higher than the expected irrigation height and stand a minimum of 2 feet wide. To construct a bund, I dig a furrow along the inside edge of the basin, piling up the soil to form the raised bund. The furrow then acts as a channel to divert water through the basin, when there's a need to move water quickly through the basin to another location. If a pass-through system is not needed, the furrow simply fills first before the rest of the basin fills with water.

Alternatively, I bring soil in from other areas to form bunds, maintaining a level basin bottom without the additional channel. Be sure to compact bund soil to prevent seepage and expect soil to settle or shrink at least 10 percent over time.

Here's how to construct a basin with bunds:

Step 1. Mark the basin dimensions. On flat land the shape of the basin can vary depending on your needs. On sloping land the basin layout should conform to the contour of the land.

Step 2. Place grading stakes along the inner edge of the basin in at least four locations.

Step 3. Using a water level or transit level, mark the grading stakes at the level contour dictating the bottom of the basin. Make another mark on the grading stakes at the desired height of the bunds. In sloped landscapes the marks on the uphill grading stakes will be belowground and the marks on the downhill stakes will be aboveground.

Step 4. Move soil to match the level marked on the grading stakes, using excess soil to form the bunds. Small basins are easily built using a shovel and hoe. If a rototiller is available to loosen the soil, the job will be easier. If you need to move a lot of soil or desire a larger basin, you may need heavy equipment.

Figure 3.3. Here's the basin by the greenhouse at the SOF under construction. Grading stakes placed around the edge were all marked at the same contour to indicate the bottom of the basin. The white lines show where twine should be run between the stakes to serve as reference points to facilitate final soil leveling.

Step 5. When you think you have the soil as level as possible, tie twine or a masonry line taut between marks on stakes along the perimeter and in an X across the middle. The twine will indicate the exact level of soil needed to achieve a perfectly flat bottom to the basin. If you see an air gap anywhere between the soil and twine, fill it with soil. Likewise, if you see the twine pushed up in any areas, move soil out of the high spot and into a low area (see figure 3.3).

Once the soil is level in the bottom of the basin, add a small amount of water; any remaining low or high areas will become obvious. It's important to make the bottom perfectly level to conserve and evenly distribute water.

Basin Water Management

What goes in must go out. A basic basin needs an intake pipe to supply water and a drainpipe to remove water. At the SOF the intake pipe for the basin by the greenhouse is connected to the supply pond up the hill. Water flows from the pond through the intake pipe and floods into the basin. After filling the basin, the water eventually finds the basin drainpipe located on the opposite side of the basin. When I need to flood the basin, I seal the drainpipe with an endcap or plug. When I don't need to flood the basin or if I want to move water to the next downhill destination, I leave the endcap off to allow water to pass through the pipe. When installing a drainpipe in a bund (a process similar to installing a drainpipe in a pond, as described in chapter 2 under Installing an Adjustable Drainpipe), it's important to pack soil carefully around the pipe where it penetrates the bund, removing all organic matter to prevent water seepage around the pipe.

Keep in mind that basin water management relies on gravity. If the elevation of the basin is higher than the floor of the supply pond, only a portion of the pond water will drain freely into the basin. Once the level of the water in the basin is at the same elevation as the water level of the pond, flow will cease. I slope intake pipes at a grade of 1.5 percent or steeper to prevent debris from settling in the bottom of the pipe and clogging the flow of water.

Flooding a Basin

Flooding a basin with water follows three predictable phases. Water advances across the basin, it ponds or wets the surface, then it recesses back into the soil.

The advance occurs as water spreads toward the opposite side of the basin. I like to shut off the inflow before the advance is completed. Water will still spread throughout the entire basin, but ponding and overwatering is avoided. However, not supplying enough water will render inadequate irrigation to the far end of the basin.

For basin irrigation to work, a large volume of water must quickly enter the basin and fill the entire basin evenly before water has a chance to sink into the ground. If a slow trickle of water enters the basin, the water simply sinks into the ground before it has a chance to spread throughout, wasting precious water. A 4-inch intake pipe is an ideal size for transmitting water from a small pond to a small basin to allow sufficient volume for rapid flooding.

Flooding a basin reminds me of making pie pastry from scratch. Too much water makes the pastry sticky and soggy, and not enough water causes it to crack and fall apart. The recipe cannot dictate the exact measurement of water, because other factors such as humidity are at play. Only by trial and error does a successful baker learn how much water is necessary to achieve the desired texture of dough.

Similarly, you can learn how to determine the optimum irrigation rate by watching the advance and recession of flood cycles in a basin. Over the course of time, you'll know when enough is enough.

Channeling Water across a Basin

Digging a ditch or channel a few inches deep and 1 foot wide across the floor of the basin between the intake and the drainpipe lets water bypass the basin without filling it. This provides efficiency in managing water. For example, another pond on the far side of the basin may need water. Instead of wasting water filling the basin when it doesn't need irrigating, you can send water directly through the channel into the next pond, in effect bypassing the basin. Besides acting as a bypass, the minicanal also prevents the basin from flooding during excessive rainfall.

I may dig a ditch in a basin on contour so it will hold water. Or I slope the channel slightly — at a 0.25 to 0.5 percent slope — to prevent water from pooling in the channel (see figure 3.5). If a basin has a sloped channel, position the drainpipe an inch lower than the bottom of the sloped channel so it will readily accept the water flowing through the channel. Placing an endcap on the drainpipe will allow you to plug it when the goal is to fill up the basin with water.

Managing Plants in Basins

Once a basin is connected to a water supply, it's time for planting. I've grown plants in pots, on raised mounds, and even on tables to create a multistory effect in basins. How you plant will ultimately depend on what you want to grow.

Irrigating Potted Plants

By using basins in my California nursery, I saved a lot of time, energy, and water. Nursery plants find a niche in a basin irrigation system. Planted in portable pots, they wick up water through their roots while they await a transplant into a new permanent home.

When I keep potted plants in a basin, I prefer to place them on a long-lasting weed barrier such as woven polypropylene landscape fabric. The fabric allows water to pass through easily while keeping weeds at bay. In sandy soils the landscape fabric also slows infiltration, improving the basin's performance. I have also created completely impermeable basins by using thick pond liners made out of material such as EPDM. However, I had to use a specialized fitting to install a pipe to drain the basin, similar to a drainpipe for a pond (see chapter 2). It's best to use 4-inch pots or larger, placed directly on top of the landscape fabric. I raise smaller flats of seedlings up slightly on bricks or trays, allowing the roots to contact the water while keeping floodwaters from pummeling the plants.

Because of the daily water demand of containerized plants, a small pond and basin irrigation system will only supplement normal irrigation regimes. But pond water stocked with nutrients furnishes free fertilizer, and as the old saying goes, every little bit helps.

Tables for Plant Production

For years I used basins only to irrigate potted plants or start seedlings in raised mounds. I guess you could say my ideas were at the ground level. But then I started thinking above the basin. By adding benches and tables, I could create multiple uses for basins.

It's like building a multiuse, multiple-story apartment complex rather than a single suburban one-story home. The benches and tables provide a protected elevated surface for propagating and hardening off plants, while adding another level of usable space within the basin.

Putting plants on tables can provide protection from ants, rodents, and other pests. If ants are eating my seeds, I simply smear a sticky barrier of Tanglefoot around the bench legs to prevent the hungry insects from accessing my seed flats. Putting flats on raised benches also prevent rodents from rummaging through them for seedlings to eat. We attach automated misters to the sides of the bench to irrigate seed flats on the bench top.

To keep costs low, salvage old dilapidated benches from greenhouse and nursery operations. Look for benches made from galvanized or painted metal with replaceable wire mesh tops. The wire mesh is usually the first to go from rust, but it's easy to replace it with diamond pattern mesh, fencing, or hardware cloth.

Figure 3.4. Misters automatically irrigate plants set on benches in the basin next to the greenhouse.

Raising Mushrooms

Besides propagation and seedling protection, the best use I've found for basin benches is as a shelter for shiitake mushroom production.

To grow mushrooms, logs are inoculated with edible fungi, mimicking the natural decomposition in a forest. The logs must remain moist most of the time to stimulate mushroom growth. Therefore, basins are a prime location for mushroom logs.

First, harvest fresh logs from dormant hardwoods with thick bark. Trees such as red and white oak make ideal candidates. I also salvage logs from tree-pruning operations, working with contractors to obtain logs with a diameter and length appropriate for our system. Ideally, I use 3- to 6-inch-diameter logs cut into 4-foot sections. However, tree pruners are more likely to part with 6-inch-diameter logs because branches this size are difficult to load into chippers. Smaller 3-inch-diameter logs are easy to chip, so tree pruners are less likely to leave material this size.

Next, drill holes into the logs every 3 inches in a diamond pattern. Then inject mushroom spawn into the holes. Spawn may be purchased from online suppliers, but I'm fortunate to buy ours from a local producer, Mushroom Mountain. The spawn comes in a sealed bag of sterilized sawdust, with the roots of the innocuous mycelium spread throughout the bag. Using a palm injector — an enlarged syringelike tool — I push the spawn into the drilled holes. Alternatively, mycelium growing on wooden dowels or plugs can be hammered into the holes. I coat the holes with hot wax using a small paintbrush. This seals in moisture and prevents invasion from insects and competing fungi.

After inoculation, the mycelium of the fungus spreads throughout the log in a process known as the spawn run. During the spawn run, water the logs frequently. The inside of the logs should remain moist, and the logs should feel saturated. However, between irrigations the outer edges of the logs should dry completely to prevent competitive fungi from gaining a foothold. Additionally, I elevate my logs on concrete blocks to prevent invasion of soil fungi.

The north side of a building creates the ideal microclimate for the spawn run. The moist, dark atmosphere provides a perfect spawning ground for fungi. If rain pours from a gutterless roof into this area, the additional watershed generates more moisture needed for good fungal growth. I have stacked my logs directly under the roof overhang to catch the additional runoff. During the long days of summer, I drape shade cloth over the logs for protection from the sunlight.

I know the spawn run is complete when the ends of the logs show the white growth of the spreading fungus. Once the fungus fully colonizes the logs, I move the logs to the basins for flood irrigation without fear of infection from other fungi. I elevate one end of our logs on a rock or brick, leaving the other end submerged in a channel. The elevated side prevents the entire log from saturating during floods. Saturated areas of logs decompose quickly, so I extend the lives of the logs by keeping a portion high and dry.

For optimum growth shiitake mushrooms need 60 to 80 percent shade. I cover the benches with shade cloth, which not only hides the worn look of old benches but also creates the perfect, shaded, protected microclimate for mushroom logs. I guesstimate the amount of shade the bench itself casts and add this to the amount of shade provided by the shade cloth. Where the shade cloth contacts the ground on the front of the bench, I wrap the cloth around a galvanized pipe and secure the cloth with cable ties. The weight of the pipe holds the shade cloth down and makes lifting up the cloth a cinch.

Figure 3.5. The channel in the basin is filled with water, ready to receive mushroom logs.

Figure 3.6. The inoculated logs rest in place with one end on concrete blocks and the other end in the water in the channel.

Figure 3.7. Recycled benches are the first stage in creating shade for the mushroom logs.

Figure 3.8. Greenhouse plastic spread across the benches serves as a roof to protect logs from excessive moisture when mist irrigation is running.

Figure 3.9. Shade cloth hides the rusting benches from view and provides additional shade.

Figure 3.10. This crop of shiitakes is ready for harvest.

The automated misters on the bench tops are great for the seedlings but bad for the logs. The excessive water leads to volunteer mushrooms competing with the cultivated variety. To prevent the logs from becoming too moist, we put a single layer of greenhouse plastic in place on the bench before spreading the shade cloth. The plastic serves as an impermeable roof over the logs. Holes in the plastic direct water to specific areas, keeping the majority of the logs dry. Before I cut holes into the plastic, I flood the tops of the benches with water and note any low areas where water pools. Next I cut a hole, about 2 inches in diameter, in the plastic at the low point to let water drain. Moving logs a foot away from drain holes will prevent oversaturation.

Once I control the excessive moisture from the misters, I can easily initiate mushroom fruiting by flooding the basins with pond water. Note that greenhouse plastic adds at least 15 percent more shade when dirty, so be sure to factor in that shade when you're figuring out the total amount of shading. Also, if mushrooms are grown under benches, avoid using the top of the bench as a permanent place for plants, which would then create a permanently dark area under the bench.

I've tried stacking logs in the hope that the fungi inserted into the uppermost logs will fuse into the lower logs, seeking necessary moisture. However, fruiting mainly occurs on lower logs in contact with water. If you stack your logs, be sure to leave enough room around each log for harvesting, and discard mushrooms submerged by floodwaters.

Ideally, we harvest the mushrooms between floods. But if I happen to walk by and see a luscious shiitake fruiting from a log, I can't resist the temptation to bring it home and sauté it in a stir-fry.

We disburse the shiitake mushrooms through our Community Supported Agriculture program. Our subscribing members look forward to receiving shiitakes in their vegetable shares, since these valuable fungi sell for ten to twelve dollars per pound in gourmet markets.

SUMMING UP THE FUNCTIONS

1. Flooded basins irrigate mushroom logs and nursery plants.
2. Pond water, containing fish waste, floods nutrient-rich water into basins to fertilize nursery plants.
3. Basins convey pond water into greenhouse ponds.
4. Basins convey runoff water from greenhouse roof to greenhouse ponds.
5. Solar reflection from basin creates micro-climates for seedlings.
6. Basin benches protect seedlings from pests and shade mushroom logs.
7. Basin retains and infiltrates stormwater while protecting lower ponds.
8. Solar reflection from basin modifies micro-climate of greenhouse and reflects light into greenhouse.
9. Basins built along pathways provide easy visibility and access for mushroom harvesting.

CHAPTER 4

Greenhouse with an "Outie"

The Bio-Integrated Greenhouse with an Exterior Pond

By harvesting rainwater from roofs and roads, gardeners and growers can create chinampa-style systems almost anywhere. At the Clemson Organic Farm I created a pattern to build a chinampa in a humid climate and supercharged it with solar energy to create the bio-integrated greenhouse with an exterior pond. The pattern has three parts:

1. A canal or pond
2. A south-facing slope constructed using soil dug out during canal/pond excavation
3. A greenhouse built on the south-facing slope

The system offers several advantages. By sloping greenhouses toward the winter sun, more sunlight penetrates the greenhouse during the season when you need it most. A properly sloped greenhouse also harvests rainwater into reflecting ponds. The ponds boost greenhouse light during winter, facilitate heating and cooling, and provide habitat for pest-eating predators such as fish, frogs, and toads. Applicable to any climate, sloped greenhouses with reflecting ponds work well with a variety of greenhouse orientations and existing slope profiles.

Leaving Level Ground

When I first started building greenhouses I always followed the standard recommendations to properly site the house: Pick a level spot, and carefully grade the land so water flows off and away from the greenhouse. After all, if water flows back into a greenhouse, it creates a boggy mess. If water flows through a greenhouse, it removes precious heat as it passes from inside to outside.

My levelheaded thoughts shifted when my father requested I build him a greenhouse. I agreed to the venture, picking out the perfect spot for the greenhouse near his home. It was close enough to the house to run electricity and plumbing easily. Using an angle finder, I checked to make sure the distant trees wouldn't shade the greenhouse in the dead of winter when it would most need the sun. Certain I'd selected the perfect spot, I happily announced the results of my site analysis to my

Figure 4.1. Here is the framework of the first greenhouse I built on a sloping site—a steep garden on my father's property. The greenhouse slopes in two directions: 8 percent from back to front and 2 percent from right to left. Water runs from the highest corner (*top right of the greenhouse*) to the lowest corner (*bottom left*). Terraces outside the greenhouse and raised beds inside placed on contour slow the flow of runoff and allow water to infiltrate.

father. However, like many design problems, the perfect spot presented me with a perplexing problem that forced me to shift my perspective on the "right way" to do things.

Sometimes it's bureaucratic red tape that compels a change of mind, such as a required setback from the edge of the property. Or the problem might be a safety issue — for instance, an overhead power line. Whatever problem arises, solving it pushes you into unfamiliar territory. With my father's greenhouse this unfamiliar territory happened to be a steeply terraced garden. How could I build a greenhouse on a steep slope?

To make matters worse, the site sloped in two directions, dipping overall to a single point at one greenhouse corner. It seemed like an impossible obstacle, but a viable solution slowly emerged. With a little extra effort we figured out how to lay the base of the greenhouse all in one plane (albeit a sloping plane).

With that worked out, the next problem to solve was how to deal with rainfall. After water flowed off the roof and hit the ground, it would naturally flow down the slope, which could cause erosion. We compensated by connecting terraces to the sloped greenhouse. The terraces slowed the flow

Elements of the System

In this chapter I'll explain how the bio-integrated greenhouse pattern with an external pond transforms a greenhouse into a living machine. This chapter shows you how to:

▸ Create a greenhouse platform sloped toward the winter sun to capture more solar energy

▸ Use slope to reduce cooling costs with passive and active multistage convective systems

▸ Use slope to reduce heat-pumping costs by creating convection loops

▸ Use sloped greenhouses and ground gutters to harvest rainwater into reflecting ponds, creating free light and heat in the winter

▸ Reduce energy consumption by capturing reflected and direct solar energy in 55-gallon drums filled with water (each drum stores up to 9,000 Btus of energy)

▸ Design evaporative cooling systems

▸ Design and build hydronic heating systems to reduce heating costs by at least 20 percent in comparison to conventional forced-air heating

▸ Use a central hydronic heating system to heat multiple greenhouses and benches on demand, reducing the need to purchase multiple heaters

▸ Benefit from Clemson University research by combining hydronic heating with row covers to increase performance of both

of the runoff and allowed water to infiltrate rather than erode the landscape.

While the system was not ideal, it worked. And it brought to my attention an important question. If I could build a greenhouse on sloped terrain where water runoff created a problem, could I somehow use the slope and runoff to my advantage? In short, could the problem become the solution?

Installing a Sloped Greenhouse System

My job at Clemson University posed the perfect scenario to test greenhouse systems. We wanted to build a new greenhouse and move several of the older existing greenhouses into a suitable area closer to other farm buildings. The lessons I'd learned building my father's greenhouse convinced me it was possible to build greenhouses on a slope to catch more winter sunlight and use the slope to harvest rainwater.

The chosen site sloped gently to the west and connected to the parking lot, office, postharvest area, and market building. The site needed a complete grading and a water-management system to keep the parking area and buildings dry. As I worked on the greenhouse plans, I spent a few extra days developing a comprehensive plan to catch and use the water shedding from the buildings and parking area as well.

Sloping the land toward the south catches sunlight and water, but too much slope causes erosion. So I chose a slope of 1 percent. This gentle slope — descending 1 foot over a length of 100 feet — catches a little more sunlight, as shown in figure 4.2. Yet it's not steep enough to cause seeds and soil to be washed away.

To create the slope, we harvested soil from areas slated to become water-holding ponds. Because of

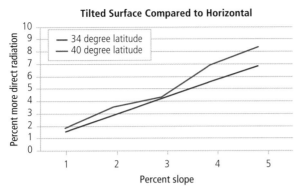

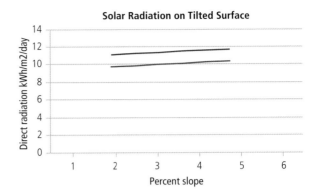

Figure 4.2. Tilted Surface Compared to Horizontal depicts the percentage increase in sunlight received on a tilted surface in comparison to a horizontal surface. Solar Radiation on Tilted Surface shows that as slope increases, the amount of direct solar radiation a site receives also increases. Analysis performed using calculator on www.pveducation.org.

the land's natural western slope, we would have had to move an enormous amount of soil to create a perfect south-facing slope. To compromise, I chose to create a two-directional sloped platform, grading the land to the southwest and maintaining the existing western slope. Because we were building five greenhouses over a large area, we built cascading ponds to accommodate the two directions of the slope. The dam of each pond became a pathway for walking into or between greenhouses.

We positioned the ponds on the south end of the sloped land to capture reflected light in the greenhouses during winter. Rainwater running off the greenhouses flows downhill into the ponds. The ponds are sized appropriately for our climate; thus, rainwater alone is sufficient to replace natural evaporation and keep the ponds full.

Setting the Orientation

The angle of sunlight becomes lower in the winter, and this effect is greater the farther north a property is located. An east-west orientation is always best for any size or shape of greenhouse because it maximizes solar energy capture, as shown in table 4.1. The east-west orientation prevents low-angle

solar rays from reflecting off the greenhouse plastic instead of penetrating into the interior.

However, placing multiple greenhouses in an east-west orientation is difficult because each greenhouse can cast a shadow on the adjacent greenhouse. The effect of winter shadow increases as you travel north, necessitating even greater distances between greenhouses.

The 40-degree latitude is the point at which an east-west orientation is the only choice that makes economic sense. If you're building multiple greenhouses on a site that is south of 40 degrees latitude, you can position them in a north-south orientation without losing an economically significant quantity of solar radiation. The north-south orientation allows much closer spacing without worries about shadows, and that's what we opted to do at Clemson.

When building greenhouses on a sloped platform, orient the platform toward the south. That way, the greenhouses will absorb more solar energy, and the rainwater collected on the south side will reflect more light into the greenhouse during winter.

The orientation of a greenhouse in regard to slope has an impact on how water will flow toward a pond. If your greenhouse is located in an

Table 4.1. Solar Radiation Transmittance (%) on December 21 with Different Roof Slopes and Orientations at Different Latitudes

	Latitude	30 degrees		37 degrees		45 degrees	
	Orientation	E-W	N-S	E-W	N-S	E-W	N-S
Roof Slope South/North (degrees)	10/10	65–70	60–65	60–65	55–60	60–65	50–55
	20/20	70–75	60–65	70–75	55–60	65–70	50–55
	30/30	75–80	60–65	70–75	55–60	65–70	50–55
	35/35	75–80	60–65	70–75	55–60	65–70	50–55
	25/30	70–75	60–65	70–75	55–60	65–70	50–55
	35/30	75–80	60–65	70–75	55–60	65–70	50–55
	30/35	70–75	60–65	70–75	55–60	65–70	50–55

Source: C. von Zabeltitz. *Integrated Greenhouse Systems for Mild Climates: Climate Conditions, Design, Construction, Maintenance, Climate Control.* Heidelberg: Springer, 2011.

east-west orientation on a south slope, then you'll need to channel water flowing off the north side of the greenhouse around the greenhouse and into the pond. If your greenhouse is in a north-south orientation, water flows off the greenhouse sides and straight downhill into the pond.

Building the Platform

Building a sloped greenhouse platform requires planning and grading. In a nutshell, the process is as easy as determining the direction and percentage of the land's slope, calculating how much soil is necessary to alter the slope, then setting grading stakes and actually moving the soil.

PLANNING

Determine the slope. You'll need to know both the direction and the percent incline of the existing slope on the land. You can quickly complete this task by obtaining a detailed elevation survey of the site. The contours on the survey show the direction and the degree of slope. In the absence of a detailed survey, take the following steps:

Step 1. Using a water level or transit level (see chapter 2) and wire flags, mark a perfect contour across the landscape on the edge of the area slated for greenhouse construction.

Step 2. Mark another contour line approximately 25 feet uphill or downhill of the first contour line.

Step 3. Continue marking contour lines at 25-foot intervals until you reach the opposite end of the area slated for greenhouse construction.

Step 4. Draw a line perpendicular to the contour lines to show the direction of the slope.

Determine the percent incline of the slope using the following formula:

Vertical elevation between contour lines ÷ horizontal distance × 100 = % slope

If the footprint of the greenhouse slopes in two directions, determine the percent slope of both directions. In figure 4.3 the vertical elevation between B and D determines the percent slope to

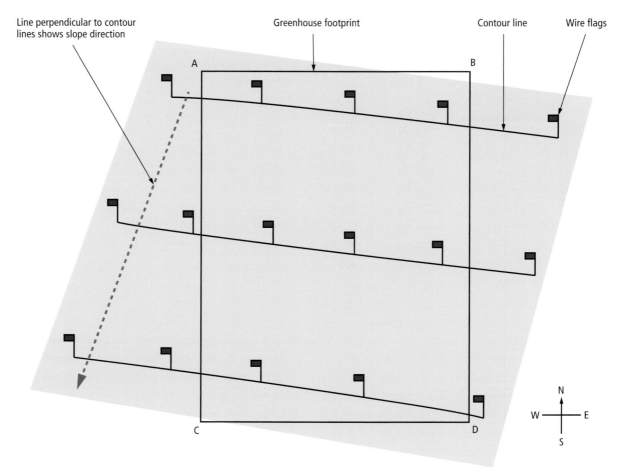

Figure 4.3. In this illustration, the contour lines show that the greenhouse is on a slope. The green line, which is perpendicular to the contour lines, shows the direction of the slope. The greenhouse footprint runs on a north-south axis, and the land slopes to the southwest.

the south and the vertical elevation between B and A determines the percent slope to the west.

Calculate your soil amount. You'll need to determine how much soil you'll produce by digging your pond and the amount of soil you'll need to alter your slope. It seems counterintuitive, but to build a sloped greenhouse platform with a reflecting pond on the south side, you have to start with level or gently sloped land. An existing slope facing south, east, or west is ideal. If the slope faces north more than a few inches, you probably

won't have enough soil to change the slope to the south unless you bring in additional soil from elsewhere. A southern slope causes water to flow into a pond on the south side of the greenhouse. Located on the south side, the pond only reflects light into the greenhouse during the winter (see chapter 2).

To figure out how much you'll be able to change the slope with excavated soil, start by determining the size of the pond the greenhouse catchment area will support based on your climate, the depth of the pond, and the size of the greenhouse. (See chapter 2 for instructions for making these calculations.)

Then use the following formula to approximate the change in slope you can achieve:

$$\text{Area of pond} \div \text{area of greenhouse} \times \text{average depth of pond} \times 0.75 = \text{height change of platform}$$

For example, if I have a greenhouse measuring 1,575 square feet, how big a pond should I build, and how much will the excavated soil contribute to changing the slope of the platform? Following the formula in on page 25, I determine that my climate supports a pond 2 feet deep and one-fifth the size of the catchment area. Since the greenhouse surface area is the catchment, the pond size is one-fifth the greenhouse size.

$$\frac{1}{5} \times 1,\!575 \text{ square feet} = 315 \text{ square feet}$$
$$315 \div 1,\!575 \times 2 \text{ feet} \times 0.75 = 0.3 \text{ feet}$$

From these calculations, I determine the greenhouse supports a pond around 315 square feet, and digging the pond an average depth of 2 feet would produce enough soil to raise any side of the sloped greenhouse platform 0.3 feet (3.6 inches).

Determine placement of soil. Next, you'll figure out where to apply your excess soil. I use excess soil from excavating ponds for the following functions: adjusting the greenhouse platform to create an even plane; filling low spots in the greenhouse platform; and exaggerating or altering an existing slope direction.

Sometimes slope is barely discernable to the naked eye. I often tell the students who work at the farm that even in the Lowcountry of South Carolina, where the land looks as flat as a board, water still flows downriver because of an imperceptible slope.

Grading the greenhouse platform so the base of the greenhouse sits on an even plane ensures that the greenhouse film will sit snugly against the framework. A twisted or distorted frame causes the greenhouse film to rest against the framework in some areas while leaving gaps in other places. If the film does not evenly touch the framework, the film moves and rubs, shortening its life span, and the greenhouse ends up looking distorted. To keep the greenhouse platform in the same plane and avoid twisting the framework, take the following steps (see figure 4.4):

Slope between A and B = slope between C and D
Slope between A and C = slope between B and D

Low spots on the greenhouse platform cause water to pool, creating wet areas contributing to poor plant growth and disease. Using soil excavated from the pond to fill in these low areas improves drainage and plant health.

The greenhouse platform must slope toward the south to harvest rainwater into a reflecting pond on the south side. You'll need a minimum slope of 0.25 percent to move water without pooling. Additionally, if the greenhouse sits on an east-west axis, water sheds off the greenhouse and accumulates on the north side. To move water around the greenhouse and into a pond on the south side, you'll need to slope the platform a minimum of 0.25 percent to the east or west.

GRADING

Locate corners of greenhouse and pond. Position your grading stakes to denote all four corners of your greenhouse and all four corners of your pond. To maximize solar reflection, place your pond on the south side and within 1 foot of your greenhouse.

Mark soil level at corners of greenhouse. Use a water or transit level to determine an even plane for the slope of your platform. Mark the level of the even plane on the grading stakes located at the four corners of your greenhouse. The marks indicate the level of the soil necessary to slope the platform. To prevent excess backbreaking labor and conserve soil, match the existing slope of the land as much as possible and respect minimum slopes. Excavating the pond will generate soil available for adjusting or changing the slope of the greenhouse platform if necessary.

Mark soil level at corners of pond. Using the technique mentioned in chapter 2, mark a level rim on the four grading stakes on the corners of your pond. The marks for the level rim should be a few inches below the marks on the grading stakes for the greenhouse platform, ensuring the pond sits lower than the greenhouse platform.

Move the soil. This probably goes without saying, but plants don't grow well in subsoil. So if you're planning on growing your plants in the ground inside your greenhouse, remove and stockpile your topsoil. I note the depth of my removed topsoil in relation to my grading stake marks.

Next, dig the pond. Use the subsoil from digging your pond to raise the soil level where needed. When you reach your desired pond depth, smooth the subsoil on the greenhouse platform. Ideally, the subsoil should be the same distance below all the original marks on the grading stakes. Check the final grade by running twine from stake to stake to form a giant rectangle. Then run two more lengths of twine to make an X across the middle of the platform. The twine indicates the exact level of soil necessary to achieve the perfect

Figure 4.4. A finished greenhouse platform slopes 1 percent to the south from B to D and A to C and 1 percent to the west from B to A and D to C. Water flows into a reflecting pond on the south side. Greenhouse bows will attach to the metal stakes along the edge. Long-lasting landscape fabric installed on the beds prevents weed intrusion.

grade for the platform. If you see excessive distance anywhere between the soil and the twine, fill it with soil. Likewise, if you see soil pushing up the twine in any areas, move soil out of that high spot and into a low area.

The gap between the twine lines and the subsoil should not be greater than the depth of the topsoil removed earlier. If the distance is greater, you'll need to lessen your desired slope. Lower the grading marks accordingly to maintain the minimum slope angles mentioned earlier, and move subsoil evenly below the new marks. You can also bring in soil from off-site to raise the platform if necessary.

I use a box scraper with scarifiers mounted behind a small 40-horsepower tractor to dig ponds and move soil. By shortening the top link of the three-point hitch, the box scraper will dig into the soil. Loosening the soil with a tiller, disc harrow, chisel plow, or scarifiers before using the box scraper quickens the process and more than makes up for itself in the time you save working with the scraper. Alternatively, a rear blade on a tractor, a small skid steer, a track loader, an excavator, or a bulldozer can do the job as well. You can use almost any piece of equipment typically used to move soil, but the rougher the finished product, the more handwork required to smooth out the platform. A small greenhouse platform can easily be built with the help of a case of beer and a bunch of friends.

Replace topsoil. Remove the twine from the grading stakes, and replace your topsoil over the greenhouse platform. Smooth the topsoil out evenly over the entire platform, and replace the twine to check and adjust the grade as needed.

Construct the pond. Construct your pond according to the directions in chapter 2. Make sure the overflow for your pond is below the level of your greenhouse floor to prevent the greenhouse

from flooding and ensure the soil will not become saturated inside the greenhouse.

Installing Ground Gutters

Conventional gutters don't fit well onto hoop houses, so harvesting rainwater from greenhouses requires a different approach: gutters at ground level. I create channels alongside the greenhouses and line them with pond liner. These ground gutters help prevent soil erosion on steeper slopes and ensure that all rainwater is captured and transferred into the ponds, rather than flooding the planting beds inside the greenhouse. The gutters also reduce rodent activity by creating a barrier the animals can't dig through.

Figure 4.5. Ground gutters harvest rainwater from the greenhouses and move it to the pond on the south side. A layer of gravel underneath the gutter liner and a piece of hardware cloth covering the mouth of the perforated pipe discourage rodents.

Ground gutters, like the one shown in figure 4.5, are similar to French drains. A French drain is basically a gravel-filled channel or ditch with a perforated pipe buried in the gravel. As rainwater fills the ditch, water streams through the holes in the pipe and flows to its final destination. In a ground gutter a pond liner covers the bottom and sides of the ditch, thereby capturing all the rainwater. The perforated pipe then directs the runoff straight into the pond. For extra protection from rodents, I sandwich the pond liner between layers of gravel.

The perforated pipe in the gutter enables the trench to fill to ground level and still provide drainage. I install ground gutters on both sides of my greenhouses. It's important to ensure that the gutters have a continual downhill slope so they will drain into the pond.

In addition to ground gutters, flood irrigation basins (chapter 3) and worm basins (chapter 13) also work well on the edges of greenhouses to direct water into ponds.

Once the gutters are complete, it's time to install the greenhouse on the carefully graded platform. With a little extra planning and work, you can sculpt the land to transform your greenhouse into a sustainable, climate-modifying, rainwater-harvesting system.

Cooling and Heating

Hot air rises, and cool air falls — this process is known as natural convection. You can feel this easily in your home by climbing a ladder. Near the ceiling the temperature is always warmer.

Envision cold air as if it were water; it travels downhill and pools in low places. The steeper the slope, the faster the cool air flows down. Hot air is the opposite. It travels upward. Or in the case of a sloped greenhouse, uphill. Since one end of a sloped greenhouse is higher than the other end,

hot air flows uphill as it ascends to the highest point of the greenhouse. The steeper the slope, the faster the ascent. By simply sloping a greenhouse, you can control the direction and speed that hot and cool air travel. This provides ventilation and reduces the need for fans to cool the greenhouse.

I've implemented both passive and active cooling systems for greenhouses. Let's explore passive systems first, then work up to more complex active systems.

Doors and Vents

The simplest, most passive way to cool a greenhouse is by installing doors at both ends. Open the doors, and heat escapes. The size of the opening controls the amount of cooling. In a sloped greenhouse, which door is opened first also has an impact. If you only need a small amount of cooling, open the uphill door first. Hot air will naturally flow up and out. If you desire more cooling, open the downhill door also. This enables cross-ventilation and assists the natural convective process. The net effect is that smaller openings in a sloped greenhouse create the same cooling effect as larger openings in a level greenhouse.

Manually opening and closing doors to cool a greenhouse requires a lot of attention and is not very effective. Although it may seem as simple as opening the doors in the morning and closing them in the late afternoon, weather and temperatures fluctuate throughout the day. If the doors remain open all day, the greenhouse may become too cool for ideal plant growth. Also, it takes only one day of forgetting to open the greenhouse to create a disaster — all the plants may die from heat stress. Manually controlled greenhouses work best in close proximity or attached to a house for careful monitoring.

The next step up involves solar-powered devices that open and close vents based on the temperature

Figure 4.6. Two venting options: A door for manual venting and a long-lasting aluminum louver-style vent.

Figure 4.7. Wax inside the black cylinder of the solar-powered vent opener expands when heated, automatically opening the vent.

inside the greenhouse. Solar-powered vents eliminate the expense of running electrical wires to the greenhouse, saving a considerable amount of money. Available by mail order from farm supply catalogues, solar-powered vent openers contain wax-filled cylinders with pistons. When the greenhouse heats up, the wax inside the cylinder expands and pushes a piston out of the cylinder, which opens the vents. When the temperature goes down, the wax contracts, and mechanical springs pull the piston back into the cylinder, closing the vents. You can adjust the cylinders to control the temperature, which allows you some flexibility.

As with doors, smaller vents in a sloped greenhouse have as much cooling effect as larger vents in a level greenhouse, because natural convection pushes hot air out the vents at the uphill end of the greenhouse. You may still want to manually open doors on the uphill end of the greenhouse to assist the convective process on hot days.

You can construct vent openings yourself, but I've never been satisfied with homemade vent openings. I'm not a welder, and I've only built vents out of wood. The wood eventually decayed or warped, leaving me with an ongoing maintenance problem.

On my latest greenhouse project I spent a few extra bucks on lifetime aluminum louver-style vents (see figure 4.6) and adapted them to solar-powered vent openers. The louver-style vents allow more airflow than homemade vents, making them not only longer lasting but also more effective.

Roll-Up Sides

To access additional passive cooling, purchase or build roll-up sides; this allows you to increase airflow through the sides of the greenhouse. Building manually controlled roll-up sides is simple and adds little expense to the total cost of the greenhouse. First, I attach an additional purlin made from galvanized tubing or treated wood about 4 feet above the ground on the side of the greenhouse. Next, I install Kwik Wire or another type of fastening system to the purlin. Then, I drape the plastic over the greenhouse bows and the new purlin and using the Kwik Wire fastener, I attach the plastic to the new purlin. Finally, I attach the plastic to a new

Figure 4.8. This roll-up side handle is made from bent 1.5-inch-diameter galvanized tubing. A slide lock keeps the handle in place.

Figure 4.9. Galvanized aircraft wire prevents the roll-up side from being banged about by the wind, as does the length of galvanized tubing.

length of galvanized tubing at ground level using Kwik Wire fasteners attached to the tubing. Alternatively, you may fold over the plastic several times and secure it with a tek screw. If you choose the latter technique, leave enough greenhouse plastic to roll the galvanized tubing up several times after securing it with screws. Rolling up the plastic several times prevents the wind from pulling the plastic at a single point of contact where the screw connects to the galvanized tubing and distributes the contact over the entire length of tubing.

Next, construct a simple handle by bending a short piece of galvanized tubing. A hydraulic press works best to bend the tubing, but a masonry chisel and hammer also work well, or you can weld the bends into the correct orientation. With the handle inserted on one end of the galvanized tubing, you can easily roll up the entire length of plastic.

Once the plastic is rolled up, a simple slide lock created out of ½-inch galvanized conduit secures the rolled-up plastic in place (see figure 4.8). Be extremely careful to control the handle when lowering the greenhouse side. If you suddenly release the handle, the weight of the roll causes the side to quickly drop, and the handle spins wildly, which could cause injury.

To prevent wind from jostling the extended roll-up side, I string galvanized aircraft wire or rope between eyebolts in the purlin and greenhouse framework. Alternatively, inserting permanent stakes made of galvanized tubing into the ground just outside the roll-up side and 10 feet apart will secure the plastic against wind. Both methods are shown in figure 4.9.

Finally, create an airtight seal to prevent wind from entering the ends of the roll-up sides by

Figure 4.10. An extra sheet of greenhouse plastic between the first two bows helps seal the corner of the roll-up side.

adding a piece of greenhouse plastic stretched taut between the first two end bows, as shown in figure 4.10.

Shading

Another easy way to cool a greenhouse is simply by covering it with shade cloth. Different densities of shade, expressed as percentages of sunlight blocked, create different microclimates. For growing plants, 20 to 50 percent shade is recommended. I roll up the greenhouse sides, then use a 30 percent shade cloth to grow lettuce in our

cool greenhouses that have internal ponds (this is described in chapter 5). The shade cloth lets enough light through for leafy vegetables to grow well, and the long days of summer help compensate for the reduced sunlight.

I secure the shade cloth using the same Kwik Wire connectors used to secure the greenhouse plastic. The thick connectors are designed to hold multiple layers of material by stacking the wire connectors on top of each other. On our largest greenhouse the shade cloth doesn't quite reach the connector on one side. To secure the shade cloth on

Figure 4.11. A 30 percent shade cloth draped over a greenhouse provides passive cooling, as do the rolled-up sides. This greenhouse has an interior pond, which also helps with cooling.

Figure 4.12. We drape this large greenhouse with 60 percent shade cloth to cut down on the need to run the electric fan. The shade cloth is pulled over the top using ropes.

Ginger and Garlic in the Greenhouse

Normally, it's not a good idea to mix production and propagation inside greenhouses because of the risk of disease. When you grow a crop for harvest, it usually suffers from disease at some point, especially in organic production. When the crop being grown to completion is grown in the same house where young plants of the same type are being propagated, disease organisms from the production crop can infect the young plants, and a vicious disease cycle ensues.

Ginger as a production crop is an exception. It's so different from all the plants we propagate, it doesn't seem to share diseases or pests with them. Ginger also requires a long growing season. Even in our southern climate, if we plant ginger outdoors in the spring, it reaches only the size of baby ginger by fall. The baby ginger tastes great, but unfortunately it's not tough enough to make it through the winter for replanting in spring. If the season is long enough, afforded by greenhouse protection, the roots reach a mature size capable of being divided and replanted. By growing and propagating ginger in the greenhouse, we save tons of money on ginger seed stock.

I keep our largest greenhouse cool by covering it with a 60 percent shade cloth. This reduces the usage of the electric fan and thus saves energy. The deep 60 percent shade on our larger greenhouse also helps in curing garlic. For good garlic flavor it's important to keep garlic out of the sun while it's drying. The deep shade of the large greenhouse, with raised benches and fan ventilation, is perfect for drying garlic, onions, seeds, sunflower heads, peppers, mushrooms, or anything else needing dehydration.

The 60 percent shade cloth also works well with cuttings and mist systems to asexually propagate nursery plants. You can use different degrees of shade to keep greenhouses productive through the heat of summer.

Figure 4.13. Ginger growing in 60 percent shade of greenhouse cover.

Figure 4.14. Curing garlic on top of greenhouse benches under 60 percent shade cloth.

this side, I string galvanized aircraft wire between the grommets in the shade cloth and eyebolts in the greenhouse frame. Then I secure the aircraft wire at the ends using wire rope clips.

Active Cooling Systems

You achieve ultimate greenhouse temperature control with active cooling systems. To automatically control temperatures in greenhouses, most active systems use temperature sensors connected to electrical relays or a computer. Active systems include fans as well as vents, and the electrical relays or computer controls the fans and vents, determining when vents open and close and when fans turn on and off. Most systems work in stages, operating low-cost devices first, then employing higher-energy devices as needed. Here is a typical four-stage system to control greenhouse cooling:

Stage 1. A small vent opens at the top of the greenhouse. On the opposing side a fan turns on low speed to pull hot air out of the greenhouse, or vents located low on the greenhouse wall open to allow natural convective cooling.

Stage 2. Larger vents open, and the small upper vent closes. If a fan did not turn on during the previous stage, the fan now runs on a low speed.

Stage 3. The fan increases to a high speed, or an additional fan turns on.

Stage 4. A sump pump connected to a tank filled with water wets an evaporative cooling surface.

In a sloped greenhouse it's possible to improve the efficiency of the system by modifying the staged cooling system to work with the convective advantage. Here are the five stages of the modified cooling system (as shown in figure 4.15):

Stage 1. A small vent opens on the uppermost portion of the sloped greenhouse. During cool weather this is the only venting necessary, since natural convection causes heat to escape.

Stage 2. Large vents open on the opposing lower side, and convection pulls air into the greenhouse from over the pond.

Stage 3. The small vent at the upper end of the greenhouse closes, and the fan turns on low speed.

Stage 4. The fan's speed increases.

Stage 5. A pond stores water instead of a tank. A pump turns on inside the pond to wet evaporative cooling surfaces.

Air cools as water evaporates because water pulls heat from the air to transform itself from a liquid into a vapor. Evaporative cooling systems for greenhouses take advantage of water's cooling effect by spreading water thinly over a material to promote evaporation. The effect works best in dry air, maximizing water evaporation. Commonly, water is pumped through a perforated pipe so it falls onto a corrugated cellulose pad. The pad spreads the water into a thin layer, catching excess water in a gutter and tank for reuse. As air enters the greenhouse, it passes through the saturated pad and cools by evaporation. For every pound of water evaporated, 1,060 Btus of heat is absorbed out of the air.

Recommendations call for 10 square feet of pad for every 215 to 322 square feet of greenhouse floor area.[1]

I hoped the pond on the south side of our greenhouse would provide a cooling effect through evaporation similar to the way an evaporative cooling system works. To test the theory I conducted a simple experiment: I monitored the temperature inside the greenhouse through data loggers. To prevent evaporation I placed a large plastic sheet over the pond and kept the cover in place for thirty minutes while the fan was running, pulling air

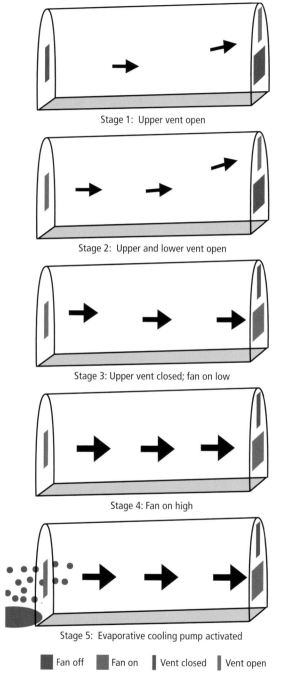

Stage 1: Upper vent open

Stage 2: Upper and lower vent open

Stage 3: Upper vent closed; fan on low

Stage 4: Fan on high

Stage 5: Evaporative cooling pump activated

■ Fan off ■ Fan on ▮ Vent closed ▮ Vent open

Figure 4.15. In a sloped greenhouse there are five stages to an active cooling system, beginning with the least energy-intensive stage of opening a single vent. In stage five a pump inside the pond wets an evaporative cooling surface.

across the pond and into the greenhouse. I then removed the cover and allowed the fan to pull in air for thirty minutes. I repeated the process several times, checking temperature changes each time. To my surprise, I detected little to no temperature difference between treatments. I repeated the treatment with a fountain running in the pond but still found no difference.

Why didn't this work? In an evaporative cooling system the evaporative pad must completely seal the opening on the greenhouse to force air through the pads without leaks to bypass the evaporation process. Delving into the matter more thoroughly, I learned that several different types of homemade evaporative cooling systems are used in other countries. These systems pass water over layers of palm fronds or mats of organic matter. Some of the best systems pass water over hanging ropes in front of greenhouse vents, like a beaded curtain hanging in a doorway. A single layer of rope is not effective, so two layers are necessary.[2] You should not be able to see through the evaporative pad. Some systems use a metal framework to hold the ropes in place, separated by a few inches. Water distributed through a perforated pipe at the top passes down the ropes and into a collection gutter that returns it to a reservoir. The reservoir is equipped with a float valve to add more water, replacing water lost from evaporation. Rope evaporative coolers are also easy to clean; simply shake or knock the ropes.[3]

Based on this research and my experiments, I think ponds on the south sides of greenhouses are not effective in cooling greenhouses. However, the ponds could serve as reservoirs for an evaporative cooling system cheaply built with ropes and gutter. It would be necessary to occasionally replenish the pond to compensate for the increased evaporation. Using a float valve connected to an additional water source would work well.

Smothering Greenhouses in Summer

I don't really need greenhouses for food production during the summer, since we barely have time to harvest everything that's growing in the fields. However, if I abandon the greenhouses, they become a weedy mess. Rather than letting that happen, I use a technique I learned from Eliot Coleman. I leave these greenhouses unshaded and plant a heat-tolerant smother crop such as sweet potatoes. Sweet potato vines grow quickly, covering every square inch of the greenhouse and preventing any weeds from growing. The plant roots also exude natural chemicals that purge weeds such as yellow nutsedge in a process called allelopathy. The greenhouses then provide a nice dry place where the workforce can stay busy on rainy fall days, digging giant sweet potatoes by hand.

Cowpeas make another good smother crop that quickly takes over and suppresses any weed growth. Before they set seed, I mow them to the ground and till them into the soil. The residue provides plenty of nitrogen for winter crops as it decomposes.

Figure 4.16. Smother crop of sweet potatoes planted in an unshaded greenhouse.

Passive Greenhouse Heating

Simply by its design a sloped greenhouse assists with heating. Because the greenhouse slopes toward the winter sun, more light penetrates the greenhouse plastic and more heat gets stored in earth or gravel floors. Additionally, if you place a heat source at the downhill end of a sloped greenhouse, hot air will rise from the source of heat and travel to the uppermost part of the greenhouse. Upon reaching the top and cooling, the air then falls back down to the heat source, creating a natural convective loop that moves heat and air throughout the greenhouse.

Even in cold climates, greenhouse temperatures can rise above 90°F on a sunny day. Heat is usually excessive during the day but deficient at night. The trick to passive heating is storing the excess heat generated during daylight hours and using it at night. A passive system stores heat in what's called a thermal mass. The thermal mass "flattens out" the temperature inside the greenhouse, creating better growing conditions for plants.

Some materials have a high volumetric heat capacity, which means they are better at absorbing and storing heat than other materials are.

Of all the common materials, water has the highest volumetric heat capacity. A given volume of water can store more than twice the amount of heat as the same volume of brick or concrete, as shown in table 4.2. The fluid property of water allows heat to transfer deep into its volume through natural convection.

Chapter 5 discusses using water as thermal mass in ponds. Fifty-five-gallon drums make excellent, cheap containers for storing water as thermal mass. Finding drums may require a little detective work, but you can usually purchase them for five to fifteen dollars per barrel. Juice manufacturers, wholesale food distributers, honey producers, food co-ops, and biodiesel manufacturers are some good places

Table 4.2. Volumetric Heat Capacity of Different Types of Thermal Mass

Material	Thermal Mass (volumetric heat capacity kJ/m³ degree C)
Water	4186
Brick	1656
Iron	3613
Concrete	1502
Stone (loose)	~1300
Granite	1896
Sand	1306
Pine	1400

Source: R. Dunlap. *Sustainable Energy.* Canada: Cengage Learning, 2014.

to start the search. Metal drums transfer heat better than plastic drums and so work better for storing thermal mass. However, the metal corrodes with time. Industrial plastic drum liners are available to protect the metal from water contact, thereby preventing corrosion. You'll need to carefully prime and paint metal drums wherever bare metal shows. At Clemson we painted our drums with flat black since it absorbs the most solar energy. However, we painted the tops of our drums white to reflect light. That's because during the summer the sun sits at a higher angle and strikes the drum tops, but in the summertime we don't want the barrels to absorb solar energy — the greenhouses don't need any extra heat then.

Plastic drums also work well and outlast metal drums. Since they have lids with watertight seals, they won't leak even if laid on their sides. However, finding black plastic drums is difficult, and most paint won't stick to plastic. I've found paint suitable for plastic in spray cans, but if you're using a lot of barrels, spray painting is costly in both money and time. A good primer may help. To save money you could paint only the portions of the barrels that will face the winter sun.

Another recent innovation uses water-filled plastic walls, an inexpensive option for thermal mass storage. The walls are composed of two layers of durable plastic and come in various sizes, with each square foot holding 7.5 gallons of water. You can also store water in thin (6 mil) plastic tubes. These inexpensive plastic tubes are sold in rolls and are normally used to transfer forced-air heat in greenhouses. The tubes usually come with prepunched holes for air and heat to escape, but you can also order them without holes. Lay the plastic tubes on the ground, tie one end closed, and elevate the closed end above the ground using a stake. Fill the other open end with water from a hose, then tie it shut and elevate it with a stake to prevent leaks. The downfall with all thin-membrane plastic systems is the fragility of the membrane. The warmth attracts cats, rodents, and other animals, and it only takes one curious claw to create a leak.

Once you've decided which type of thermal mass material to use in the greenhouse, the next step is figuring out where to place the thermal mass for best performance. In a bio-integrated greenhouse, sunlight reflecting off the pond in the winter

Figure 4.17. A number of 55-gallon drums filled with water capture the solar energy reflecting off the ponds and store the energy for nighttime heating.

supersaturates the south side with heat. With water barrels located on the south side, we've recorded a single 55-gallon drum releasing over 9,000 Btus of heat. The average Btus per day over December and January measured 4,447. A set of twenty barrels stores and releases more energy than burning a gallon of propane every night.

Since our bio-integrated greenhouse slopes to the south, the southern end is also the lowest point inside the greenhouse. Having the thermal mass heat source at the lowest point creates a natural convective loop that circulates heat throughout the greenhouse. The drawback of locating thermal mass on the south side is the shadow it casts in the

China's Thermal Mass Greenhouses

Most of the greenhouses in China utilize north-wall thermal mass systems. With an estimated 6,820,098 acres in greenhouse production, China dwarfs the 20,818 acres of greenhouse production in the United States. Recently, farmers in Manitoba brought the Chinese technology to North America, achieving an astonishing 18° to 27°F

of protection on cold nights. The Chinese greenhouse style uses passive solar techniques, storing solar energy in soil and rock walls. Insulating blankets roll down over the top of the greenhouse membrane to help keep the stored heat inside the greenhouse. Coupled with a reflecting pond, the Chinese system has huge potential.

Figure 4.18. This Chinese greenhouse uses soil for thermal mass. The insulating blanket is rolled up at the peak of the roof. Photo by Dr. Desmond Layne

Figure 4.19. Dr. Desmond Layne, Tree Fruit Extension Program Leader for Washington State University, and his Chinese hosts are inside the same greenhouse next to the thermal mass wall, admiring peach trees. The vertical support posts are made of concrete, and the roof supports are bamboo. Photo by Dr. Desmond Layne

greenhouse. To compensate for this I place the plants on benches to raise them closer to the height of the thermal mass.

Traditionally, thermal mass is located at the north side of a greenhouse. You can stack thermal mass all the way to the ceiling on the north side without blocking the winter sun. Thermal mass on the north end also helps insulate the greenhouse from cold northern winds in winter. The thermal mass material should not contact the greenhouse plastic; otherwise, heat can escape through the point of contact.

In our propagation greenhouse we elevate all our plants on benches or place them in pots. This leaves the ground available for thermal mass heat storage. To take advantage of this area, I laid 4 inches of gravel at the base and covered the gravel with a long-lasting, black, woven polypropylene landscape fabric. The black fabric captures heat and transfers it to the gravel during the day, and the gravel radiates the heat back into the greenhouse at night.

Active Heating Systems

Traditional greenhouses require active heating, and it takes enormous amounts of fossil fuel energy to maintain heat. With passive solar heating techniques, the amount of additional heat needed is minimal. However, on cloudy and extremely cold days, you may need supplemental heat to keep plants alive, depending on the crop. Extra heat will also boost seed germination times and speed plant growth.

Forced-air heating of the entire greenhouse has quickly made greenhouse production uneconomical as fuel prices rise. Also, forced-air whole-house heating is not effective for warming the soil. Soil temperature is 10°F lower than air temperature in a greenhouse heated by forced air. Hydronic heating systems offer a great alternative

to forced-air heating, and a quick payback in fuel savings. Hydronic heating systems pump heated water through a closed-loop pipe system to wherever heat is needed. I like to compare hydronic systems to the vascular system in humans. In humans, blood travels out from the heart through arteries, into the organs, then back to the heart through veins. Likewise, hydronic systems have pipes that bring water out from the heat source to where it's needed, then back to the heat source and pump.

The pipe system bringing heat to plants can run inside the soil of planting beds, embedded in concrete or gravel floors or simply placed on top of benches and planting beds. If the greenhouse has a pond or water barrel inside, the hydronic heating system can heat the pond or barrel using a heat exchanger, turning the pond into a heat-storage battery. After the water makes its delivery of heat to plants or ponds, it returns to the heater to pick up more heat for delivery. Hydronic heat has the following advantages:

- Heat is applied under the plants and rises up naturally, reducing fuel requirements by at least 20 percent.
- Heating systems are easily divided into zones or individual benches. Rather than heating the entire greenhouse for one bench of plants, only the bench receives heat.
- A variety of fuel sources can be used to heat the water, including compost (as described in chapter 8).
- A single central heater can serve many greenhouses, eliminating the need to purchase several heaters.
- Heat can be stored in insulated tanks and ponds during the daytime, to be used by greenhouses at night, which enables a smaller heater to serve a larger area.

Hydronic heating systems come in kits to fit any size greenhouse. Alternatively, you can piece together component parts to save money. While hydronic systems look complicated at first glance, once broken down into components the system is simple.

- Heater. An electric, compost, or solar water heater (or any combination thereof) provides heat for the system.
- Expansion tank. Water expands as it heats; the expansion tank gives the water a destination.
- Circulation pump. A small circulation pump moves heated water through the distribution system.

- Pressure regulator. The system runs on low water pressure; the pressure regulator reduces the incoming pressure to around 12 pounds per square inch (psi). The pressure relief valve on the hot water heater also needs to be exchanged with a lower pressure relief valve to accommodate the new pressure.
- Controller. A temperature sensor placed in the soil or pond connects to the controller. The controller determines when the circulation pump turns on based on the preset temperature.
- Pressure switch. If the water pressure drops, the pressure switch engages and adds water to the system.

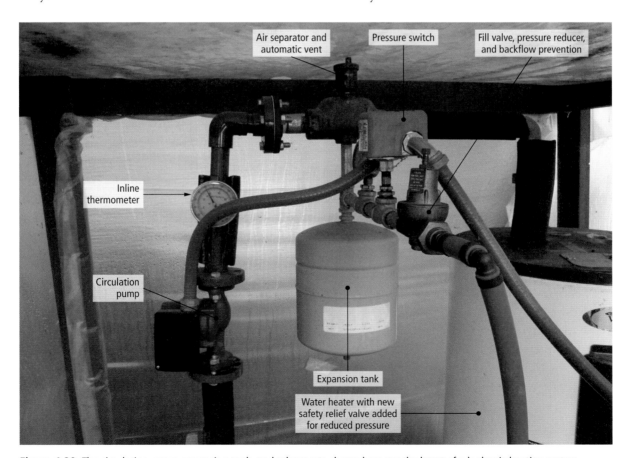

Figure 4.20. The circulation pump, expansion tank, and other parts shown here are the heart of a hydronic heating system.

- Air scoop. Since air trapped in the system reduces the flow and causes corrosion, the air scoop placed at a high point helps remove the trapped air.
- Thermometers. The thermometer monitors water temperature as water enters and leaves the water heater.

The water-heating device determines the total amount of heat a system can produce. Each heating system is rated on output, usually defined as British thermal units or Btus. One Btu is the amount of energy needed to heat or cool 1 pound of water 1°F. Heating needs depend on the crop, climate, and insulation of the greenhouse. Use online calculators to determine required Btus based on your needs.

Heaters vary by type in the amount of output. For example, electric water heaters commonly produce around 20,000 Btus of heat, while a propane water heater may produce 40,000 Btus.

Water from heating devices travels through distribution pipes to benches and greenhouse beds, where smaller pipes convey the water to plants. For distribution larger pipes move more heat, as shown in table 4.3. If the distribution pipe runs aboveground, I use a thick, tough schedule 80 PVC pipe. When the pipe runs belowground, protected from sunlight and damage, I use flexible PEX pipe with an oxygen barrier designed for hydronic heating.

To retain heat and prevent insect problems, insulate all hot water distribution pipes. I made the mistake of not insulating the heat transfer pipes that travel between the greenhouses at the Clemson Student Organic Farm (SOF). Since 90 percent of these pipes traverse inside the greenhouses, I assumed the greenhouses would capture any lost heat. Though logical, this method didn't let us control where we applied the heat.

After we installed noninsulated distribution pipes in our greenhouses, I determined we were

Table 4.3. Distribution Pipe Size

Pipe Size	Btus Moved
½"	15000
¾"	40000
1"	75000
1¼"	160000

losing 5,000 Btus per hour from the 240 feet of pipes. For every month of operation, we were wasting heat equivalent to 40 gallons of propane. The realization prompted us to rip up all the pipes and reinstall with insulation.

Normal pipe insulation used for hot water pipes is inadequate and too expensive for all but the shortest runs. Preinsulated pipes are available but cost an astronomical amount of around a thousand dollars per 100 feet. I believe the best way to insulate the pipes is to dig a trench a few inches wider than the bundle of distribution pipes. Lay the pipes inside the trench on top of wire supports typically used to hold rebar off the ground when pouring concrete. Next, hire a spray foam insulation contractor to place insulation around the buried pipe. Then cover the top of the trench with soil.

To insulate our pipe system at the SOF, we had to rebuild it in sections, so we didn't have the option of using spray foam insulation. Instead we dug a perfectly square trench slightly wider than the bundle of heat distribution pipes. Next, we used 1-inch-thick extruded polystyrene foam insulation to line the square trench, forming an insulated box. Then we laid pipes inside the box and placed perlite around the pipes to seal gaps and provide additional insulation. The top of the box was slightly wider than the bottom to shed water away from the pipes and perlite inside the box. Finally, we covered the trench with soil to conceal everything. I tested the system out and had less

Figure 4.21. Blue distribution pipes for hydronic heat are laid inside a trench insulated with extruded polystyrene foam. Perlite is added to seal the gaps, and the trench is topped with a wider sheet of insulation to shed water. The white pipe carries compost gases (rich in carbon dioxide) extracted from an exterior compost windrow.

than 1°F temperature drop compared to a 5°F drop without the insulation.

Another benefit of insulation was fire ant control. The warm pipes without insulation traversing the greenhouses caused an explosion in fire ants. Attracted to the heat, the ants thrived all through the winter, eating lettuce and chard seedlings and stinging workers. Now that the pipes are insulated, the ant attack has abated.

Two lines distribute the heat, a feed line and a return line. The feed line simply extends from the heater to all the greenhouses or benches in need of heat. The return line brings the water back to the heater post–heat extraction. To ensure heat is distributed evenly, I start the return line at the closest bench or greenhouse, then extend the return line toward the farthest bench or greenhouse before returning it back to the heater (see figure 4.22).

Heat transfer lines convey heat from the distribution line to the plants. One common technique for heat transfer is to bury PEX piping in the soil below plants. Instead of PEX, I prefer tough strands of EPDM rubber tubing laid on top of the soil or bench surface. Extremely flexible, this tubing easily rolls up for removal.

I use two manifolds to connect the distribution pipes to the small rubber heat transfer tubing, a feed manifold connected to the feed line, and a return manifold connected to the return line. The small rubber tubing then connects between the

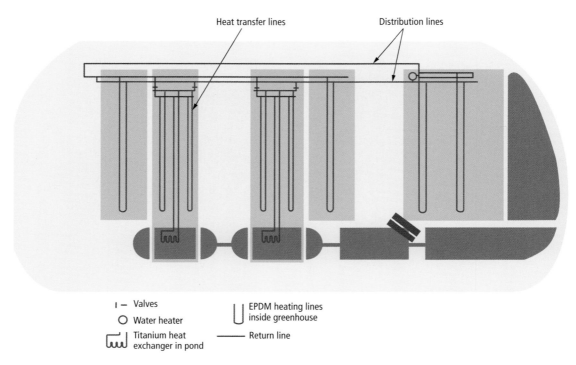

Figure 4.22. Hydronic heating pipe distribution system at the Student Organic Farm. Note how the return line in blue starts at the closest greenhouse, then extends to the farthest greenhouse before returning back to the heater, ensuring the heat is evenly distributed between all the greenhouses without "short circuiting."

two manifolds, forming a loop. A valve on the feed manifold line and return manifold line turns the manifold on or off.

The manifolds are made of schedule 80 PVC piping with fittings inserted into the pipe to accept the small rubber tubing. Purchasing the manifolds is convenient but expensive; you can save a lot of money by building them yourself. Simply drill holes in schedule 80 PVC piping and use tongue-and-groove pliers (Channellocks) to press the fittings into the pipe. I lubricate each fitting with a little soap to ease the insertion. Once built, paint all exposed PVC piping to protect it from sun damage.

Two sizes of rubber tubing move heat to plants. The smaller tubing, with an outside diameter roughly ¼ inch, extends a maximum length from the manifold of 60 feet. If the tubing runs farther, the small pipe is unable to carry the heat that distance, and the tubing is no longer effective. The larger tubing with an outside diameter of roughly ½ inch extends almost 100 feet from the manifold.[4]

The amount of heat radiated from the tubing depends on the temperature of the water and how much space is left between the lengths of rubber tubing on the bench.

After you assemble the distribution lines, it's important to flush all systems of debris before installing the rubber tubing. I've found pipe shavings lodged into circulation pumps, preventing flow. The shavings could also block flow in the small rubber heating tubes. Once you've assembled and flushed the system with

Figure 4.23. In this closed-loop hydronic heat system, rubber tubing on top of the bench carries heated water that warms the soil and roots of plants growing in flats placed on top of the tubing. Tubing is attached to a feed-and-return manifold controlled by valves.

Figure 4.25. To make a manifold, use tongue-and-groove pliers to push a lubricated fitting into a 7/32-inch hole drilled into PVC pipe. The fitting is sized for rubber tubing that circulates heated water to greenhouse benches.

Figure 4.24. For inground plants the rubber tubing is coiled when not in use. The manifold pipe, made of schedule 80 PVC, runs in a shallow trench underground, wrapped in insulation (not shown).

water, install the rubber tubing, chase air out of the system, then turn it on. After running the system for a few days to ensure no leaks are present, I add anticorrosive chemicals to prevent metal components, such as the circulation pump, air scoop, and metal pipes, from corroding and blocking the flow of water. Three options are available for applying the chemicals: through the air vent using a funnel, pumping chemicals into the drain in the water heater, or pouring chemicals through the port supporting the anode rod in electric water heaters.

In cold climates, adding nontoxic antifreeze to the water in a hydronic system prevents pipes from freezing and breaking. It takes a lot of antifreeze to achieve a good level of protection, though, and the material isn't cheap. Any leak in the system could cause a large financial loss. Antifreeze also doesn't carry as much heat, reducing the efficiency of the system. In our mild South Carolina climate, rather than adding antifreeze to the system, I have opted to insulate all pipes and manifolds exposed to cold temperatures. When the forecast predicts

Table 4.4. Heat Output of EPDM Rubber Tubing Used for Hydronic Heating (Btus/Square Foot)

Tube Spacing	100°F	110°F	120°F	130°F	140°F	150°F	160°F	170°F
2″	45	56	68	79	90	102	113	124
3″	30	37	45	53	60	68	75	83

Source: Delta T Solutions: www.deltatsolutions.com

cold nights, I circulate warm water through all noninsulated zones to prevent freeze-ups.

The Row Cover Bonus

After seeing the hydronic heating systems in our greenhouses, Dr. Geoff Zehnder (coordinator of Clemson's Integrated Pest Management (IPM) and Sustainable Agriculture Programs) posed the question, "What happens when hydronic tubing is combined with row covers?" Traditionally, growers use row covers to protect plants in fields and greenhouses from late or early cold spells that would otherwise wipe out a crop. More recently, growers started using them to speed up plant growth in greenhouses during winter. Wires strung a few feet above the ground allow you to quickly pull the covers over plants at the end of the day, then pull the covers back off the plants in the morning.

Surprisingly, I found no research on the topic of combining row covers with hydronic heating systems. As a result, Dr. Zehnder and I decided to research the topic. Our preliminary results are interesting: In separate experiments we tested thick- and medium-grade row covers with and without hydronic heat under the row cover. To conduct the tests, we placed temperature sensors connected to data loggers at different distances from the hydronic rubber tube at a height of 4 inches above the soil. The water in the hydronic system was around 125°F.

The preliminary results of the experiment showed that hydronic heating more than doubles the performance of medium-grade row covers and drastically improves the performance of thick row covers directly above the heating tube. Less but

Figure 4.26. Here are our research plots for the comparison of the effect of row covers with and without hydronic heating underneath.

Table 4.5. Average Temperature Boost (in °F) above Nighttime Minimum inside Greenhouse Using Row Cover with and without Hydronic Heat

Type of Row Cover	Temperature under Row Cover with Hydronic Heat Tubing		Temperature under Row Cover Alone (No Hydronic Heat)
	12″ from Tubing	Directly above Tubing	
Medium Agribon AG-19	12	20	8
Thick Agribon AG-20	12	22	10

still substantial heat is gained 12 inches from the hydronic tube. Table 4.5 presents results based on the average of a few of the coldest nights that occurred during the experiment.

While I did see improved germination and growth with hydronic tubes alone, the tubes provided little to no cold protection on the cold nights. We ran only a single loop of hydronic tubing down a 5-foot-wide bed. Our heat transfer would have improved had we placed more hydronic tubes beneath the plants. However, even a single loop of tubing used in conjunction with row covers has a drastic effect. On our coldest night, when it was 12°F outside the greenhouse, the temperature under the medium-grade row cover directly above the hydronic tube was a balmy 41°F.

SUMMING UP THE FUNCTIONS

1. South-facing sloped platform allows greenhouse film to catch more sunlight than a traditional greenhouse.
2. Gutters on sloped platform harvest rainwater.
3. Rainwater collected in pond reflects sunlight into greenhouse in winter.
4. Sloped platform, filled with gravel and topped with long-lasting landscape fabric, maximizes the absorption and storage of heat.
5. Water-filled barrels store direct and reflected sunlight.
6. Automated irrigation system irrigates plants, saving time and money.
7. Hydronic heating heats soil and reduces energy consumption by at least 20 percent, compared to forced-air heating of the whole greenhouse.
8. Hydronic systems allow spot heating and connection of other greenhouses with a central hydronic system, eliminating the need to purchase separate heaters for every greenhouse.
9. Greenhouse provides space for storing and mixing potting soil ingredients and tables for seedling production.
10. Benches with Tanglefoot applied to bench legs protect seedlings from pests.
11. Misting system under benches provides a shady area for asexual propagation.
12. Predators of plant pests find habitat in and around the pond.
13. Convective loops facilitate distribution of heat stored in thermal mass at nighttime.
14. Sloped platform allows for convective removal of heat through vents, which facilitates cooling.
15. Insulated walls and roof trap nighttime heat.
16. Shade cloth over greenhouse provides shade for dehydrating garlic and vegetables.
17. Greenhouse located on the west side of hardening-off tables protects the tables from intense afternoon sun.
18. Greenhouse blocks wind from hardening-off table.

CHAPTER 5

Greenhouse with an "Innie"

The Bio-Integrated Greenhouse with an Interior Pond

Growers spend thousands of dollars, sometimes hundreds of thousands, to heat greenhouses during the winter. That's a problem because, as finite resources, fossil fuels aren't getting any cheaper. Luckily, there's a logical solution — building ponds inside greenhouses.

Over a six-month period, a pond as small as 12 feet by 15 feet can store solar heat equivalent to the heat produced by 195 gallons of propane. At $4 per gallon that's a total of $780 worth of nighttime heat during the colder part of the year. For less than $400 you can build a pond that will pay for itself in a few months' time. What other investment can yield that kind of return?

Some growers argue that ponds take up precious space in a greenhouse, that vegetables planted in the same amount of space would produce more profit than the savings due to the pond. I agree — unless you grow a marketable product in the pond.

At the Student Organic Farm (SOF), we're always experimenting. The two greenhouse-interior ponds we built have provided fertile ground for intriguing projects, some of which

worked out better than others. For a few years we grew tilapia in the ponds, but that venture wasn't very profitable. We did save money on heating costs for the greenhouses and on fertilizer costs for plants in our hydroponic systems. Currently, we are using the ponds inside the greenhouses as nurseries to overwinter high-value freshwater prawns. We then stock the prawns into larger outdoor ponds during the growing season, where they grow larger with a minimum input of energy. We sell the prawns to our community-supported agriculture (CSA) members for ten dollars per pound. And as I describe in chapter 8, we've also integrated the interior ponds into our hydronic heating system for the greenhouses. Our hydronic heating pipes run through our compost piles; the water in the pipes warms as it travels through the hot compost, then transfers heat directly or through heat exchangers into the interior ponds. The ponds become heat storage batteries, releasing and recharging as the water flows through.

Water stores heat better than soil. When solar energy research was advancing during the early

Figure 5.1. The sides of this greenhouse at the SOF are rolled up, revealing the pond inside. The pond collects solar energy during the day, and at night it releases the heat to warm the greenhouse. We also raise prawns in the pond.

1970s, shallow solar ponds were proposed as a cheap means to collect and store solar energy.[1] Another popular approach for storing thermal energy in water is using 55-gallon barrels or water tanks. I've tried this technique, and though it did warm the greenhouse, I wasn't satisfied, because I couldn't use the hot water in the tanks for any purpose other than heating. I prefer ponds because I can grow fish in them, too, which in turn gives the water an additional function: It provides a nutrient-rich fertilizer for plants.

About Aquaponics

I remember going snorkeling for the first time when I was a kid. We were in Hanauma Bay on the island of Oahu in Hawai'i. I nearly missed the experience of a lifetime because I didn't want to wear a snorkel and mask. I'm not sure if my resistance arose out of sheer embarrassment or fear of suffocation, but boy, I'm glad I got over it.

There's a whole other world beneath the surface of the ocean. Fish dart in and out of brightly colored coral, pausing to nibble algae off the reef. Sea turtles slowly drift along, occasionally chomping on sea grasses swaying in the current. Underwater life has always intrigued me, and so has horticulture. So it was a natural progression to explore aquaponics.

Aquaponics combines aquatic animals with hydroponics (growing plants in a nutrient solution rather than in soil) in a symbiotic system. It sounds complex, but it's fairly simple — fish waste feeds

Elements of the System

Building a pond inside a greenhouse opens the potential to store free solar heat during the day, then pump that heat at night through the greenhouse using natural convective loops. The pond can be connected to the hydronic heating system using a simple heat exchanger that transfers heat from an exterior compost pile into the pond water (as described in chapter 8). The pond water can be filtered through an aquaponic system, which allows for rearing fish or prawns during cold weather in the interior pond to stock large outdoor ponds in the spring for commercial production. In this chapter you'll learn how to:

- ► Construct a rainwater-harvesting pond inside a sloped greenhouse, which can result in 6.5°F of additional cold protection beyond that provided by a standard greenhouse
- ► Connect an interior pond to a hydronic heating system for heat storage, enabling higher greenhouse and pond temperatures if desired
- ► Use an interior pond to grow fish or prawns, which saves money on expensive tanks
- ► Filter pond water using several different types of productive hydroponic systems
- ► Transfer and raise fish and prawns in outdoor ponds

plants rooted in the water, the plant roots clean and filter the water so the fish stay healthy, then we eat the fish *and* the plants. The basic components of an aquaponics system include a greenhouse pond, fish, waste, biofiltration, and hydroponically grown plants.

Aquaponics is a natural use of an interior greenhouse pond. Not only does the pond warm the greenhouse, saving us money and energy, it becomes its own bountiful cornucopia, diversifying our overall production. I've grown baby lettuce year-round in our aquaponics system. The pond water moderates the temperatures for the plants, keeping them warm in the colder months and cool in the warmer months. Since it tolerates fluctuating nutrients and flourishes in flotation, lettuce is perfect for growing in a hydroponic system. This leafy green commands a high price per square foot and is always in demand. And as I described above, beneath the surface of the water, we've grown tilapia fingerlings and are experimenting with freshwater prawns.

Aquaponics balances the growth of these aquatic animals with the cultivation of plants in a recirculating system. With recirculating soilless techniques developed for hydroponic systems, plants thrive from the nutrient-rich waste of the water. In trade the plants remove toxic waste from the water and allow a larger population of fish to inhabit the pond. Though the plants get the most credit for this biofiltration process, it's really the bacteria we should thank, as I explain later in this chapter.

Building the Pond

Before we dive into the details of managing an aquaponics system, let's discuss how to build interior ponds. Building a pond inside a greenhouse is not as difficult as you might think. The pond is primarily constructed belowground; it's similar in some ways to a pool or water garden. The pond is formed with a heavy pond liner that attaches to wooden baseboards or concrete blocks, and it has

Walking in Water

Getting in and out of a pond is sometimes an ungraceful experience. Unfortunately, I can't count on one hand the number of times I've clumsily fallen into a pond, then had to race across the farm to change into dry clothes — not so bad when it's 90°F outside but breathtaking when it's below freezing. It's not always necessary to work in the water, but getting your feet wet is a part of pond management when it's time to harvest or move the fish. I've found that vertical walls ease this venture. Sloped sides become slippery when wet, but a vertical wall allows a safer transition. I've cast the sides of ponds in concrete and used cinder blocks to create vertical walls. The extra space also provides more room for water storage. If I have to negotiate the slippery sides of a pond, I place a ladder or net inside the pond on the steep slope to supply traction.

---

insulation underneath the liner and a hardware cloth edging (to prevent rodents from taking up residence under the liner). What's more complex is the mechanism to *fill* the pond once you've built the structure.

If you build the greenhouse on a sloping platform as described in chapter 4, the slope of the platform in conjunction with modified ground gutters will direct water into the interior pond. If you dig the pond prior to greenhouse construction, you can use the soil from digging the pond to create the slope of the greenhouse platform. Alternatively, if you build the pond post–greenhouse construction, you may not be able to harvest rainwater from the greenhouse and a sloped platform, but you could fill the pond from another water source, such as a well, a municipal source, or an uphill pond. Building an interior pond after greenhouse construction will require disposing of or using the soil from digging the pond in other areas around the greenhouse instead of modifying the slope of the greenhouse platform.

Here's an overview of the construction process.

Step 1. Figure out how big you want the pond to be. Our ponds occupy a space 12 feet wide and 15 feet long, spanning the full width of the houses at the low end (remember the greenhouses are built on a gentle slope). Occupying about one-fourth of the total greenhouse area, each pond has an average depth of 2 feet and holds approximately 1,400 gallons of water.

Step 2. Procure a liner of the proper size, along with wood for the baseboard and the polystyrene insulation and the hardware cloth. I prefer to use a thick, 45-mil EPDM pond liner.

Unlike the outdoor ponds I build, I don't cover the liner of an interior pond with soil, because the black pond liner absorbs more solar energy than soil would. Also, if there is soil in a pond, the soil becomes suspended in the water when the water circulates, and water circulation is essential for raising fish and prawns in high densities. So soil would cause multiple problems: It would block sunlight, it would clog up the hydroponic system through which we filter the pond water, and it would make it difficult to remove solids (fish manure, etc.) from the water; not to mention, it would inhibit your view of the fish.

Step 3. Install a wooden baseboard at the junction of the pond wall and the greenhouse wall, using treated or engineered wood. This wooden border forms the upper edge of the pond. I place this border inside the metal hoop framework of the greenhouse, which makes it easier to attach the pond liner. On the side of the pond that doesn't border a greenhouse wall, I pound in galvanized tubing stakes and secure the wooden baseboard to the stakes.

Step 4. Create the slope on the sides and bottom of the pond. The sides can be steep, because there's no worry about soil slumping, and the pond will hold more water because there's no soil taking up space. A two-to-one slope ratio is sufficient for a pond with an uncovered liner.

Step 5. Uncovered liners and ponds inside greenhouses create the perfect habitat for rats. If rats take up residence under the liner, they will eventually chew through, creating a leak in the pond. To protect the liner, I fasten a 2-foot-wide strip of galvanized hardware cloth to the baseboard all the way around the rim of the pond, as shown in figure 5.2. Since applying the hardware cloth a year ago, I haven't seen any rats or rat damage. While the hardware cloth does provide permanent insurance, I attribute the absence of rats to ground gutters (described in chapter 4) and a few friendly rat snakes who call the greenhouses home.

Step 6. I line the bottom of the pond site with 1-inch foam insulation to prevent heat loss into the ground. This type of extruded polystyrene insulation is typically used under roofs and on walls. I cut the insulation into sections that fit together tightly like pieces in a puzzle: the goal is no gaps. The foam should lie flat against the ground to prevent cracking and should extend to the farthest edge of the pond.

Once the insulation is in place, I cut out a 2 × 2-foot square of insulation at the low point, creating a sump, as shown in figure 5.3. Since

Figure 5.2. A 2-foot strip of hardware cloth around the rim of a pond protects the insulation and pond liner from rodent damage.

Figure 5.3. Insulation laid over the pond floor prevents heat from escaping into the soil. The square cutout at the center creates a sump, which is useful when the pond needs to be drained.

the last inch of water is difficult to remove when pumping or siphoning water out of a pond, the sump creates a confined low point so that almost all of the water can be removed easily. The square of insulation cut out for the sump is then applied to the bottom of the sump after removing 4 to 6 inches of soil to create the low area.

Step 7. I spread out the liner over the insulation and the baseboard and smooth it to prevent folds. Before securing the edges of the liner, add enough water to partially fill the pond. The weight of the water removes any slack and ensures that the edges of the liner won't shift after filling.

Step 8. Once I fill the pond about halfway, I use furring strips and screws to secure the edges of the liner to the wooden greenhouse base. If you use pressure-treated wood for the baseboard, wrap it with extra pond liner so there is no contact with the soil, or chemicals may leach out of the wood.

Figure 5.4. The interior edge of the pond is fastened to galvanized stakes and the support posts for the roll-up cover. The treated wood is wrapped in excess pond liner to prevent ground contact. The liner of this pond is pulled back to show the wooden baseboard and the galvanized stake.

Filling the Pond

After the construction is complete and I fill the pond, I can suddenly see the slope of the greenhouse. The subtle slant is usually undetectable, but the level water surface reveals the illusion like logic exposing a magician's trick.

As I mentioned above, filling an interior pond is not a simple matter, because we don't simply turn on a tap to do so. Instead, we create a mechanism that allows us to fill the pond with harvested rainwater. Since that water is harvested at ground level, we have to position an interior pond at the lowest point of a sloping greenhouse. We also connected our greenhouse pond to the turkey nest pond located uphill from the greenhouses using a siphon. The uphill pond is a stored water source we can rely on to fill the greenhouse pond when needed (see chapter 6 for how to make a siphon). The siphon is connected to the greenhouse pond with a float valve to replenish water automatically as soon as the water level in the greenhouse pond drops below the preset point.

Harvesting rainwater is cheap and easy using a ground gutter system like the one described in chapter 4. The challenge lies in how to direct the rainwater into the interior pond. One approach is to allow the water to drain into an exterior basin, where it flows into a feeder pipe that leads into the pond. Or the rainwater can flow from the gutters into a distribution box. In that case, simply moving a plug between pipes inside the box controls the direction of flow.

If the greenhouse platform slopes in two directions, as presented in chapter 4, a low point exists at one corner of the greenhouse. Water in ground gutters or ditches alongside the greenhouse flows downhill to this single low point. For the pond inside the greenhouse to harvest water from the gutters or ditches, the surface water level in the pond should be equal in height to the lowest

Figure 5.5. No, the camera wasn't crooked. But the water in this pond naturally forms a perfectly level surface, which reveals how the entire greenhouse sits on a slight two-directional slope.

point on the sloping platform. At this low point a feeder pipe penetrates through to the interior pond, as shown in figure 5.6. (I recommend using a specialized pipe fitting suitable for the pond liner material to seal the pipe penetration.) A cap, sewage cleanout, or valve can be placed on the feeder pipe to control when water is allowed to flow into the interior pond (see chapter 3 and figure 2.40 for an overview of water flow for the entire system).

Covering the Pond

I started covering ponds once I realized the enormous expense it took to keep pond water warm. Luckily, ponds are like pools, and keeping pool

Figure 5.6. The basin collects rainwater flowing off the sloped greenhouses. When the overflow pipe of the basin is closed and the valve on the pipe feeding into the greenhouse is opened, the force of gravity will send water through the pipe into the pond inside the greenhouse. Water from the uphill side of the greenhouse flows into the ditch and toward the feeder pipe as well. Water from upper ponds can also be channeled through this same ditch.

water warm is an advanced science. Solar pool covers provide a cheap, easy way to heat ponds and prevent evaporation. At thirty cents per square foot, covering our greenhouse pond cost a whopping fifty-four dollars. Evaporation accounts for most of the heat lost from ponds, so simply placing a plastic sheet over the pond will help. However, high-tech pool covers are designed to prevent evaporation, insulate, and last for many years. Experiments with "bubble wrap"–type covers have proven their insulation capabilities. By trapping heat gained from sunlight, the covers increased water temperatures by 10° to 16°F.[2]

Covers come in different levels of opacity. Opaque black covers help convert sunlight into heat and also prevent algae from growing by blocking sunlight from entering the water. Translucent covers allow sunlight to enter the pond, and that light strikes the black bottom liner, converting to thermal radiation and heating the water. If I exclude fish and other aquatic organisms, the pond will stay relatively clear. However, when I add life to the pond, an accelerated ecosystem forms, powered by the heat and light trapped by the clear pool cover. Warm water, fish manure, and plenty of sunlight combined with a solar pool cover creates an ideal cocktail for algae growth. If I stock algae feeders such as tilapia or prawns in the pond, the increased algae may provide all the feed necessary to grow the fish. Without algae feeders,

Figure 5.7. A solar pool cover rolls up for easy access to the pond.

clear covers turn ponds into a thick, green algae soup. I advise caution, because too much algae can clog pipes and pumps.

If you want to be able to access the pond without having to remove and fold up the cover, you can install a roll-up system that's similar to the mechanism for a roll-up side on a greenhouse wall (described in chapter 4). Here's how to install a cover with a roll-up system.

Step 1. Figure out what size cover you need. The cover should be the same width as the pond and 4 feet longer.

Step 2. Procure the cover, a 1.5-inch galvanized tube, a piece of 1-inch PVC pipe with endcaps attached to both ends, tek screws, cable ties, a handle, two 4 × 4 posts 4 feet long, and four 2 × 2 pieces 1 foot long.

Step 3. Attach the handle to the galvanized tube, as shown in figure 5.8.

Step 4. Dig holes for the 4 × 4 posts at the inside corners of the pond, and set the posts in place.

Step 5. Attach the 2 × 2 pieces to the 4 × 4 posts to form a U at the top of each post. (An alternative is to cut notches into the tops of the 4 × 4 posts.)

Step 6. Seal the PVC pipe ends using the caps to ensure the pipe will float (this makes it easy to pull the cover across the pond). Attach one edge of the pond cover to the galvanized tube using tek screws. Attach the opposite end of the pond cover to the PVC pipe using cable ties to prevent puncturing the PVC pipe. I perform this task in an open area outside the greenhouse where it's easy to unfold the cover completely.

Step 7. Roll up the cover on the galvanized tube, and insert the ends of the tube on top of the 4 × 4 posts.

Step 8. Attach a rope to the PVC pipe, and run the rope through a pulley attached to the baseboard at the far end of the pond.

Step 9. Use the rope to pull the PVC pipe and cover across the pond. The handle rolls the cover back up, removing it from the pond.

Figure 5.8. Handle and support for the roll-up cover.

Figure 5.9. A pulley is used to pull the cover over the pond.

Though pond covers are beneficial, they limit access for feeding fish and watching breeding and feeding behaviors. I cut a 2 × 2-foot flap in the cover that I can lift with a long pole, which allows me to feed the fish easily. One problem is that fish may jump through the opening and become stranded on top of the cover, so I make sure to close the flap before I leave. Or I use plastic netting to cover permanent openings in covers. Pond covers also limit the amount of oxygen naturally being diffused into the water through the pond surface. You can compensate for this by using high

circulation rates and a blower with diffusers (this is described later in this chapter). But in a pond with low water circulation, I reduce the population of fish to compensate for the lower oxygen level caused by the cover.

Water Quality: Biofiltration and Aeration

In an aquaponic system fish waste inevitably accumulates in the water, affecting the water quality.

Preventing Evaporation and Condensation

Condensation is a natural result of evaporation. Ponds inside greenhouses create evaporation when the warm water condenses on the cool greenhouse roof, generating a constant rain of water. This extra moisture and humidity can wreak havoc on plants, forming an ideal environment for fungal disease. The condensation can also block sunlight, reducing photosynthesis and solar heat gain. Additionally, evaporation leads to heat loss in the pond, reducing water temperatures and plant and fish growth.

Prevent evaporation and condensation by covering all exposed areas of the pond with floating hydroponic rafts or floating solar pool covers. The covers keep heat inside the pond and prevent evaporating water from condensing on roofs. Running fans can also help reduce condensation.

Consider the roof design of your greenhouse. Steep, angled roofs allow condensed

water to roll down the sides and back into the pond. Roofs with flat tops permit condensed water to form large droplets that rain down on top of plants and create disease problems.

Figure 5.10. Condensation on the roof of a greenhouse with a pond inside can block sunlight and cause disease problems.

Overall there are three main water quality parameters to manage in intensive production systems: oxygen, pH, and nitrogen wastes. To maintain high oxygen levels for the fish, the solids from the waste must be removed from the water. If too many solids remain, decomposition depletes the water of needed oxygen and produces carbon dioxide, a toxic compound for fish. The pH of the water drops, because of the conversion of feed into acidic compounds, and the pH must be adjusted using basic compounds. Also, harmful nitrogen wastes are excreted through fish gills and must be converted into less toxic compounds using the help of bacteria.

To filter the water we pump it from the pond into three different filtration locations, splitting the flow using tees and valves:

- Flood-and-drain hydroponic system
- Floating raft hydroponic system, preceded by swirl separators to remove solids from the water
- Packed column aerator, a large column filled with aggregate to encourage bacterial growth

In the aquaponics system I designed at the SOF, I plant vegetable plants rooted in perlite or aggregate directly into rafts of expanded polystyrene insulation and a flood-and-drain bed full of aggregate. Bacteria colonize the outer edges of the perlite and aggregate and convert nitrogen fish waste, such as ammonia, into nitrates, which the plant roots can absorb. The bacteria are the unsung heroes. While the veggies are hamming it up on center stage, the bacteria are busy below the surface keeping everything working properly.

This is biofiltration — symbiosis at its best: The fish have clean water, the plants have nutrients, and we reap healthy food to eat. We use two separate systems to filter nitrogen wastes: a flood-and-drain system and a floating system.

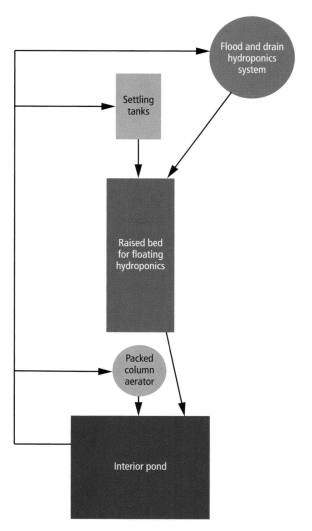

Figure 5.11. Illustration showing the flow between components used to filter water from the interior pond. Arrows show the direction of water flow.

Flood-and-Drain System

A flood-and-drain system mimics the ebb and flow of tides in a marsh. At high tide the salty seawater floods the marsh — at low tide it recedes back into the ocean. Marsh grasses and organisms such as oysters depend on this life-sustaining ebb and flow. At SOF there's no ocean, but a similar process occurs.

We fill a round waterproof tub, 5 feet in diameter, with 1 foot of aggregate or gravel. Then we pump pond water into the tub, enough to nearly cover the gravel but leaving the top 2 inches dry to prevent algae growth. (I talk about pumps in detail later in the chapter, under Pumps.) We place a special valve known as a bell siphon in the middle of the tub. The pump fills the tub, and the siphon drains the water as soon as the tub is full. The water then drains into the floating hydroponic system, which then flows into the pond, creating a perpetual ebb and flow of water in the tub.

The bell siphon drains water out of the tub faster than the pump fills it, and a valve preceding the tub ensures that the water enters at the correct flow rate; too fast, and the bell siphon won't work.

Three different mechanisms work together to filter the water: The aggregate in the tub physically filters the solids, earthworms living in the aggregate eat the solids, and microbes attached to the outer edges of the aggregate convert the solids into nitrates. The constant influx of oxygen in every drain cycle enhances the microbial and earthworm activity.

Figure 5.12. An ebb-and-flow hydroponic bed with red lettuce growing in the aggregate. The white top of a bell siphon is visible amid the lettuce; pipes on the right feed water from the pond into the tub.

We've made the flood-and-drain tub multipurpose by planting lettuce and basil directly in the tubs. The plants use the aggregate as a rooting structure and absorb the nitrates produced by the bacteria. In essence, the flood-and-drain technique is an all-in-one ecosystem accomplishing multiple functions in a compact environment.

Flood-and-drain systems are the simplest to operate but require the most expense per square foot of grow area. They perform well in low-nutrient situations because the accumulating solids steadily release nutrients for plants. However, the solids need periodic flushing. We accomplish this by stirring the aggregate with a shovel, then flushing out the solids with a hose through a drain in the bottom of the tub. To drain the water for cleaning, we remove the middle riser of the bell siphon. Beneath the tub we rotate an elbow that usually directs water into the floating system. We direct it instead so the solids drain into adjacent soil beds during flushing to fertilize the soil. We do this after harvest, before planting a new succession of crops in the tub. Attaching a short length of PVC to the end of the hose creates a stiff rod to stir and plunge into the gravel, easing the cleaning process.

Flood-and-drain systems rely on pumps properly working. If the electricity shuts off or a pump breaks while the basin is full of water, the plants will be fine. However, if the basin is empty during a drain cycle and the power fails, the plants will quickly perish without water.

Floating Systems

Floating hydroponic systems offer the lowest cost per square foot of grow area. They make use of shallow raised beds, but instead of filling the beds with soil, we fill them with water. Polystyrene sheets float like rafts in the water. We cut holes directly into the polystyrene for the plants, and their roots dangle in the water below. Water from the ponds circulates

How a Bell Siphon Works

The bell portion of a bell siphon usually sits over a standpipe, surrounded by a perforated pipe. As the water level rises in the tub, water is forced through the teeth in the bottom of the bell and up between the walls of the bell and standpipe. As the water level breaches the lip of the standpipe, a siphon is created, which then drains the tub. Once the water drains out of the tub the snorkel on the side of the bell breaks the siphon, and the tub begins to fill again. The University of Hawai'i College of Tropical Agriculture and Human Resources has a great fact sheet on building bell siphons.[3]

Figure 5.13. Here's the bell part of the bell siphon, which has been pulled up and out of the standpipe at the center of this flood-and-drain tub.

through the bed, nourishing the plants with a continuous stream of nutrients. Plants are also not vulnerable during power failures, because the roots are always suspended in water; there is no drain cycle.

Every system has its disadvantages. Suspended solids in floating systems may accumulate on plant roots, creating anaerobic zones that prevent nutrient uptake. Therefore, floating systems require the addition of a solids-removal component to keep roots clean. Floating systems also require higher feeding rates to generate enough nutrients for good plant growth. A good guideline is ¾ pound of fish feed per day for every 4 × 8-foot sheet of floating grow area. Feeding rates this high require adequate aeration and stocking rates.

Both techniques work well in an aquaponic system but fulfill different needs. I decide which technique to use based on my stocking rates. I choose ebb-and-flow tubs if my pond has fewer fish, and floating rafts if my pond has more fish.

BUILDING THE BEDS

The foam insulation sheets I use to grow plants in these beds are usually sold in 4-foot widths, so I design my bed to accommodate this standard width. It's common to use treated lumber to build aquaponic beds. A typical wooden frame stands approximately 16 inches tall and has support posts on the exterior of the bed, keeping the interior free of protrusions. A pond liner made from high-density polyethylene (HDPE) laid inside the bed forms a watertight barrier. Commonly, a white liner is used so solids settling on the bottom of the basin are easily visible for removal with a siphon.

Because treated lumber harbors many toxic chemicals, I use it sparingly at the SOF to retain our organic certification. I am not concerned about the plants growing hydroponically in the floating beds because the pond liner will prevent their roots from contacting the treated wood.

Figure 5.14. When constructing a raised bed for hydroponics, filling trenches with compacted gravel before putting cinder block walls in place is essential to prevent the bed from sinking.

Figure 5.15. The plumbing installation in the central trench in this raised bed is complete. The blue pipe moves heat from the hydronic heating system into a heat exchanger inside the pond; the white pipes move water from the pond into the floating and flood/drain hydroponic systems.

Figure 5.16. The basin is lined with insulation to limit heat transfer out of the water. The insulation is cut to fit snugly, limiting gaps between sections. Multiple overlapping layers would provide even more insulation in colder climates.

However, plants rooted in the ground and ponds near the treated lumber may absorb the harsh chemicals leaching from the lumber. That's why I'm a big fan of cinder block beds. Since cinder blocks don't contaminate the ground, I can grow a mixed production of hydroponic and soil-grown plants in the greenhouse.

To build a bed with cinder block walls, I first dig a level trench to a depth that is 22 inches below the final desired height of the beds. I fill the trench with 6 inches of compacted gravel.

Next I build the walls, using two layers of cinder blocks. I use grading stakes and masonry lines to achieve a perfectly level top. (Leveling is imperative for beds built with treated lumber, too, or water will be wasted.) The bottom of the bed itself can slope to a low point for easy solids collection during draining and cleaning.

Figure 5.17. I cap the cinder blocks with paving stones to hide and hold the edge of the pond liner in place. The polystyrene rafts in the water are ready to receive plants.

I also dig a trench down the center of the bed for plumbing: pipes that pump water from the pond into the floating beds and the flood-and-drain beds. By placing the pipes directly under the bed, I ensure that any heat lost from the pipes will transfer directly into the grow bed for reuse.

Before the bed can be lined, it has to be insulated to limit heat transfer out of the water. I fill the cinder blocks with sand for extra thermal mass and cover the floor and walls of the bed with polystyrene insulation.

The final stage of construction is installing the liner. I lay a pond liner (Permalon) inside the trench and up the walls, with the liner edges cut to rest on the top face of the cinder blocks.

DEALING WITH OVERFLOW

Once we've built the floating bed, we fill it by pumping water from the greenhouse pond into the bed. Water siphoned from ponds uphill replaces the water removed from the greenhouse pond. Alternatively, we fill the bed with city or well water.

Since water from the pond constantly pumps into the floating basin, we install an overflow to bring the water back to the pond. There are two ways to do it: The overflow can be piped out through a standpipe or allowed to flow over the edge of the wall like a waterfall.

To pipe out the overflow, attach a removable standpipe to a bulkhead fitting in the bottom. With a standpipe attached, surface water flows over the horizontal lip of the pipe, down through a drainpipe, and into the pond. With the standpipe removed, the water exits through the drain in the bottom of the bed. It's similar to a bathtub. Bathwater can exit through the overflow drain on the side of the tub or through the drain on the bottom of the tub. But unlike bathtubs, the drainpipe is located inside the hydroponic bed. (Thank

goodness bathtubs aren't designed this way, or that would make for one uncomfortable bath!)

At the SOF I let the water flow over a low point in the wall of the bed, creating a cascade of water over the side and into the pond. The low point is about 1 inch lower than the rest of the bed and has a breadth of 4 feet. Make sure the rest of the bed is perfectly level so the only low point is the desired overflow area. This low-tech method is an easy way to move water from the raised bed to the pond, and the waterfall effect aerates and filters the water. At the farm we connect the overflow with the packed column aerator. Water spills over the edge of the bed, and the same pond liner connecting the bed to the pond forms the impermeable base for the packed column aerator (see figure 5.23). An aesthetic alternative to a packed column aerator in this location is a waterfall of rocks, gravel, or biofiltration media on top of the liner as it slopes into the pond.

THE FLOATING RAFTS

Modern aquaponic rafts are made from extruded polystyrene (XPS). A cheaper alternative for a floating surface for plants is expanded polystyrene (EPS); however, the open-cell form of EPS absorbs water and degrades quickly. I use polystyrene sheets that are 1.5 inches thick so the rafts will not break when they're full of plants. Edges of the foam should be square because tongue-and-groove edges will break. I have also fused together two ¾-inch sheets with food-grade adhesive, since 1.5-inch foam is difficult to find. Carefully select your foam material by checking the chemical composition from the manufacturer because some foams harbor fire- or mold-preventing chemicals that are toxic to fish.

The foam sheets are usually 4 × 8 feet but can be shortened to 4 feet to facilitate easier handling and lifting. Using a hole saw, I drill holes into the foam sheets at a spacing appropriate for the type

Figure 5.18. A floating raft made from extruded polystyrene (EPS) insulation supports the growth of romaine lettuce. Roots from lettuce transplanted into the rafts six weeks earlier hang in the water below, absorbing nutrients from fish waste.

of plant. We space holes 8 inches apart, measuring from the center of one hole to the center of the next, and make the holes 2 inches in diameter to accommodate lettuce and other closely spaced plants. Net pots are placed inside the holes to hold the aggregate and plants. Alternatively, we simply stuff well-rooted plants into the holes if they're rooted in a lightweight potting soil mix. We add enough perlite to the potting soil mix so the ratio is 3:1 (75 percent perlite to 25 percent peat and compost ingredients), then place the well-rooted transplants directly into the floating rafts, saving time and money.

I've noticed iron deficiencies in some plants in our floating systems. The symptoms appear as interveinal chlorosis, or a yellowing between the veins of the leaves. We correct the deficiency by adding about 5 tablespoons of 10 percent chelated iron to every 1,000 gallons of water to bring the concentration of iron to 2 to 3 ppm. Iron testing kits are available to check iron levels.

Removing Solids

Unlike flood-and-drain biofiltration systems, floating systems don't benefit from solids in the water. A solids-removal component is essential to prevent nutrient-blocking debris from accumulating on roots. Accumulating solids also deplete the water of oxygen as they decompose, and that's bad for both plant roots and fish.

Since our raised bed lacks a drain, I periodically pump or siphon out the water, then scoop the sediment from the bottom of the bed. I have also vacuumed the solids with a wet vac, pool vac, or siphon. The solids and nutrient-rich water from the bed is pumped or dumped onto our compost pile or fields and treated like raw manure for safety reasons.

To avoid having to clean solids from the floating hydroponic bed, it's best to remove the solid material from the water before it enters the bed. I use two techniques for solids removal—settling tanks and swirl separators. The water from the pond continuously flows into the settling tanks before it enters the floating grow bed to help remove solids.

SETTLING TANKS

Settling tanks or basins are cheap options for the removal of solids. Water from the pond is passed through a basin or tank, which slows the flow of water. Since solids in the water have a higher density than the water, they sink to the bottom and accumulate. Recommended depths for basins are 3 feet with retention times (how long the water stays

Figure 5.19. Settling tanks help remove solids before water flows into a floating-raft hydroponic system.

in the tank) ranging between fifteen and forty minutes.[4] Longer retention times will allow smaller, less dense particles to settle out of the water, too, which improves the water quality. Once the solids settle out of the water, the water flows out of the tanks and into the raised bed for floating hydroponics.

We use two 55-gallon drums in series as settling tanks for our floating beds. Recycled food-grade drums are readily available and cheap. Some recycled drums may have been used to store toxic chemicals, so it's important to know the history of any drum you plan to use as a settling tank. I insert a pipe to serve as the inflow a little lower than the midpoint of one drum. (If water enters any lower it will disturb the sediment settled on the bottom.) Water exits the drum through a standpipe at its top center. The standpipe connects to the second drum's midpoint to repeat the process. Both drums are connected to a single drain line with ball valves to manually remove accumulated solids. When the valves are opened water quickly exits the bottom of the drum, and the solids siphon out. Water from the drums drains into a retention basin below our compost pile, which allows for reuse of the nutrients in our compost (see chapter 8).

The drains need to be opened every few days depending on the amount of feed given to the fish. Flushing out about half the water in each barrel removes most of the accumulated solids in the bottom. A float valve inside the pond automatically siphons water from an upper pond to replace the lost water. The influx of fresh water helps improve overall water quality.

Be careful when you insert pipes through the plastic 55-gallon drum. I use several techniques to make a tight seal. For the cheapest, I simply drill a hole through the barrel the size of the inner threads on a male pipe coupler. I use a threaded metal pipe to cut threads into the plastic on the side of the barrel. The male coupler is then threaded into the

Figure 5.20. Plumbing for settling tanks. Both valves on the left are opened to drain solids. The valves in front are for two inflow lines (see "Installation and Plumbing" section on page 128).

barrel and sealed with silicone or marine-grade silicone for a longer-lasting seal. Another, slightly more expensive, option uses a Uniseal rubber ring. The rubber ring slips into a hole cut in the barrel, and the pipe pushes the rubber out, sealing the hole. The final and most expensive technique uses a bulkhead fitting, with a rubber washer that presses down on the inside of the barrel and creates a seal.

The two settling tanks also make a great table for placing the floating rafts when installing plants.

You can calculate retention times by recording how long it takes to fill the settling tank. Then adjust a valve on the inflow pipe until the desired time is achieved. The easiest way to measure retention time is to capture the water leaving the settling tank using a 5-gallon bucket. I use the following formula:

$$\text{Volume of tank} \div 5 \times$$
$$(\text{minutes to fill 5-gallon bucket}) = \text{retention time}$$

For example, if the settling tank is a 55-gallon drum and it takes two minutes for the water flowing

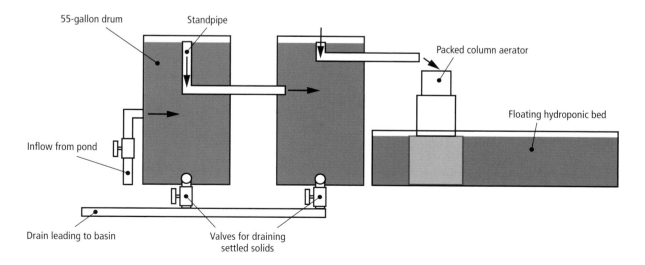

Figure 5.21. This diagram shows the flow of water through the plumbing of the settling tanks.

from the settling tank to fill a 5-gallon bucket, the retention time is twenty-two minutes.

$$55 \div 5 \times 2 = 22 \text{ minutes}$$

This is within the recommended time interval of fifteen to forty minutes.

Another option for settling solids is to use an expensive cone-bottom tank, which concentrates the solids at the drain valve opening, improving removal. However, with the flat-bottom 55-gallon drum, the drainpipes penetrate the tank on the side, and an elbow facing down siphons solids from the bottom. The efficiency is poor but might be improved by installing the tank on a tilt.

I also use the drain system of the settling tanks as a way to drain the whole pond, since I don't have a drain in the bottom of my pond. To use them as a pond drain, I pump water out of the pond and into the settling tanks using the existing pump and pipe system. I then leave the drain valves open, and all the water flows through the settling tanks and into the retention basin, thereby draining the pond.

SWIRL SEPARATORS

Centrifugal forces are greater than gravitational forces. That's why swirl separators can remove the same amount of solids as a settling tank in a much smaller container. Swirl separators operate like settling tanks with the addition of a whirlpool current. They, too, take advantage of the dense, heavy nature of pollutants in relation to water, but they also put a "spin" on things.

It's easy to convert round 55-gallon drums into swirl separators by adding a few pipe fittings. Simply placing a 90-degree elbow on the intake to the drum, as shown in figure 5.22, sets up the swirling motion. The incoming water is directed toward the outside periphery of the drum, creating a whirlpool effect. The swirling motion pushes solids

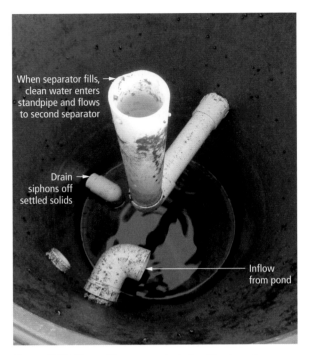

When separator fills, clean water enters standpipe and flows to second separator

Drain siphons off settled solids

Inflow from pond

Figure 5.22. The swirl separator is drained to show the components inside the drum.

toward the outer edge of the tank and concentrates the cleaner water in the center of the drum. Since water exits the tank through a standpipe in the center, the swirling action helps keep the outflow clean. Swirl separators allow a loading rate up to five times higher than traditional settling tanks, making efficient use of small areas.[5]

Aeration

Aeration of pond water in an aquaponic system is important, because as fish production increases, oxygen becomes a limiting factor, and nitrogen and carbon dioxide may build to toxic levels. Aeration removes carbon dioxide and nitrogen from water while increasing the concentration of oxygen dissolved in the water. Oxygen goes in — carbon dioxide and nitrogen go out. Dry air is composed of approximately 78 percent nitrogen and 20 percent oxygen by volume, and water is composed of 60 percent nitrogen and 40 percent oxygen when in equilibrium with the air. The air constantly exerts pressure on water, and the water takes up the oxygen and releases nitrogen until an equilibrium is reached and the water becomes fully saturated. The temperature of the water will determine how much oxygen it can hold. Cooler temperatures hold more oxygen.

The goal of aeration is to optimize the conditions for the exchange of gas between air and water by creating a large contact area between them. The thinner the layer of water, the more effective gas transfer becomes. I use two basic techniques to aerate water: cascading water through air and pumping air into water.

PACKED COLUMN AERATOR

A packed column aerator cascades water through air. Despite its complex name, it's a simple design. Basically, water runs through a column filled with aggregate. The water thins as it spreads over the many pieces of aggregate. This increases the surface area of the water, exposing it to more oxygen while ridding it of nitrogen and carbon dioxide.

Ideally, a packed column measures 6 feet in height. The taller the aerator, the greater the amount of time for air and water to exchange gases as the water cascades through the column. The porosity of the medium inside the aerator determines the width of the column. Smaller particles have more surface area but clog more easily, requiring wider columns. Higher flow rates will also require wider columns. Most columns measure 1 to 2 feet in diameter.

For even water distribution the column is topped with a perforated plate (a piece of Plexiglas with holes or a filter media pad). Water pours onto the perforated plate and cascades evenly over the aggregate. Aggregates are rated based on their

ratio of surface area to volume (A:V). Recommendations call for a surface area to volume ratio between 100 and $200 m^2/m3$.[6]

Alternatively, I've used a type of aggregate called bio balls. Bio balls are plastic balls with high surface areas. Since bio balls weigh much less than aggregate, it's easier to move a packed column aerator filled with bio balls.

Packed column aerators are easy to build and inexpensive. We built our own out of on-hand materials. Ours looks like a tall wedding cake with multiple tiers. Each tier is a black plastic pot filled

Figure 5.23. A homemade packed column aerator filters and oxygenates the water. The wooden bridge spans the low point in the floating hydroponic bed. Overflow from the floating hydroponic bed cascades over the low point (underneath the bridge). The liner of the raised bed extends all the way under the packed column aerator, forming an impermeable base.

Figure 5.24. Construction of a packed column aerator in progress. *Left*, a pond liner is laid on top of the sand and a support pipe with rebar is attached to the roof support. The two white pipes are inflow lines that will bring water from the pond to the top of the packed column aerator. *Right*, an "L" shape at the bottom of the pipe "keys" the pipe into the aggregate that will fill the pot.

with aggregate, and holes in the bottom of the pots allow water to pass from one pot to the next. The largest pot forms the base tier, and the smaller pots are stacked one on top of the next, with the tiniest pot at the top. Similarly to the way a fountain works, the pond water is pumped to the top, then cascades through the pots and aggregate, flowing to its next destination.

The packed column aerator also functions as a solar heat exchanger to help warm the pond. Situated on the north side of the pond, the stack of black pots collects solar energy reflecting off the pond and transfers the heat into the aggregate and water inside the pots before it falls back into the pond.

Before assembling the fountain, I lay a sturdy, level base of gravel with sand excavated into the soil to support the column on the greenhouse floor next to the pond.

As you can imagine, 6 feet of stacked pots seems like a tower on the verge of collapse. To prevent this balancing act from toppling over, I first place a ½-inch UV resistant PVC pipe down the center of the column and insert a piece of rebar through the length of the pipe. The middles of the pots are then strung through the PVC pipe like beads on a necklace. I secure the top of the pipe to the galvanized purlin running the length of the roof on the greenhouse. The bottom of the pipe forms an L shape to anchor the pipe into the aggregate, like the roots of a tree. To fill the pots, you must tie or hold up out of the way all but the lowest pot while you fill the lowest pot with aggregate. Once that's full, the next pot is lowered into place and filled, and so on until you have filled the top pot. Be sure to measure the total height of the stacked pots to make sure it matches the height of the pipe that will pump water into the pots. It's easier to cut out an extra pot than add another pot once the whole column is in place.

DIFFUSERS AND BLOWERS

Another way to efficiently add oxygen and remove carbon dioxide is by pumping air into water using a blower with diffusers. The blower is the first mechanism to move the air; the diffusers create small bubbles of air. The transfer of gas occurs as the bubbles move throughout the water. The rising of the bubbles also creates currents that move water to the pond surface for additional gas exchange and move solids toward a drain. If solids removal is poor, the diffusers act to resuspend solids, creating a biofloc system (see "Surface Aerators" on page 125).

There are three main types of diffusers — perforated hoses, silica glass airstones, and ceramic airstones — and each type varies in efficiency. The two types of airstones create smaller bubbles that ascend slowly, maximizing the gas transfer rate. However, diffuser hoses are easy to clean by hand. Airstone material must be cleaned yearly in a 5 percent muriatic acid solution, which takes more time.

Diffusers also create a pressure drop as air is forced through the material to create the bubbles. For example, a diffuser may cause a pressure drop equivalent to 10 inches of water depth. So when the diffuser is placed 20 inches under water, the total pressure drop will be 30 inches.

Diffusers also work well in conjunction with solar pond covers. Placed under the pond cover, the diffusers create the needed bubbles for good gas exchange. However, air pockets between the water and solar cover insulate the water from the sun's energy and may reduce total heat gain. I recommend placing diffusers on the edge of covers or making holes in the cover to help release the air to prevent air pockets.

A blower moves air by trapping a small quantity, then pushing it forward using blades enclosed

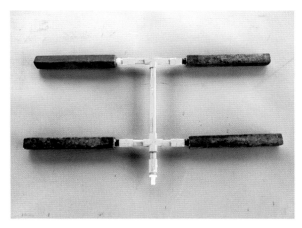

Figure 5.25. Connected to the blower, silica glass diffusers placed underwater create tiny bubbles for aeration.

in housing. Not all blowers are created equal; they are rated on how many cubic feet of air is moved in a minute (cfm), with efficiency decreasing based on the depth of the diffuser. Blowers that move large volumes of air at low pressure are ideal for aquaculture. Regenerative blowers are energy efficient and perform well. The blades in a regenerative blower allow some air to slip past the blades, where it's picked up by the next turning blade, in essence moving more air for the same turn.

As a general principle, the more fish in a given volume of water, the more aeration will be needed. To translate that principle into practical terms, such as determining the size of blower needed, I use a general rule of 3 cubic feet per minute of air for every pound of daily food ration.[7] Next, I determine the number of diffusers I need based on the feeding rate and the stated cubic feet per minute of the diffuser material. Last, I compare performance graphs for different-size blowers based on the pressure drop determined by the depth of the diffuser added to the pressure drop created by the diffuser material. To reduce pressure drop from friction created by the pipe distribution network, I ensure the distribution pipe is larger than the output

Figure 5.26. *Left*, the filtered intake for a regenerative blower must be placed outside the greenhouse because the air inside the greenhouse is too rich in CO_2 for use in aerating water. *Right*, an overturned plastic bucket serves as a protective cover for the intake filter.

orifice on the blower. (For a discussion of the effect of friction created by pipes, see chapter 8.)

I ran our air distribution pipes underground inside insulated trenches alongside our hydronic heating lines. The insulation prevents condensation from forming inside the lines, and having the lines underground protects them from damage from people and sunlight. There are a few low points in the distribution line, and I placed a ⅛-inch seep hole in these locations to drain condensation.

The air inside the greenhouse has high concentrations of carbon dioxide caused by fish and bacterial respiration. To protect the blower, we place it inside the greenhouse with the intake outside so it will take in "regular" air.

Our blower is designed for dry locations with temperatures less than 104°F and humidity less than 80 percent. We protected the blower inside the greenhouse by attaching an internal roof made from greenhouse plastic. However, a dedicated blower house outside the greenhouse would be ideal to protect the blower from the elements, as

Figure 5.27. To save material and space, we use the wall of our greenhouse to form the back wall of our blower house. Then we fashion a sheet of greenhouse plastic as a cover over the pump, like a minigreenhouse with open sides for proper ventilation.

the heat and humidity would be lower than inside the greenhouse.

We need fresh air for aeration, but we don't want to introduce cold outdoor air into the greenhouse during the winter. We solve this problem by attaching the intake of the blower to a corrugated pipe that we've placed on top of the compost pile adjacent to the greenhouse. The intake end of the pipe sits in an area uncontaminated by compost gases. The pipe then travels along the top of the compost pile but under the compost pile cover, allowing heat produced by the compost to warm the air inside the pipe. Since this pipe isn't buried, removal and reuse is easy.

To avoid supersaturating the water with nitrogen, don't add air to the pond below a depth of 5 feet. Below 5 feet the pressure of water has time to force toxic levels of nitrogen from the bubbles into the water. Since our greenhouse ponds are only 3 feet deep we don't have to worry about this.

SURFACE AERATORS

Surface aerators use paddle wheels or propellers to push water up into a bubbling cascade. The thin films, bubbles, and additional air contact generated by the water movement produces a large surface for gas exchange. The aerating action also removes carbon dioxide from the water if the greenhouse is well vented. Surface aerators are generally used in larger ponds but also fill a niche in small greenhouse ponds and tanks. Smaller surface aerators called agitators are available for small ponds and tanks and can completely replace other forms of aeration.

The oxygenation and resuspension of solids by surface aerators allow algae and microbes in the pond to convert toxic ammonia and waste material into food for filter feeders such as tilapia. The pond itself becomes a self-contained filtration system in a practice called biofloc. At low feeding rates of less than 0.71 lb/ft^3, biofloc systems are dominated primarily by algae in green water systems.[8] The algae keeps ammonia levels low and feeds the fish. As feeding rates increase, the density of algae shades itself out, and heterotrophic bacteria become the dominant species. With bacteria-dominated systems, growers must add a carbon source such as chopped hay, grain pellets, or molasses to the pond to enable the bacteria to absorb the excess ammonia. The fish then consume the bacteria as part of their feed intake. Over time, solids may accumulate to high levels, and you may need to drain the system to remove the solids.

Figure 5.28. The small agitator at left is perfect for a small greenhouse pond. The large ¾-horsepower agitator at right is what we use in our ½-acre outdoor pond.

Growers can use Imhoff cones in biofloc systems to measure settleable solids. Imhoff cones are like clear funnels with a closed bottom. Solids settle to the bottom of the cone for measuring; the desired solids concentration is 25,000 to 50,000 ppm.[9]

When settling tanks, swirl separators, and hydroponic systems are attached to ponds or tanks, they help remove solids. This gives the system added security and allows for longer time frames between draining and cleaning out solids.

Our pond aquaponic system usually starts out as a clear water system, then converts to a green biofloc system as fish grow and solids accumulate. By the time the system is maxed out with solids, the fish or shrimp are ready to be moved to an outdoor grow-out pond.

In addition to providing aeration for a biofloc system, surface aerators make great backup systems in case pumps break, and they provide additional oxygen for peak demand periods. However, I don't recommend using surface aerators in conjunction with ponds that contain larval freshwater prawns because the aerators can suck them in and injure them. Having enough diffusers connected to blowers will resuspend the solids and provide the aeration needed for biofloc systems with prawns.

Managing Water Circulation

Aeration adds oxygen and removes carbon dioxide and nitrogen while creating currents that resuspend solids for biological filtration inside the pond. But recirculating aquaculture systems require pumps to move the water from the pond into and between exterior filtration components such as settling tanks, hydroponic beds, and packed column aerators.

Pumps

The two basic types of pumps used to move water outside the pond are centrifugal and airlift.

Figure 5.29. A submersible pump is designed for use directly inside a pond or tank.

Centrifugal pumps are the most common and use a motor to rotate an impeller inside a casing. They whirl water outward through piping by creating pressure differences. To operate properly, centrifugal pumps must always remain full of water. They come in virtually any flow range with two basic types available: submersible and external. In either case I prefer a pump with high-flow and low-lift capacity.

Submersible pumps are a good choice in small systems to simplify the installation process. Simply place the pump in the pond or tank, and attach it to an outlet pipe. Submersible pumps made for ponds generally come enclosed in a protective casing to prevent large fish and particles from entering the pump. As long as the pump is operated only while under water, the risk of running the pump dry and damaging the seals is minimal. However, submersible pumps are not energy efficient. They move a small amount of water for the amount of electricity consumed.

External, in-line centrifugal pumps are placed outside the pond and move more water for less expense. However, they do require additional

Figure 5.30. This external, in-line centrifugal pump is installed below pond level inside a pump house made from a 55-gallon barrel buried in the ground. Fittings allow the pump to be extracted easily if needed.

Figure 5.31. A corrugated roof for the pump house protects the pump from being wet by greenhouse misters.

plumbing and management. To operate properly, prime the external pump by filling it with water after connecting it to the water source with an airtight, water-filled pipe. If the pump can be placed below the water level of the pond or settling tank (outside the pond), gravity will push water into the pump without the need for a priming pot. Place the pump inside a plastic 55-gallon drum or other suitable underground box to allow more depth below the pond water level. Locate the box in an area protected from rainwater, either inside the greenhouse or outside the greenhouse, because rainwater flowing into the box will damage the pump. A deep gravel bed under the box with an underdrain also helps with drainage.

Alternatively, you may place the pump above the level of the pond and use a priming pot to form suction. Water is held in the priming pot before it proceeds into the pump. Fill the priming pot and the intake line leading into the pond with water before turning on the pump; the water in the priming pot helps the pump develop the suction needed

to start pulling water from the pond and prevents air from getting sucked into the pump (which could damage it). Since the pump and priming pot are higher than the pond, water will inevitably flow back into the pond and empty both the pump and the priming pot. To prevent this I install a one-way check valve in the line below water level. Like submersible pumps, in-line pumps must never be allowed to run dry to prevent seals from breaking. In-line pumps also require draining in winter if pumps sit idle and freezing is an issue.

Airlift pumps work when air is blown into the bottom of a submerged vertical pipe with the same type of blower used with diffusers. The bubbling water rises upward through the pipe, which creates suction because the bubbling water is less dense than the surrounding water. Rising bubbling water is then displaced into a higher outlet on the vertical pipe. At most, airlift pumps can draw water up at a distance equal to the depth of the pipe. Therefore, airlift pumps work best under low head or height conditions.

SIZING THE PUMP

Desired flow rate along with the total height, or head, the water must travel determines the size of the pump. The flow rate needs to be sufficient to supply the necessary filtration and oxygenation for the species grown in the pond. However, since filtration systems vary, a good guide for flow rate is the effectiveness of water in pushing solids toward a drain. As water flows along the wall of a tank or liner, it dislodges particles and transports them toward the drain. The flow rate should be large enough to promote such self-cleaning. The recommended velocity of water at the bottom of the pond or tank is 2.5 to 3 inches per second. When the recommended bottom velocity is achieved, the velocity toward the surface is slightly faster, at 4.5 to 6 inches per second. If water flows too fast, fish may become stressed. If flow is too slow, solids accumulate on the sides and bottom of the pond or tank. I gauge the velocity of water by the retention time or how long it takes to fill the pond or tank with water from the pump. The ideal retention time falls between thirty and one hundred minutes.[10]

To determine the retention time for the pond, first determine the pond's volume using the calculations presented in chapter 2. Next, check your pump manufacturer's recommendations; they have graphs displaying the volume of water moved based on the height, or head, the pump must move the water. More height means less volume. Once you know the maximum height needed to pump the water and the volume of the pond, select a pump that will fill the pond within the thirty-to-one-hundred-minute time frame.

For example, I have a 2,000-gallon pond, and I need to pump the water to a height of 6 feet and want a retention time of one hundred minutes. I would shop for a pump that moves 20 gallons per minute (2,000 gallons ÷ 100 minutes) to a height of 6 feet.

INSTALLATION AND PLUMBING

Secure pumps to a concrete slab with concrete screws and rubber mounting bushings. The rubber bushings prevent the vibrations of the pump from wearing down the base or creating noise. Locate direction-changing elbows an equivalent of three to four diameters of the openings on the pump from the discharge or suction points. Most importantly, don't run pumps dry! You can purchase in-line pump controllers to detect if a pond is out of water. Alternatively, you can add a float switch that floats in the pond and turns the power off when the pond runs low.

Both submersible and external pumps move water from ponds into hydroponic systems, settling tanks, and packed column aerators. I pump water out of the pond through a single pipe, then use three valves to direct the water to the three filtering components: the hydroponic systems, the settling tanks, and the packed column aerator. The ebb-and-flow hydroponic system and the settling tanks both have specific flow rates necessary for the system to perform properly (discussed earlier). To regulate the amount of water going to each component, I adjust these three valves. Any excess water flows to the highest component in the system, the packed column aerator, which readily handles a large volume of flow.

I run two separate outflow lines to all the components from the pump. If one line clogs with algae or debris, I simply shut it off and switch to the other line. With the closed line shut off, the inside of the pipe goes anaerobic and all the debris decomposes. Before I turn the anaerobic line back on, I flush all the debris out through a drain at the bottom of the system.

The size of the pipe is determined by the amount of water moved. A pipe the same size as or larger than the pump orifice usually works well. As you design the piping layout, avoid small pipes

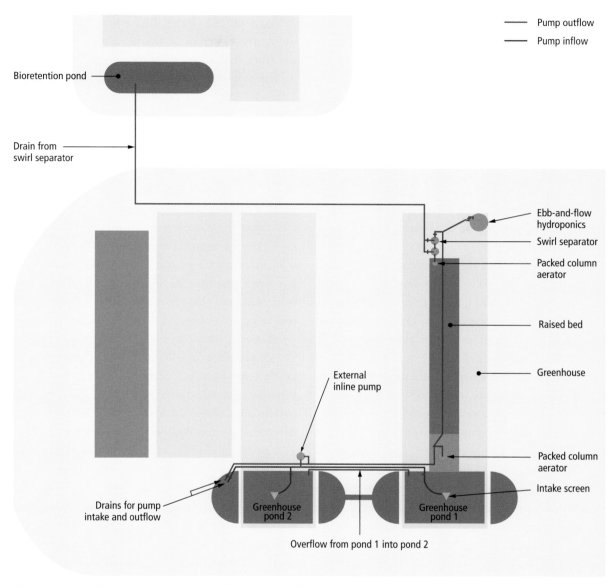

Figure 5.32. This diagram gives an overview of the plumbing for aquaponic systems connecting multiple greenhouse ponds.

and sharp curves to reduce friction loss. I've had problems with any pipe less than 2 inches in diameter, and 3-inch pipes are commonly used even in small systems to prevent clogging. I also use flexible PVC to avoid complex turns that would reduce flow.

Outlets

Think of the pond as a living organism — food goes in, and waste goes out. Since we've already established a successful flow pattern to push solids toward the pond outlet, we need to determine how to efficiently expel the waste through that outlet.

The outlet drain can connect directly to a submersible pump. But if you want to use an external pump to remove water from a pond, you have to either embed a drain in the bottom of the pond or pump out the water over the wall of the pond. I opt for pumping water over the side, since it's the cheaper, easier solution. In our square-shaped greenhouse pond we inserted the end of the water intake pipe inside a depression in the middle of the basin. To prevent fish from getting sucked into the pipe and then the pump, I placed a strainer on the end of the submerged pipe. We attached flexible PVC to the strainer and ran the piping up over the side of the pond. (During harvesting, we can easily move the flexible PVC pipe.) Improving upon this technique, we also placed diffusers on top of the strainer to force water up, then back down the sides of the pond, pushing solids toward the strainer underneath.

Embedding a bottom drain into the liner probably would remove more solids than pumping them through a pipe laid over the pond wall. The pond drains easily if it's plumbed downhill to an outlet connected to the bottom drain, and numerous types of commercial drains are available. Some dual drains separate polluted water from the main flow for treatment in settling tanks. Others have diffusers mounted on top to create currents that rise, then fall along the edges, forcing solids back down to the drain.

However, embedding a drain in a lined pond introduces the risk of leaks. If you go this route, follow the manufacturer's guidelines. In my opinion, an embedded drain involves more risk and more expense — pumping water over the side of the pond involves less risk and less expense.

Flow Patterns and Inlets

In nature a river's current pushes water downstream — over waterfalls, around bends, and through gorges. Inevitably, some water escapes the rushing current and pools along the edges of the river. Although in nature these stagnant pools probably host a habitat for living organisms, in aquaponics this undesired stagnation is referred to as a dead zone.

Our goal is to develop flow patterns absent of stagnant dead zones, where the current pushes water throughout the entire pond or tank. We achieve our desired flow patterns by carefully selecting the placement of the inlet and outlet. A circular pattern works well when the pond resembles the shape of a circle or square. To create a circular pattern we deflect the inlet water to the side and insert the outtake at a low point near the middle of the pond. The current sweeps around the edges and forces solids to pool on the center bottom of the pond. In long, narrow ponds a cross-flow pattern distributes water equally. We simply place the inlet on one side and the outlet on the opposite side, forcing the water to flow down the length of the pond. I've observed that a circular pattern in a round or square pond removes solids better than a cross-flow pattern in a long rectangle.

The design of the inlet ensures the entire volume of water is enriched with filtered water. A single point of entry primarily moves water at one level in the pond; for example, if the inlet is placed near the top of the pond, the bottom may lack current. However, having several smaller inlets at varying depths creates currents at multiple levels in a pond. We drill a series of holes in our inlet pipe and submerge it vertically on the side of our pond to distribute the water evenly. To reduce friction we prop the pipe slightly off the edge of the pond. Additionally, we design for easy removal during harvesting.

We are still experimenting with the size of the inlet pipe and the number of holes to drill. Available pressure and the head of the water play a major role in this determination. Smaller holes produce higher velocities but require more pressure to force the water through the inlets. We've used tees and removable bushings to research flow rates. The removable bushings allow us to adjust the flow easily, which is helpful in reducing the current to prevent stressing small fish. It's a process of trial and error — but we enjoy the challenge.

Emergency Backup Systems

Maintaining water quality is a delicate balance. As fish production intensifies, it's crucial that all systems function properly because the fish perish quickly without proper filtration. A broken pipe, power outage, or broken or clogged pump can be fatal to fish grown at high densities. I can't count how many times the power has failed, usually on a weekend when I'm not around. If a pipe breaks in the middle of the night, depending on the size of the leak, the pond could drain by morning.

Fortunately, alarm systems are available to monitor temperature, water level, power, and many other environmental parameters and will alert you via cell phone, Internet, or landline. Sensaphone Web600, Freeze Alarm, and HOBO data loggers are but a few common devices found online. A standby generator with an automated power transfer switch is an ideal backup. Though costly, they're worth the expense when also used to protect greenhouses and refrigeration units. However, generators do require monthly testing and maintenance.

Far cheaper and simpler, we use a standby oxygen injection system to protect against commonly

Figure 5.33. An oxygen tank connected to a regulator and solenoid valve. The solenoid valve is plugged into grid power and remains off while electric power is in operation. If the power fails, the solenoid valve opens and pure oxygen is released through the diffuser to keep fish alive.

occurring power outages. Oxygen is stored in a 244-cubic-foot tank purchased and refilled at a welding supply store. A pressure regulator and sole-noid valve are connected to the tank; we plug the solenoid valve into a power outlet, and it's triggered open if the power turns off. When it's open, oxygen flows from the tank through a diffuser and into the bottom of the pond, maintaining oxygen levels until the power returns.

Active Heating

We use pool covers over our ponds to capture solar energy for nighttime use. The pond is like a battery, recharging during the day and depleting at night. However, sometimes more heat is needed to increase either greenhouse or pond temperatures for growing tropical plants and fish. In these cases the water is charged with heat from an external source. Two basic types of heaters are available: immersion heaters and heat exchangers. With immersion heaters an electrical element sealed from water contact is heated using electricity. Either the heater sits in the water or water is passed over the elements. Typically, in-line electric heat-ers are connected to pond pump systems.

The pond is easily connected to a hydronic heating system using a heat exchanger. Similarly to storing solar heat, the pond also stores hydronic heat pumped in during the day. At night the pond heat is turned off, freeing up the hydronic heat for use in other greenhouses, or left on in colder climates to provide heat continuously. Connecting the hydronic heating system to a hot compost pile transfers compost heat into the pond. Chapters 4 and 8 discuss hydronic heating systems for green-houses and compost heat extraction.

If tropical fish are grown in the pond, a properly sized heater and insulated pond are needed to keep them alive. To determine the amount of heat needed, you'll first need to figure out the tempera-ture tolerance of the fish, the volume of the pond, and the desired temperature change of the pond water. Refer to chapter 2 to calculate the volume of the pond. Pond and greenhouse insulation can vary, which makes it difficult to predict how much heat you'll need until you monitor the pond temperature drop during a cold winter night. For example, some greenhouses have only one layer of plastic. Others have two layers, and air trapped between the layers provides extra insulation. The thickness of the insulation underneath a pond may differ. Additionally, climate, wind buffers near the greenhouse, and leaks in the greenhouse structure also affect heat exchange. Here's one example: For a pond insulated with 1 inch of EPS insulation inside a single layer plastic film greenhouse, in my climate zone 8a I may need to heat the water 14°F on a cold night to keep temperatures at 67°F.

As a general rule, it takes approximately 6,950 Btus/hour or 2 kW/hour of electricity to heat 1,000 gallons of water 10°F over a twelve-hour period. Since my pond is 2,000 gallons and I need to heat the water 14°, I would need a heater capable of generating 19,460 Btus per hour to keep the pond at 67° on a cold night. If I have to run the heater for fourteen hours, that would be 272,440 Btus of heat or about 3 gallons of propane. An uncovered pond may need twice that much heat to keep it warm. As you can see, heating a pond can become expensive, and insulation is critical to keeping costs down.

We provide heat using a titanium heat exchanger immersed inside the pond and connected to our hydronic heating system. The titanium efficiently transfers heat into the pond water. The heat exchanger is also connected to our hot compost pile and 9,000-watt electric water heater. Alterna-tively, pex pipe embedded inside a layer of concrete on the bottom of the pond could also transfer heat, but the efficiency of heat transfer with pex is not

Figure 5.34. I'm lifting the titanium heat exchanger out of the pond. The ends dipping into the pond are connected to the hydronic heating system. Hot water circulates through the exchanger to heat the pond.

as good as metal. The temperature requirements change depending on our production strategies and the time of year. Fish may survive at a lower temperature, but breeding and reproduction require higher temperatures. Heated water for fish requires aeration to avoid supersaturation of gases in the water as temperatures increase.

Tilapia Production Strategies

Our aquaponics production strategy is simple — use the greenhouse ponds as nurseries to extend the growing season for tilapia or prawns. By raising young fish or prawns in the greenhouses in the winter we can have a head start when we move the fish to outdoor ponds in the spring. The carrying capacity of our greenhouse pond is 200 pounds. The 200 pounds can either compose two hundred 1-pound fish (500 grams) or two thousand 50-gram

fish. Size is significant. If we stock our outdoor pond with 50-gram tilapia in the spring, they will reach marketable size before water temperatures decline in the fall.

The cycle is continuous, but we begin preparations for indoor production about a month before we harvest market-ready fish from the outdoor ponds. At that time we start plants in the greenhouse hydroponic system and fertilize them with a complete organic fertilizer. We have to plan ahead, because for the biofiltration system to work properly, the system needs time to build the required microbial population to filter the water. One month of fertile pond water flushing through the biofilter will prime the system for fish. You may stock fish before finishing the priming process, but I recommend withholding feed until nitrite levels drop (see Testing Water Quality).

After priming, we stock the indoor greenhouse pond with approximately 150 pounds of mixed-sex fish from the outdoor grow-out ponds — an amount less than the total carrying capacity of the pond. A few of the stocked fish become breeders for the next year's production, and we hold the rest for extended sales and growth. Nutrients in the fish manure supply the necessary fertilizer for the hydroponic plants after the initial feeding with the complete organic fertilizer. Periodically during the fall, we harvest fish to reduce the capacity and maintain water quality. Maintaining fish at high densities or at temperatures between 65° and 68°F prevents spawning and places the fish in a holding pattern until reproduction is desired.

Ten to twelve weeks before we want to stock our outdoor ponds, we drain the greenhouse pond and clean any remaining solids. We apply all the waste into our fields of cover crops or on the compost pile and treat the waste as raw manure. At this time we reduce the fish population to eight females and four males (1 fish / 15 sq ft pond surface area). Since each female will produce 250 fry, we only need eight to produce the desired 2,000 fry. The population reduction triggers spawning, and out comes a crop of evenly sized fry. The fry swim in and out of the mouths of the protective parent while young and eventually start to swim more freely. At this point the fingerlings are less than an inch long.

Once the approximate amount of desired fry is produced, we harvest the adults to prevent them from eating their young. Without the adults the carrying capacity of the pond is increased, making room for the growth of the young. Water temperature is critical during the entire spawning cycle. We maintain our pond's temperature as close to 80°F as our heating system can manage. Tilapia produce more fry at higher temperatures.

We affix small partitions in our greenhouse pond to provide protection during aggressive reproductive behaviors. Built of half cinder blocks laid on top of scrap pieces of pond liner to protect the main liner, the partitions jut out from the sides of the pond and create a wall that limits the fish's view. We then remove the cinder block partitions when we remove the adults.

The young fingerlings continue to grow, and we increase the feed accordingly. A specialized commercial feed with high protein and a fine consistency ensures quick growth. To avoid large fluctuations in water quality, a belt feeder distributes the feed evenly over the entire day.

Separating fingerlings from adults is tricky. We place a fine mesh net with ¼-inch openings inside the pond to assist in removing the adults. Gathering the net toward the corner crowds the fish and enables us to scoop them out with a dip net. Alternatively, we partially drain the pond to crowd the fish and similarly scoop them. While the fry are very young, they swim in and out of the netting on the bottom of the pond searching for "greener grass" on the other side. Some fry will get stuck on the other side because they grow too large to fit back through, becoming hopelessly too big for the small area. To prevent small fish from swimming through the net, make sure the netting lies flush on the bottom of the pond. Because the net also prevents solids from moving toward the drain and may inhibit water quality over time, it's probably best to remove the netting after separating the adults.

Figure 5.35. A net cage suspended from the roof of the greenhouse temporarily holds fish inside the hydroponic channel for live sales or until transport.

If you allow the netting to remain to facilitate removal of the fingerlings later, it's best to convert to a biofloc system to resuspend solids for treatment in the water by algae.

While the fry are young, they will obtain a major portion of their diet from algae growing on the sides of the pond. Any netting or material placed in the pond also supplies a good substrate for algae growth. As the fish age they soon exhaust all natural feeds and require supplemental feeding based on the total weight of the fry in the pond. To determine the weight, we acquire a few samples of fish with a dip net and weigh them. We can then determine an average weight per fish, which we multiply by the total approximate number of fish in the pond to figure a total weight. While young, the fish eat 8 percent daily of their total weight. We slowly taper off to 2 percent toward the end of the nursery period. If water quality declines, we reduce feeding rates.

Sometimes we partition off holding pens for fish in the floating hydroponic systems. We use the holding pens during harvesting or when we're moving the fish as temporary areas to keep the fish happy and healthy. To increase water quality we place aeration and oxygen airstones inside the pen.

I recommend the following two techniques for constructing a fish pen inside a hydroponic bed. The first uses a net cage suspended from the roof of the greenhouse; PVC pipe establishes a framework to keep the top of the net cage open. The second technique uses rigid plastic netting attached to a PVC frame to partition off part of the hydroponic channel. The same netting and framework is then used to seine fish out of the channel.

Grow-Out Ponds

Clemson University's aquaculture facility comprises about 12 acres of ponds. Located within several thousand feet of the farm and greenhouse ponds at the SOF, these large ponds range in size

Aerator Tip

We use surface aerators to increase oxygen in the water, which allows us to increase our fish feed. Surface aerators tend to stir up sediment and deposit it on the outer edges of the pond, creating a depression under the aerator. If the aerator is located on the shallow side of the pond, a depression forms that prevents water from getting to the drain. Therefore, locate aerators above drains whenever possible to simplify harvestings.

from ¼ acre up to 2 acres. A nearby creek supplies water for these ponds via a dam and pumps.

Several months prior to stocking the outdoor pond, we conduct a soil test and lime the ponds accordingly (see "pH and Alkalinity" on page 144). One week before stocking, we apply 300 pounds of cottonseed meal per acre of water, spread evenly over the pond. A dense algae bloom forms within a week, starting a chain reaction that leads to an abundance of fish food. We transfer the fish from the greenhouse pond to the grow-out pond when morning water temperatures reach over 68°F (usually late April in our area). To prevent shocking the fish, we temper the greenhouse pond water used to transport the fish with water from the grow-out pond, slowly acclimating the fish to the temperature of their new habitat. I add about 5 gallons of water every five minutes to the 100-gallon transport tank. The water inside the transport tank reaches the same temperature as the pond water within an hour.

Relocation stresses the fish, but as soon as they will eat we feed them as much as they can consume in ten minutes once or twice per day. The fish manure continues to fertilize the pond, producing an abundance of natural feeds to supplement the commercial feed. Water quality is monitored, and a surface aerator is used as needed to maintain morning dissolved-oxygen levels above 1 ppm.

We stock largemouth bass in our grow-out pond as soon as possible after stocking the tilapia to prevent the pond from becoming overpopulated with small fish from rampant tilapia reproduction. The small bass don't eat the large tilapia but do hastily consume small tilapia fry. To ensure the larger stocked tilapia survive, we stock small fingerling bass, only a few inches long, with 6-inch-long tilapia. As a general rule bass can eat fish measuring half of their own body length. Recommendations call for stocking bass at one bass for every fifteen tilapia.[11] However, at $1 per fish the bass are expensive, so

Figure 5.36. To fill the grow-out ponds we insert flashboard risers into a dam, causing a nearby creek to back up. A pipe with a screened inlet connects to a pump to pull water out of the creek. If enough gravity were available, a siphon would also work.

we stock one bass for every thirty tilapia. I haven't seen any signs of tilapia fingerlings at harvest, so I believe the lower stocking rate of bass works.

All kinds of wildlife may prey on fish in the pond. We overstock our pond to help account for losses. Maintaining a presence by feeding fish and monitoring the pond definitely deters the birds. Otters, on the other hand, hunt during the night. I've heard that an otter family can wipe out an entire pond within a week. I monitor the margins of the pond for otter slides worn into the mud so I can promptly trap the otters. Luckily, I haven't had to deal with any otters yet.

Harvest

About six months after stocking fry, the harvesting begins. We catch the first fish with a cast net, using feed to attract the fish. The fish become wary of humans as soon as harvesting begins. Therefore, cast netting only works well once a week, which gives the fish enough time to forget about the last cast. I've also tried a seine to net fish out of the pond. However, tilapia jump over the net and dig into the mud, making successful seining a nearly impossible task.

Draining the pond is the simplest way to harvest the fish. We either pump or drain water out of the pond, and the fish crowd in the shallows where we can quickly scoop them up with dip nets and place them into baskets. As oxygen in the water depletes, the fish slow down and become an easier catch. Once the baskets contain about 6 inches of fish, we move them to holding pens in adjacent ponds. Alternatively, we place the baskets into tanks or coolers filled with pond water on trucks next to the pond.

We've run into our fair share of problems growing out fish in our outdoor ponds. Over the past two years, several five-hundred-year storm events have limited our harvest by flooding our pond and washing away most of our fish. We've moved our

Figure 5.37. After draining the grow-out pond, my aquaculture mentor, Scott Davis (*left*), and I scoop out fish with dip nets to harvest.

operation to a higher pond that's less susceptible to flooding.

Another problem we've encountered with outdoor culture of tilapia is the short harvest period. Since tilapia need the entire growing period to mature, we must harvest fish over a short period of time before water temperatures cool to lethal levels. Processing and storing fish also require facilities we don't have, and most customers want processed fillets. To grow fish to a marketable size, we have to stock large tilapia along with expensive predators (i.e., the bass) to control overpopulation. Generally, fish farms use hormone treatments or hand sex their fish to quickly grow only male tilapia. However, hormone treatments go against our organic practices and hand sexing is time consuming and produces a lot of wasted females. Overall, growing tilapia has been educational but not profitable.

Freshwater Prawns

Prawns are invertebrates, so they don't fall under the same regulation hurdles as fish at universities. Freshwater prawns are also easier to sell and command a higher market price. We partially feed our prawns by fertilizing ponds to produce feed through natural pond growth. Since organic commercial feeds are cost prohibitive but organic fertilizers are not, organic production is a viable

Transporting Fish and Prawns

If you decide to raise fish or prawns, one thing to consider is how you'll get your breeding stock to your farm in the first place. Freshwater prawn postlarvae are hard to come by. I imported my first school of prawns from Houston, Texas. They flew into Atlanta's international airport on a cargo plane along with the other miscellaneous travelers that don't fit the criteria for a passenger flight — mice, dogs, and cadavers.

I veered toward the cargo terminal and pulled into a lot filled with loading docks. After parking the truck, I navigated between baggage trolleys piled high with pet carriers and boxes. An island of normalcy in a sea of chaos, two airline employees sat behind a desk similar to the setup at arrival and departure gates inside the terminal. I checked in and was directed to take a seat in a small waiting area with magazines.

The flight arrived at 8:30 p.m. By 9:30 — after a long day's work at the farm and a two-hour drive to the airport — I was anxious to pick up my packages and head home. While the employees were distracted, I meandered back into the warehouse, intent on locating my larvae. Among the many kenneled dogs and caged mice, I watched a few men carry out a long plywood box about the length of a human. That's when I decided to return to the waiting area.

A short time later the airline rep called me to the desk and pointed me toward my packages. Each of the three boxes measured about 2 feet by 2 feet, weighed approximately 20 pounds, and contained over four thousand postlarval prawns. By the time I arrived back at the farm, it was near midnight. I felt like a

thief in the night as I toiled with the combination lock of the farm gate and wound down the dirt road, mesmerized by the full moon's light reflecting off the ponds.

Opening the packages was like peeling back layers of an onion. First the cardboard boxes, then the Styrofoam insulation, finally the doubled plastic bags. The bags contained oxygen and a dash of salt for disease resistance. The postlarvae are tiny, and it felt a bit anticlimactic to dump what seemed like mostly water into my greenhouse pond. All that preparation — draining and cleaning the pond and attaining the proper temperature for my first school of prawn larvae — and I could barely see them. But by midnight the thought of a hot shower and a warm bed dissolved any possible disappointment.

Compared to that experience, picking up my first school of tilapia was a breeze. I lived in the Lowcountry of South Carolina at that time, and the closest hatchery was located in Columbia, South Carolina. Though it was still a four-hour round trip, no chaotic airport terminals or corpses were involved.

It was my first time transporting fish, so I had no idea what I was doing. I threw a large cooler into the bed of the truck and hit the road. At the hatchery they filled my cooler with pond water, dumped in a school of tilapia, and closed the lid. I didn't own an aerator then, so out of concern for the fish's oxygen level, I pulled over every forty-five minutes, opened the cooler, and splashed the water around with my bare hands. I wasn't certain it worked, but I figured it was better than nothing.

 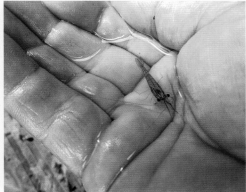

Figure 5.38. Postlarval freshwater prawns are transported via airmail. Each box and bag contains about four thousand prawns.

Now I'm a little older and a little wiser. I bring along the standby oxygen injection system normally used to protect fish from power failures at the farm. It works great as an oxygen injection system during transport. I simply strap it in like a passenger in the front seat and run the tubing out the window into a diffuser inside the 100-gallon plastic stock tank in the truck bed. It oxygenates the water while I drive. Back at the farm, I slowly add pond water to the tank to acclimate the fish to the pond's temperature. Then I cinch the net that I lined the tank with and easily transfer the fish into the pond.

option. Additionally, customers can cook whole large prawns on the grill. If any processing is desired, it's as easy as removing the head, a task most customers tolerate.

Our greenhouse pond systems also support the growth of freshwater prawns (shrimp). The giant river prawn (*Macrobrachium rosenbergii*), native to the Indo-Pacific region, reaches sizes of over 1 pound in tropical climates and resembles a lobster more than a shrimp. Freshwater prawns' life cycles are composed of three parts — hatchery, nursery, and grow-out phases. In temperate areas growing the prawns to a marketable size requires a hatchery and nursery phase in heated ponds to extend the growing season. The heated ponds also protect the prawns from predation until they reach a less vulnerable age.

Adult freshwater prawns mate in either fresh or brackish water. The newly hatched larvae can only live in freshwater for a few days. In the wild, larvae move downstream to brackish coastal waters to continue feeding in the rich marsh areas. After molting eleven times, the larvae enter into a postlarval stage in which they resemble adults. During the postlarval stage the prawns migrate back upriver to freshwater. The prawns continue to molt and grow into juveniles, then into adults in a freshwater environment.

Farmers can also purchase postlarval and juvenile prawns for maturation in grow-out ponds. However, the number of boxes needed varies drastically depending on the age of the prawns. For example, to ship twenty thousand prawns, or enough to stock a 1-acre pond, it takes 4 boxes of postlarval prawns, 25 boxes of thirty-day-old juveniles, or 114 boxes of sixty-day-old juveniles.[12] Since it cost eighty-five dollars per box for the required next-day shipping, postlarval is the only profitable choice when shipping is necessary. We raise our postlarval prawns to juveniles in our heated greenhouse pond systems to save money on shipping. The limited culture time of thirty to forty days makes heating costs reasonable.

In our recirculating aquaponic systems we stock our postlarval prawns at twelve to twenty-three postlarvae per gallon of water to grow them out to juvenile size. In nature, prawns live on the bottoms and edges of ponds, and adding a substrate to increase the surface area in a production pond greatly reduces aggressive behaviors and cannibalism. Typically, bird netting or PVC fencing is hung vertically or horizontally. One-sixth-inch openings in netting substrate are ideal.[13] We use inverted vegetable bins that have small slits in the sides, and I've heard some farmers use inverted bread trays. You can also determine stocking rates based on the surface area of the bottom and sides of the pond, plus any added substrates such as vegetable bins or netting in the pond. When basing stocking rates on surface area, use a rate of twenty

Prawn Primers

The hatchery phase of production starts with mating adults and progresses into a postlarval stage of growth. Hatchery production uses recirculating brackish water ponds and complex feeding regimes, as well as growth of natural feeds. For more information I recommend reading "Culture of Freshwater Prawns in Temperate Climates: Management Practices and Economics" by Louis D'Abramo et al., and *Freshwater Prawn Culture: The Farming of* Macrobrachium rosenbergii by Michael New and Wagner Valenti.

to forty postlarvae per square foot. Ideally, water temperature should range between 78° and 82°F during the nursery phase.

Typically, postlarvae grow for forty to sixty days to mature into juveniles with an average weight of 0.0035 to 0.01 ounces. Surface aerators may damage the young prawns, so we use diffusers and recirculation to maintain oxygen levels for good growth. To prevent the postlarval prawns from being sucked into the pump system, we place window screening over an enlarged intake for the pump. As solids accumulate in the system during the end of the culture period, the screen gets clogged frequently and needs cleaning. I like to attach the intake to a rope and pulley connected to the top of the greenhouse. This allows me to raise the intake off the floor of the pond and spray it with a hose for cleaning. As the prawns grow I remove the screen covering the intake to reveal slightly larger holes than the screen provided.

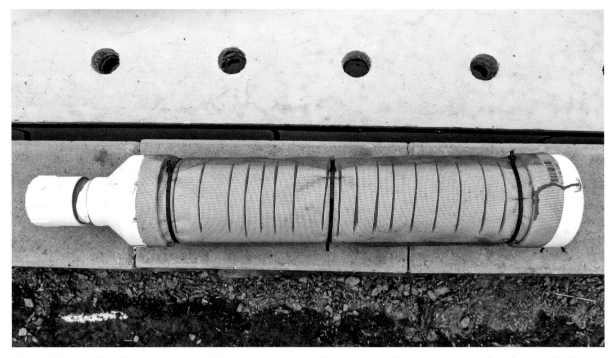

Figure 5.39. An adjustable intake screen. When prawns are small the screen is added to prevent them from being sucked into the pump.

The larger prawns won't be sucked through the slightly larger holes, and debris is less likely to clog them.

We prepare our outdoor ponds by draining them to eliminate all competition and predators of prawns. Five days before stocking, we fill the pond and inoculate it with fertile pond water. We apply cottonseed meal at a rate of 300 pounds per acre to create an algae and zooplankton bloom to feed the young prawns. If the ponds are filled too far in advance of stocking, predators of prawns, mainly dragonfly larvae, will grow, and the prawns become nothing more than expensive dragonfly feed. We continue applying cottonseed meal at a rate of 15 to 20 pounds per acre every few days to sustain natural food populations. When soldier fly larvae are available I store the frozen larvae in 5-gallon buckets in the freezer. I

Figure 5.40. After growing postlarval prawns in our greenhouse pond for forty days, I'm scooping the prawns to Hunter Fields and David Robb, Clemson students, for weighing, then transport to an outdoor pond.

Figure 5.41. Transporting juvenile prawns to grow-out pond by truck. Branches in the tank provide habitat, the dip nets scoop up the prawns, a colander and a scale weigh prawns, and the bucket tempers the water before moving the prawns to a larger pond.

Figure 5.42. A meat grinder is used to process black soldier fly larvae into feed for prawns. Yummy.

thaw them weekly and run them through a meat grinder. The ground larvae make a great feed, since it's easier to throw a wet patty into the middle of the pond, and it distributes the food more evenly over the surface.

We raise the juveniles in the grow-out ponds for 100 to 140 days. The culture period depends on the growing season for the area and is dictated by water temperatures above 68°F. The southeastern United States north to southern Ohio has a growing season long enough to grow prawns. Prawn growth rates slow below 68°F, and all prawns must be harvested before morning water temperatures drop below 60°F in the fall. We

apply continuous aeration at 1 horsepower per acre to maintain dissolved oxygen above 3 parts per million. Research indicates 1 horsepower of aeration can increase yields by 350 to 450 pounds per acre.[14] Experiments in Mississippi have shown that whole prawns weighing twelve prawns per pound can be produced at 1,200 pounds per acre in ¼-acre ponds when stocked as thirty-day-old juveniles at a rate of twenty thousand per acre. To harvest the prawns, we drain the water from the pond and place a catch net and sump on the drain outlet. The pond must be weed free and continually slope toward the drain to make the harvesting possible.

Figure 5.43. To determine how many prawns we have in a given harvest, we first weigh about fifty prawns and determine an average weight per prawn. We weigh the rest, then we use this formula to determine how many prawns we have in all: total weight of all prawns ÷ average weight of a single prawn = total number of prawns.

Testing Water Quality

Water quality in a pond fluctuates daily and depends on the total weight of fish or prawns in the pond, the amount of feed given, accumulation of solids on the bottom, and the amount of photosynthesis occurring. As waste, feed, and the weight of fish increase, more demands are placed on dissolved oxygen and filtration of nutrients.

Oxygen

When the sun shines, algae and plants photosynthesize and pump oxygen into the water. But little oxygen is produced in the pond during cloudy weather. At nighttime plants and algae remove oxygen as they respire and consume carbohydrates. Thus, oxygen concentrations are lowest during the morning, especially after cloudy days. If the pond has too much algae, extreme fluctuations may generate too little oxygen in the morning from nighttime respiration of algae.

We gauge oxygen concentration by observing algae growth, fish behavior, and by using dissolved-oxygen meters. I use a Secchi disk to monitor algae growth. It should be visible at 12 inches of depth. I make the Secchi disk using a white yogurt container lid with two black triangle pie pieces painted on it. I then affix the lid to a wooden ruler with a staple and dip the Secchi disk vertically into the pond until the Secchi disk disappears. If algae growth inhibits visibility at a shallower depth, the density of algae may cause oxygen depletion in the early morning. Tilapia also let us know when oxygen levels are low. If we see the fish piping — gulping for air at the surface of the water — then we know oxygen levels are too low.

A dissolved-oxygen (DO) meter such as the one shown in figure 5.44 will give more accurate readings than pure observation. Meters vary depending on the manufacturer, and I recommend following the manufacturer's instructions. The meter we use has to move constantly to allow the proper reaction to occur inside the probe. The probe also needs periodic cleaning and membrane replacement. I recommend taking readings several feet from the edge of the pond and 12 to 18 inches below the surface and away from any aeration devices.

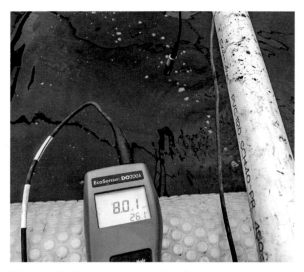

Figure 5.44. To take oxygen readings in appropriate locations, we attach the dissolved-oyxgen sensor probe to a stick, pipe, or float.

It's best to take readings at a consistent time and place in the morning and again in the evening to monitor trends in dissolved oxygen. Then you can predict morning DO conditions based on nighttime readings and use supplemental aeration devices before dissolved oxygen levels breach thresholds.

PH AND ALKALINITY

The pH of water measures its acidic or basic properties, or the amount of hydrogen ions it contains. Photosynthesis affects the pH of water. During nighttime, plants and algae consume oxygen and produce carbon dioxide. The carbon dioxide diffuses in the water, forms carbonic acid with a free hydrogen ion, and lowers the pH.

$$H_2O + CO_2 = H_2CO_3$$

During daytime the opposite occurs. The pH rises as carbon dioxide is removed from the water during photosynthesis. Intensive recirculating systems produce hydrogen ions that lower pH over time by using biofilters to convert toxic ammonia into nitrates.

Fish, plants, and biofiltration components are all affected by the pH of water. Fish tolerate a pH of 6 to 9.5; biofiltration bacteria's optimum range is 7 to 8; and plants prefer a pH of 6.5 to 7 to access nutrients for proper growth. Maintaining a pH around 7 and an alkalinity of 50 to 100 ppm will keep all organisms happy and healthy.

The alkalinity or the capacity of water to neutralize acidity also affects the pH of water. Highly alkaline water buffers against pH changes using bases such as carbonate and bicarbonate to absorb hydrogen ions that would otherwise create acid water. Hydrogen ions combine with bases, which prevents the pH from decreasing, but the bases get used up in the process. Since acid compounds are constantly produced as the fish are fed, a continuous supply of a base applied as a percentage of feed will neutralize the acid-forming compounds.

At least weekly we monitor the pH and alkalinity in greenhouse ponds with a digital tester and adjust as needed. Alternating between potassium carbonate and calcium carbonate at a rate of 2.5 teaspoons per 100 gallons of water or 25 percent by weight of the amount of feed will raise or maintain the pH and add alkalinity to the water. Adjust according to test results on subsequent days. Since pH and alkalinity change slowly over time, it's best to add a small amount of base daily corresponding to feed to maintain levels, instead of waiting until problems emerge.

Alternatively, potassium hydroxide or calcium hydroxide can be used as a base to raise the water pH, but they are not considered organic, and they don't add to the alkalinity. The potassium and calcium from either hydroxide or bicarbonate sources have the added benefit of adding calcium

Table 5.1. Water Quality Parameters for Various Organisms

	Oxygen (ppm)	pH	Ammonia (ppm)	Nitrites (ppm)	Temperature °F
Tilapia					
Range	1–7	5–10	0–1	16@pH7.9	50–105
Ideal	2–7	6–9	0.08	1	85–88
Freshwater Prawns					
Range	1–7			2@pH8.5 1@pH9	54–107
Ideal	3–7	7–8.5	0.1–0.3	1	77–91
Biofiltration					
Range	2–7	6–9	0–3		32–86
Ideal	4–7	8–9	3		86
Plants					
Range		6–7.2			65–86
Ideal		6.8–7			75

Sources: T. Popma and M. Masser. *Tilapia Life History and Biology*. Southern Regional Aquaculture Center Publication No. 283, 1999.

J. E. Rakocy and A. S. McGinty. *Pond Culture of Tilapia*. Southern Regional Aquaculture Center, Publication No. 280, July 1989.

M. B. New and W. C. Valenti. *Freshwater Prawn Culture: The Farming of* Macrobrachium rosenbergii. Osney Mead, Oxford: Blackwell Science Pub., 2000.

and potassium, nutrients commonly deficient in aquaponic systems. Sodium bicarbonate (baking soda) should not be used with aquaponic systems to avoid sodium toxicity with plants.

We adjust the pH and alkalinity of our large, soil-bottom ponds by treating the soil. First, we conduct a pond-bottom soil test through a soil testing service. We scoop pond mud from several areas in the pond, dry it, mix it, and submit it as a single sample. Test results indicate the amount of lime needed, and we then add dolomitic limestone to the pond bottom based on recommendations. I recommend adding the lime after draining the pond for harvest in fall or late summer. Watch for an opportunity when the pond bottom becomes dry enough to drive a tractor across it. We use a rotary cone spreader to apply the lime easily. Mixing lime into the soil is not necessary, but it will speed up the reaction process while also aiding

the decomposition of accumulating organic matter that may hinder water quality during production.

NITROGEN WASTES

As fish consume food they release nitrogen wastes. Approximately 2.2 pounds of ammonia nitrogen are produced from every 100 pounds of feed.[15] The fish excrete ammonia through their gills as ammonia gas. Bacteria living on the surface of biofilter media convert the ammonia into nitrite and from nitrite to nitrate. Since ammonia and nitrite are both toxic to fish, it's important to maintain levels of these substances below the threshold for the species grown. If I see fish crowding around the water inlet, I know ammonia or nitrite levels may be too high. Testing kits are available through online suppliers to accurately test ammonia, nitrite, and nitrate levels. To quickly but temporarily lower levels, we increase water exchange, add more biofiltration media, or reduce feeding rates or stocking densities. Tilapia can survive ammonia concentrations as high as 3 ppm for three to four days if they're gradually acclimated. However, food consumption is depressed at concentrations as low as 0.08 ppm. Nitrite levels for tilapia should be less than 27 ppm.[16] With freshwater prawns, tolerance varies based on pH, but larvae can tolerate levels around 1 ppm for ammonia and nitrite.

SUMMING UP THE FUNCTIONS

1. Greenhouse film catches more sunlight because of south-facing slope.
2. Sloped greenhouse platform catches rainwater.
3. Interior pond reflects sunlight in winter.
4. Interior pond encapsulates and insulates collected rainwater for use as thermal mass.
5. Insulated rainwater pond stores heat from compost pile using heat exchanger.
6. Interior pond provides nursery for tilapia fingerlings and prawns.
7. Fish and prawns grown in interior pond produce nutrients for ebb-and-flow hydroponics.
8. Fish and prawns grown in interior pond produce nutrients for floating hydroponics.
9. Fish and prawn waste in pond fertilizes adjacent beds in greenhouse.
10. Adjacent farm field is irrigated and fertilized by draining pond.
11. Interior pond cools air.
12. Sloped greenhouse platform stores heat in soil.
13. Solar energy is converted to heat using stacked pots of packed column aerator.
14. Plants in greenhouse are automatically irrigated.
15. Hydronic heating warms soil and plants.
16. Pond provides habitat for predators of plant pests, such as toads and frogs.
17. Sloped greenhouse and pond facilitate distribution of heat at night through convective loops.
18. Sloped greenhouse platform facilitates convective removal of heat through vents for cooling.
19. Greenhouse walls and roof trap nighttime heat.
20. Solar-powered wax valves control ventilation.
21. Roll-up greenhouse sides provide ventilation.

CHAPTER 6

The Big Flush

Bio-Integrated Rainwater Harvesting

The urban water cycle requires a considerable amount of energy to extract, distribute, and treat water for use in homes. Not only that, but according to the Environmental Protection Agency, the United States uses 30 percent of its potable domestic water outdoors for washing cars, watering lawns and gardens, and filling swimming pools. Of the remaining 70 percent used indoors, 30 percent is flushed down the toilet. Consequently, we use 50 percent of our potable water for purposes that don't require treated water.

A study conducted by the Electric Power Research Institute (EPRI) concluded that our nation uses 3 to 4 percent of its electricity to move and treat drinking water and wastewater. However, a more recent report by the Government Accountability Office (GAO) determined that the highly cited EPRI report was outdated, and energy use to manage the urban water cycle is probably much greater. If the United States reduced potable water demand by 10 percent, we would save approximately 300 billion kilowatt-hours of energy each year.[1]

Given our energy-guzzling water use habits, using less water not only conserves water but also prevents pollution and conserves the fossil fuel resources we would use to treat and move the water. Harvesting rainwater for irrigation and flushing toilets offers an easy solution for reducing water consumption. And by reducing runoff, rainwater harvesting not only conserves natural resources and energy but also helps prevent erosion, flooding, and the contamination of rivers, lakes, and oceans from fertilizers and other pollutants.

Rainwater harvesting is a lot like banking. We deposit money into our accounts, then direct withdrawals to pay our bills, fill our fridge, and purchase products. Our finances are a constant recycling of resources. We retain and expend depending on our current needs.

Rainwater-harvesting systems convey water from roofs and land surfaces into a pond or storage container for later use. Gutters collect the rainwater and transfer it into downspouts. Piping then conveys the water into storage tanks or ponds. This chapter isn't about installing a simple rain barrel at the bottom of a downspout. The system is creatively and carefully designed to make use of the power of gravity. The unique fluid properties of water allow transfer to high locations to provide gravity-powered flush toilets in a home and drip irrigation systems for a garden.

The Parts of the System

The basic parts of rainwater-harvesting systems are divided into conveyance, roofs, gutters, downspouts, filtration, storage, and use. Making a dent in the massive amount of water consumed in the average home requires careful planning and design of a system to capture water in larger storage areas based on needs and rainfall potential. Always base rainwater-harvesting conveyance systems on one-hour, one-hundred-year rain events, and place your water storage high in the landscape to enlist the help of gravity (see the "Designing a Gravity-Driven System" section). Keep in mind that the easiest low-pressure, nonpotable uses for harvested rainwater are for toilets and irrigation systems. Pumping offers more pressure if necessary but comes at a cost.

Conveyance

Two basic configurations convey the water: dry systems and wet systems. In dry systems all the water drains out of the conveyance pipes between rain events. In wet systems, also called reverse siphons, part of the pipes stay full between rain events. Both systems have their benefits and disadvantages.

DRY SYSTEMS

In dry systems, pipes primarily remain empty. They only contain water during a rain event, when water is running from the roof, through the pipes, and into the pond or storage device. The pipes of a dry system must continuously slope downhill toward the storage area without forming any dips or low points where water can pool. The minimum slope is 1 percent, or approximately a ⅛-inch drop per foot. The advantages of this system are that, because water doesn't pool in pipes, the water stays cleaner, and hard freezes do not damage the piping.

Dry systems do have their limitations, though. If you want to connect two downspouts from

Figure 6.1. A dry rainwater harvesting system conveys water from gutters through the black pipe on top and into the tank on the left. The overflow from the tank travels through the white pipe under the building, then up into an elevated pond, forming a "wet" system. A pump just above the white pipe pressurizes the water in the tank for use.

opposite ends of a house to a single storage device, you must run the pipes on a continuous slope along the side of the house, high enough above the ground to empty into the top of the storage device. A pipe running across the side of your house can be an unsightly nightmare, especially if windows are in the way — the pipes might have to cut across the windows.

Dry systems also require locating a pond or storage container close to the gutters. If you locate storage areas far from the building, you will need to build a long, aboveground aqueduct-like system to convey water into the top of the storage area.

Elements of the System

Inexpensively capture rainwater from your roof and store it at the highest point on your property to maximize the force of gravity. This chapter teaches you how to:

▸ Position ponds or tanks to provide free water for flushing toilets
▸ Save money on water bills by using harvested rainwater to flood-irrigate hilly terrain and small gardens
▸ Engineer both wet and dry rainwater-harvesting systems and size systems for your climate
▸ Reduce evaporation in drier climates with floating covers for ponds, turning them into cheap tanks

Figure 6.2. Installing a wet rainwater-harvesting system to bring water to a pond at the top of the property 200 feet from the house. The surface of the pond sits 3 feet lower than the lower edge of the roof.

WET SYSTEMS

Wet conveyance systems are shaped like a large horseshoe. They continuously contain water inside the piping at the bottom of the horseshoe. The side of the horseshoe that connects to the gutters is taller than the side of the horseshoe that connects to the storage device. So when water enters from the gutters, it pushes water through the horseshoe and out the lower end of the system. In spite of this movement, water always remains in the bottom of the horseshoe unless it is drained. Since water constantly stays inside the pipe, you have to control mosquitoes and debris, manage for freeze damage, and prevent stagnation. A few design considerations can solve these problems and free up the advantages of wet rainwater-harvesting systems.

Wet rainwater-harvesting systems seamlessly connect downspouts to underground piping and convey water from multiple locations into central storage areas without littering the landscape. Since pipes are "out of sight — out of mind," you can locate central storage areas a great distance from your home or building at high elevations. Catching and storing water as high as possible is the golden rule for rainwater harvesting. Elevated storage allows gravity to move water where it's needed without expensive pumps and ongoing use of electricity. The "Designing a Gravity-Driven System" section shows you how to maximize gravity in wet rainwater-harvesting systems and turn any encountered problems into insightful solutions.

Roofing

A roof is an obvious choice for an elevated area to catch water from. Sitting higher than the rest of the landscape, a roof can easily catch and move water to elevated storage areas by taking maximum advantage of the power of gravity. The nature of the catchment surface is an important consideration. If you plan to use the collected water for potable purposes, your roof should be metal with baked-on enamel, galvanized steel, or Galvalume steel. The smooth surfaces of metal roofs are ideal because they don't trap debris or hold water after a rain. (Debris reduces water quality, and biological contaminants can grow in trapped water.) Asphalt shingles and some older metal roofs may contain toxic compounds and thus aren't suitable for harvesting potable water. However, it's fine to use water from such roofs for irrigation and flushing toilets, especially if you integrate filtration devices into your system.

You can also modify your roof to improve your water quality. I wouldn't make a change to a new roof just to enhance my rainwater-harvesting system. But when it's time for roof replacement, water quality would be a factor in my roofing material decisions. After all, replacing an old worn-out asphalt shingle roof with a longer-lasting metal roof is logical.

Applying an approved potable-water roof coating is another way to increase the life of an aging roof while also improving water quality. When our eighty-year-old toxic metal roof started leaking, we coated the roof with an approved elastomeric paint approved for potable water harvesting. The elastomeric paint, from Scott Paint Company in Florida, can be used to coat asphalt, metal, or thermoplastic olefin (TPO) roofs. The coating lasts seven years before requiring reapplication and is considerably cheaper than roof replacement. The paint costs less than thirteen hundred dollars in materials for a 1,700-square-foot house. The light-colored material also reflects light, reducing the amount of energy needed to cool the building in summer.[2]

Gutters and Downspouts

Gutters collect rainwater, and downspouts channel the water into pipes that lead to the pond. For potable water, gutters made from galvanized steel or aluminum are the safest. If you're working with

Figure 6.3. A long, seamless gutter (seen here from underneath) slopes to one end of the house, facilitating centralized harvesting.

copper gutters, coat them with an NSF-approved paint, and avoid gutters made of zinc or bamboo. Installing gutters correctly is difficult, and I recommend hiring a professional service capable of building long, seamless gutters on-site using specialized equipment. Seamless gutters last longer and require less slope than sectional gutters typically bought by homeowners. Seamless gutters require a minimum of 1/16-inch drop every 10 feet, and sectional gutters require a minimum of 1/16-inch drop every foot to handle the heavy load of water.[3] The slope and size of gutters also dictates how much water the gutters can move between downspouts.

Designers should work with gutter installers to ensure rainwater-harvesting goals are incorporated into gutter design. Gutter installers typically install more downspouts than necessary, making central-ized harvesting difficult. Normally, distributing roof runoff to multiple points helps prevent con-solidated flows that may cause erosion. However, with rainwater harvesting, distributed flow defeats the goal of centralized harvesting. With careful design you can modify your gutters to distribute water to a limited number of downspouts, which will save you money on filtration and conveyance systems. For example, sloping all the gutters to the side of your house closest to a storage tank or pond may eliminate the need to install expensive underground pipes to convey the water.

Much like sizing gutters, sizing downspouts correlates with the volume of water entering the downspout. Using PVC material for the down-spouts facilitates installation of filtering equipment (as described later in this chapter). I coat the PVC with a paint suitable for plastic or prime and paint it with latex paint to protect the PVC from UV degradation. See the "Designing a Gravity-Driven System" section for sizing gutters and downspouts.

Filtration

Roofs are leaf and dirt magnets, collecting every possible airborne contaminant. I commonly find bird and rodent feces, and even the occasional dead frog, in gutters or on roofs. To make matters worse, pollutants settle on surfaces, waiting to be washed into gutters with the next rainfall. Luckily, filtering contaminants out of the rainwater before it enters the storage container is easy. Filtration involves not only filtering out large debris but also making sure that fine debris, such as pollutants that stick to the roof itself, don't get washed through the system and into the storage container. Two different devices help with this: downspout filters and first flush diverters.

Ideally, trees don't overhang your house. This limits the amount of leaf debris falling on your roof. Designing a landscape with medium-size

trees placed a good distance away from your house prevents leaf problems. I learned this valuable lesson through trial and error. Our first house was located in an older, historic district of a small town. Taking down a tree involved more than an arborist or a landscaping crew; it required a compelling argument to a board of tree committee members dedicated to saving old trees. I love trees as much as the next guy, but when they're littering my gutters weekly I become a little less loving.

So for all those homeowners with overhanging trees, here are a few possible solutions. The first is gutter guards. I've tried three different types and found all of them lacking. Some work on the prin-ciple of water tension, with a solid plate covering the top of the gutter and hanging over the edge. While it appears that the plate will prevent water from entering the gutter, the edge is curved, which allows water to flow around and under it into the gutter while the leaves are shed off the gutter com-pletely. Though these gutter guards are successful at preventing leaf intrusion, I've noticed that with steep roofs and heavy downpours water overflows the guards and misses the gutters.

Other systems use filters placed inside the gut-ter or screens placed over the gutter. With foam filters placed inside the gutter, debris ultimately clogs the filter and water eventually overflows the gutter. *Consumer Reports* rates fine-mesh screens placed on top of the gutter as the best-performing gutter guard.

The next line of defense is a downspout filter. These filters are small boxes with sloped screens on top. They can be attached directly below the gutters to direct flow into a downspout or placed midway along a downspout. As water exits the gutters it falls on the sloped screen. Leaves get pushed off the screen in a self-cleaning fashion, and water penetrates the screen and enters the downspout below, free of debris. The screens also keep mosquitoes out of "wet" rainwater-harvesting systems, a necessity, especially in the southeastern states. I prefer a downspout filter called Leaf Eater Advanced. The system is compact, is unobtrusive, and adapts to various configurations with a swiveling outlet. The Leaf Eater literature doesn't specify a gallons-per-minute rating for the filters, and I've found they can overflow during extreme rain events, not from pipes overflowing but from the screen's acting as a diverter. Smaller pieces of debris eventually accumulate at the bottom of the screen, and cleaning is required about twice a year. Since I don't have a lot of leaves falling on my roof, I use a downspout filter only, and no gutter guards.

After roof runoff flows through the downspout filters, first flush diverters remove the first flush of dirty water. First flush systems divert the initial rainfall into a separate holding area, enabling the harvest of cleaner water for later use. Recommendations vary regarding the amount of rainwater to divert through the first flush. Pollution in rainwater runoff likely depends on the amount of time

Table 6.1. Capacity of Schedule 40 PVC Diverters

Length of Pipe (ft)	2" Capacity (gal)	4" Capacity (gal)	6" Capacity (gal)
1	0.2	0.7	1.5
3	0.5	2.0	4.5
5	0.8	3.3	7.5
10	1.6	6.7	15.0
15	2.4	10.0	22.5

Downspout filter

Downspout to send rainwater to storage

First-flush diverter

Figure 6.4. Downspout filter prevents debris from entering harvesting systems and keeps mosquitoes out of "wet" rainwater-harvesting systems.

Figure 6.5. First flush diverters and leaf guards clean the water and prevent mosquito reproduction.

between rain events and the amount of pollution falling onto the roof. Roofs located in high-pollution areas and next to dirt roads require more first flush diversion. A good rule of thumb is, "For each mm of first flush the contaminate load will halve."[4]

In Australia, where rainwater harvesting is more common, the government has established a guide that states the first 5.3 to 6.6 gallons of first flush should be diverted for the average-size house.[5] I easily build first flush diverters by adding a PVC tee anywhere in the water piping system, as shown in figure 6.5. The first flush of water falls straight down through the tee, and fills the diverter pipe with dirty water. Once the diverter is full, water then flows through the horizontal projection of the tee and continues to flow through the system toward storage.

Some commercially available first flush systems come equipped with a floating ball that rises and blocks incoming water from mixing with the first flush water once the diverter is full. I equip each downspout of a first flush diverter with a sewage cleanout fitting at the low end. I leave the threaded fitting on the cleanout a little bit loose to allow water to slowly leak out of the diverter and automatically reset the system between rain events. The capacity of the diverter depends on the size and length of the pipe, as specified in table 6.1.

Storage Ponds and Tanks

Rainwater storage is the most expensive part of a harvesting system. Strive for a storage amount large enough to support your needs and maximize collection, yet small enough to maintain reasonable costs. Storage ultimately depends on usage patterns and rainfall patterns of your specific location, as well as the type of storage container you use. Recommendations call for starting storage volume calculations based on the sum of the four highest-demand months.[6] See the section on

"Designing a Gravity-Driven System" for information on sizing storage devices.

PONDS

Ponds inexpensively store harvested rainwater, as discussed in previous chapters. However, ponds lose water to evaporation, especially in the heat of summer, when it's needed most for irrigation. Pan evaporation rate measures the amount of evaporation based on the time of the year and the site's location. You can find information about pan evaporation rates through your state's climate office. Once you know your pan evaporation rate, you can determine the amount of water lost per square foot of pond surface area on a monthly basis and apply this figure to the water demand charts presented later in this chapter. In Clemson, South Carolina, the highest amount of evaporation occurs in July, with 0.22 inch of water lost per day. Since 0.62 gallon is equivalent to 1 inch of water over an area of 1 square foot, we can use the following formula to determine gallons lost from a pond from evaporation. This calculation helps in determining whether building a pond in a particular location will be worthwhile and whether a cover will be needed to prevent excessive evaporation:

$$\text{(inches water lost)} \times (0.62) \times \text{(square feet of pond surface)} = \text{total gallons water lost}$$

Based on a pan evaporation rate of 0.22 inch per day, let's calculate how much water I lose to evaporation during the month of July from my 625-square-foot pond. The first step is to calculate gallons lost per day:

$$0.22 \times 0.62 \times 625 = 85$$

My pond loses 85 gallons per day from evaporation during the month of July. That's 2,557 gallons per month. About 37 percent of all the water I harvest in July is lost to evaporation. If you live in a climate that has plentiful amounts of rainfall in comparison to your evaporation rates, open ponds are good storage containers for harvested rainwater. However, for an accurate water budget you should subtract the amount of evaporation at your site from the total supply available. Digging your pond as deep as stable slopes will allow reduces evaporation and maximizes your water storage. A pond full of plants holds up to 40 percent less water since the plants take up space. A planted pond also loses more water to evaporation from plants transpiring. Chapter 2 contains a formula for estimating pond volume.

Adding a floating pool cover to a pond prevents evaporation. To maintain aesthetics easily, surround the pond with screening vegetation — this definitely looks more appealing than a large plastic or metal tank plopped in the middle of the landscape. The drawback of ponds is algae growth. I've unclogged a lot of algae from filters and pipes that convey harvested rainwater for toilet flushing and irrigation. Speaking from experience, I recommend taking the following steps to prevent algae growth when planning on using ponds for drip irrigation or toilet flushing (see chapter 2 for more information):

- Build a pond with raised edges or build low berms around an existing pond to prevent overland flow and associated nutrients from entering the pond. Overland flow may also compromise the quality of the harvested water, limiting use to flood irrigation.
- Use dark-colored pool covers to prevent light from entering the water.
- Stock algae-eating fish such as sterile grass carp or tilapia to eat algae and keep the water clear.
- When constructing a pond, discourage pond life by not covering the pond liner with soil.

Although not covering the pond liner with soil does discourage algae growth, it may also decrease the life expectancy of your liner. If you cover your pond liner with soil to protect it from damaging sunlight, its life expectancy is indefinite unless it's punctured. Without soil to protect your pond liner, its life expectancy depends on on the type of pond liner you use, with EPDM liners lasting more than fifteen years.[7]

TYPES OF TANKS

Tanks provide a protected area for water, free from the vagaries of birds, rodents, and other defecating animals. However, tanks cost more in materials, shipping, or time to build. Site preparation and grading is also needed to build concrete platforms or graveled foundations to support the heavy weight of a tankful of water. Some tanks are difficult to conceal, and neighbors may frown on the idea of a large tank in the yard next door.

Concrete. Known for their strength and durability, concrete tanks can last a lifetime. Installed above- or belowground, the tanks come precast or cast in place and are usually made with reusable forms in a single monolithic pour. Because of the weight of concrete, all but the smallest tanks are immovable. Concrete tanks are common for large municipal projects. I lived in a town that, for about a hundred years, used a 100,000-gallon concrete tank elevated 130 feet above the ground on a concrete foundation. Life expectancy of concrete tanks exceeds thirty years.[8]

Ferro-cement. Ferro-cement tanks are a combination of steel and concrete. To build a ferro-cement tank, you construct a grid of steel, then cover the grid with a mixture of sand and cement concrete. You spray the concrete onto the frame to form a thin but durable tank. Because the framework can be shaped into unlimited forms, you can sculpt it to resemble rocks or other structures.

Ferro-cement tanks are durable, like concrete tanks, but require fewer materials and offer a wider range of sizes. And like concrete, they are pretty permanent—once in place they are immovable. The life expectancy of ferro-cement tanks is over forty years.[9]

Galvanized metal. Multiple types of galvanized metal tanks are available. Some tanks are made of smooth pieces of metal bolted or welded together, and others are made from thin, corrugated metal sealed with a plastic membrane. Both welded and bolted metal tanks usually contain thicker metal with a galvanized coating protecting the steel from corrosion. Because of the zinc in the galvanized coating, the water is toxic to fish, something to consider if you operate or are planning an aquaculture system.

Corrugated metal tanks with plastic membranes are considerably less expensive. Unlike for plastic tanks, large sizes up to 50,000 gallons are available, and since the tank ships in stackable pieces and is built on-site, shipping costs are reduced. The plastic membrane protects the metal from water contact, increasing the longevity of the metal. Because the liner complicates adding inlets and outlets in the plastic-lined metal tanks, I recommend having them installed by manufacturers. The life expectancy for metal tanks is over twenty years.[10]

Wood. Redwood tanks were popular when old-growth redwood was available to supply the rot-resistant heartwood required to build the tanks. Since wood shrinks when it's dry, the tanks must be kept full to prevent water from leaking out, making them useful only in humid climates and at sites that have a consistent water source.

Polyethylene. Widely available, lightweight, and portable, high-density polyethylene (HDPE) tanks block sunlight and subsequent algae growth in the water because of their dark colors. You can drill and cut the plastic to retrofit inlets and outlets

Figure 6.6. I wanted to install this HDPE tank in a remote location that's inaccessible by vehicle, so I'm rolling it to the site.

using bulkhead fittings. Part of the fitting must be installed through the inside of the tank and part through the outside—a difficult task through the small access opening on the top of the tank. To accomplish this feat, first cut the appropriate size hole for the bulkhead fitting recommended by the manufacturer of the fitting. Next, use a long, stiff wire to feed a rope through the access opening on the top of the tank into the cut hole. Pull the rope through the hole, then slide the inner fitting down the rope, pulling it through the hole. Finally thread and tighten the outer piece of the bulkhead fitting onto the inner piece.

Tanks are available in a variety of sizes up to 12,500 gallons. Since tanks consume a large amount of space, they cost a lot to ship. Tank suppliers usually ship from multiple locations, so I suggest finding a tank size that corresponds with a nearby shipping center; you'll save a considerable amount of money. Shippers may also require forklifts on-site to unload large tanks. However, you can easily roll smaller tanks off trucks. HDPE tanks last fifteen years or more if shaded by a building or tree. Placing tanks on the north side of buildings will provide shade for part of the year.

TANK FOUNDATIONS

Your tank is only as good as your foundation. I recommend locating tanks in elevated locations away from the flow of water, thereby preventing erosion of the tank's foundation. If you use fill to level the site, the fill dirt will likely sink and cause the tank to tilt with time, breaking pipe connections and possibly the tank. Flexible connections between tanks and pipes are required in some municipalities to prevent broken pipes if tanks move. Water weighs 8.34 pounds per gallon, and a tank full of water exerts a considerable amount of force on the foundation. Tanks less than 500 gallons are probably suitable for soil foundations.[11] However, larger tanks require careful consideration of the safe loading capacity of the soil. You may need pea gravel, sand, or a concrete base to provide the appropriate support for your tank. Do not use gravel because the sharp edges can puncture tanks.

Empty tanks weigh little and can easily blow over in high winds. Tanks usually come equipped with areas for attaching support straps and securing tanks to foundations or the ground. You can also use anchors, traditionally used to tie down mobile homes, to secure tanks in place.

BURYING A TANK

Burying a tank hides it from view and protects the tank from UV degradation and freezing winter temperatures. However, burying a tank is expensive, and an underground shift may crack the tank or connecting pipes, causing hard to find leaks. Cracks may also allow contaminants to enter the tank, affecting water quality. Plastic tank materials are also permeable to chemicals, pesticides, and wastewater that could affect water inside tanks. Belowground tanks should have an opening at least 8 inches above the surrounding grade, with the soil sloping away from the opening. In addition, locate underground tanks uphill from septic leach fields and prevent vehicles from crossing over the top of tanks. Not all tanks can be buried, so use specially designed tanks for underground burial that can withstand the external forces of soil when the tank is empty. When underground tanks are empty and the surrounding soil is saturated with water, the tank becomes buoyant and may rise out of the ground. Codes call for securing the tanks to meet or exceed the buoyancy effect of the buried empty tank. According to Archimedes's principle, an object is buoyed up by a force equal to the weight of fluid displaced. Therefore, a 1,000-gallon tank may have an upward force greater than 8,000 pounds in saturated soil — enough to turn the tank into a surfacing submarine.

FITTING UP A TANK

Equip all tanks with inlets, outlets, vents, and inspection ports. A bulkhead fitting creates a tight seal when pipes must protrude through tanks. I design all openings to prevent rodent and mosquito intrusion, as well as limit the stirring of sediments on the bottom of the tank that leads to increased turbidity. Turbidity is a measure of the cloudiness of water or how difficult it is to see through the water. Suspended silt, mud, dead plants, algae, and aquatic organisms all contribute to turbidity. Inlet water should cascade through a basket filter or enter at the bottom of the storage container through an upturned pipe. This helps maintain water quality by preventing the disturbance of sediment on the bottom of the tank. This technique is also useful for discharging into ponds to avoid stirring the pot.

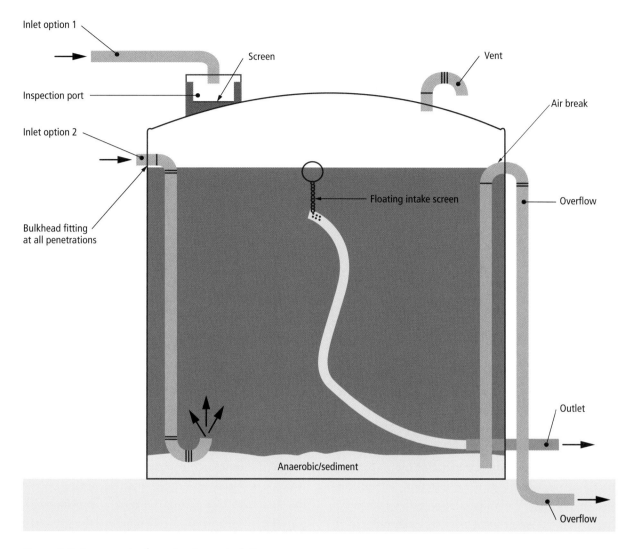

Figure 6.7. Components of a water storage tank.[12]

Ideally, seal a tank at all inlets and outlets. As the tank fills with water, air becomes displaced and needs to escape through a vent. Commercial vents screw into the top of the tank and come equipped with shielding to protect against contaminants and mosquitoes. Alternatively, I've attached a pipe to the top of a tank with the open arm facing downward, then protected it with metal hardware cloth and mosquito screens. I've also drilled a hole in the top of the tank and sealed an aluminum mosquito screen over the hole with silicone, but this allows contaminants into the tank so only use this technique with containers that store nonpotable water.

Situate the outlet pipe at least 4 inches above the base of the tank to prevent the outlet from pulling sediment and dirty water settled at the bottom of the tank. The cleanest water in a tank or pond sits about 4 inches below the surface. I use a floating intake

Figure 6.8. A floating intake screen is built using PVC parts. It's used to siphon water from the cleanest top 4 inches of a pond.

Figure 6.9. The intake screen is assembled, painted, and attached to a float and flexible pipe.

to pull water from this area, with a stainless steel chain and coarse filter made from PVC, as shown in figure 6.8. I attach flexible tubing to the filter, long enough to pull the filter toward the inspection port for cleaning, as shown in figure 6.9. Building this type of floating intake is simple and saves over a hundred dollars compared to the commercially available options. For outdoor ponds I use flexible PVC as the flexible pipe for the floating system to prevent UV degradation. Since flexible PVC is not rated for potable use, I recommend substituting HDPE piping if the water is for potable purposes.

To prevent water from backing up into the inlet, make tank overflow outlets at least the same diameter as the inlet pipe. I design the overflow pipe to siphon water and sediment from the bottom of the tank, automatically cleaning the tank with every rain. To do this, locate the intake for the overflow a few inches from the bottom of the tank. The overflow pipe then extends up to the top of the tank and exits the tank at the maximum water level height, forming the shape of a horseshoe. The overflow from this type of system could create a siphon that would suction all the water out of the tank. To prevent water in the tank from siphoning

out as it exits the other side of the tank, drill a hole, one-quarter the diameter of the pipe, at the very top of the horseshoe on the interior of the tank to create an air break. I've also designed overflow pipes to skim debris from the top of tanks by locating the overflow inlet at water level and cutting a 45-degree angle into the intake.

I always make sure the overflow pipe of a tank discharges water into an appropriate area. Tanks usually fill quickly during a rain event, and excess water can flood or erode sensitive areas. Locate the discharge away from the tank to prevent tank foundations from eroding. A flapper valve attached to the end of the overflow pipe prevents rodents and mosquitoes from entering the storage tank. Ponds, rain gardens, diversion channels, and basins all make great outlets for tank overflow.

Various tank-level indicators are available to show the water level inside the tank. Devices that use a float and counterweight are easy to install and not susceptible to freezing like site glass devices are. You can easily make one by running a copper pipe through the tank to allow the cord to move easily through the hole as the float and counterweight move. Remote wireless systems are also available so

Combining Inlet Pipes with Siphons

Inlet pipes can also serve as siphons to pull water out of a tank or pond. This is useful in wet rainwater-harvesting systems, which need a drain at the lowest point of the pipe system to remove water during a hard freeze and to flush the system of any debris that may accumulate. If you locate the inlet pipe below water level, when you open the drain all the water will siphon out of the tank or pond unless you provide an air break at the top of the wet harvesting system pipework. If you install valves as air vents at all the high points in a wet rainwater harvesting system, then you can close the valves to allow siphoning or open them to prevent siphoning, as shown in figure 6.10. Some of the high points may be downspout drains, and you'll need to affix them with valves to prevent unwanted siphoning. I use this technique when I want nonpotable water in the following situations:

- When I'm using a wet rainwater-harvesting system to fill a pond that doesn't have a drain
- When I need water at the inlet of the tank or pond and the outlet drain is on the opposite side

Figure 6.10. A combined inlet and outlet pipe can send water into this pond or siphon it out. Closing the valve at the downspout and opening a valve at a lower point in the system siphons water from the pond.

- When a long pipe is needed in a wet rainwater-harvesting system to deliver the water from the catchment area to a storage tank or pond; if I also want to pipe the water back close to the source, I conserve a considerable amount of money and pipe material if I use a single pipe system to fill and drain the pond

you can monitor the water level from the comfort of your home.

POND AND TANK SAFETY

Water storage devices such as tanks and ponds inevitably attract the curiosity of children. Tanks also pose

a danger since children and adults could fall in and drown. Lock access ports to prevent entry to unqualified persons, and install a ladder inside the tank if the port is large enough for entry. Also, remove external ladders or install a removable lower section to prevent children from climbing on top of the tank.

Using Harvested Rainwater

Your local jurisdiction may have regulations regarding the use of harvested rainwater. If not, you can consult the International Association of Plumbing and Mechanical Officials' *Green Plumbing and Mechanical Code Supplement*, which provides guidelines for using harvested rainwater.[13] If local codes don't cover rainwater harvesting, they will probably adopt the Green Plumbing codes eventually. Using rainwater for potable uses such as showers, dishwashing, and drinking is complex and outside the scope of this book. I recommend reading the book *Rainwater Harvesting: System Planning* by Texas AgriLife Extension Service.[14]

This section covers my use of harvested rainwater for nonpotable uses: toilet flushing and drip irrigation systems. I've tried to follow the *Green Plumbing and Mechanical Code Supplement* for the following sections and refer to them as "Code." See chapters 3 and 7 for instructions on using harvested rainwater for flood irrigation and manure flushing.

Water Pressure and Flow

Using harvested rainwater requires understanding principles of water pressure and flow. Municipal water suppliers usually pump water up into water towers high above the ground; then gravity generates the needed pressure to run water utilities. Locating water storage as high as possible in the landscape serves the same purpose but probably won't provide the same amount of pressure as a municipal water system. The difference in height between the level at which water is stored and the level at which it is used determines water pressure. For every foot above the point of use, gravity generates 0.43 pound per square inch (psi) of pressure. Typically, cities provide water at 40 to 70 pounds per square inch by locating water towers at the

Table 6.2. Minimum and Optimum Water Pressure

Application	Pressure (psi)	
	Minimum	Optimum
Fire Hose	40	100
Drip Irrigation	15	25
Drip Tape	4	10
Shower	5	50
Washing Machine	10–15	50
Toilet	1	Depends on valve and tolerance to long refills
Tub	0.5	Not critical
Kitchen Sink	0.5	20
Reverse Osmosis Filter	30–60	40–80

Source: A. Ludwig. *Water Storage: Tanks, Cisterns, Aquifers, and Ponds.* Santa Barbara, Calif.: Oasis Design, 2007.

highest point in the landscape, usually elevated 100 feet above the ground.

However, not all water applications require high pressure, and you can modify some applications to function with low-pressure systems. With a little ingenuity you can store roof rainwater high in your landscape and use gravity to convey it. Pumps are sometimes necessary to supply the needed pressure when gravity pressure is insufficient. I present information on pumps under "Adding a Pump to the System," later in this chapter.

There's a common misconception about water pressure as it relates to the volume of storage. Many people assume a large tank full of water exerts more pressure on a drainpipe than does a small tank of water. But volume is not the determining factor for pressure; only height of the stored water is significant. For example, a 1,000-gallon tank storing water at a height of 10 feet exerts 4.3 pounds of pressure on the outlet pipe at the bottom of the tank. A 10,000-gallon tank storing water at a height of 10 feet still only exerts 4.3 pounds of pressure on the outlet pipe at the bottom of the tank.

$$0.43 \text{ psi/foot of height} \times 10 \text{ feet}$$
$$= 4.3 \text{ pounds of pressure}$$

Static pressure is the amount of water pressure generated by the height of the water storage. Once water starts moving inside the pipe, friction and turbulence against the sides of the pipe reduces the amount of pressure — this is known as dynamic pressure. Four factors affect dynamic pressure in piping systems.

- Velocity of water (measured in feet per second or gallons per minute)
- Inside diameter of pipe
- Roughness of the inside wall of the pipe
- Change in direction and fittings such as tees, elbows, valves, and filters

Every device that uses water — every sink, toilet, shower, and sprinkler — not only has a minimum pressure to operate properly but also has a flow or gallons per minute necessary for the application to determine the dynamic pressure. The flow for some applications isn't a significant factor. For example, water flow into a toilet is not significant unless you want the toilet to fill up quickly. Other applications, such as drip irrigation systems, require the flow to be balanced with the number of plants requiring irrigation. Otherwise you won't have enough volume of water to fill the irrigation system regardless of the pressure. Luckily, increasing the size of the pipe or limiting the amount of applications running off a single line increases the available flow.

Using harvested rainwater requires elevating rainwater high enough to provide the needed pressure for the specific application (see table 6.2)

and determining the size of pipe to reduce friction and ensure the flow capacity. I use two methods to determine pipe size, the velocity limit method and the friction loss tables. The velocity limit method mostly applies to high-pressure systems using city water or pumps and ensures that the velocity of the water stays below 5 feet per second to avoid damaging systems from pressure surges. High velocities also apply with gravity-fed systems if storage is elevated high above the point of use. Use the following equation to compute the velocity of flow:

$$V = .408 \, (Q \div d^2)$$

where

V = velocity in feet per second; Q = flow volume (gpm); and d = inside diameter of pipe (inches)

The velocity should remain less than 5 feet per second. If the velocity becomes greater than 5 feet per second, enlarge the pipe diameter to reduce the velocity.

Here's an example from our farm: I want to run 20 gallons per minute (gpm) to an irrigation system in our field. I'm considering using a 1.5-inch schedule 40 PVC pipe to run the water, and I want to make sure I'm not going to exceed the recommended velocity. The inside diameter of the pipe averages 1.59 inches.

$$V = .408 \, (20 \div 1.59^2)$$
$$V = .408 \, (20 \div 2.53)$$
$$V = .408 \, (7.9)$$
$$V = 3.22$$

Therefore, at the needed flow volume the velocity of the water will be 3.22 feet per second or less than the 5 feet per second limit. However, if I use a 1-inch pipe, the velocity would be over 8 feet per second and would exceed the limit.

Table 6.3. Friction Loss Chart for Schedule 40 Pipe

Pipe Size	½"		¾"		1"		1¼"		1½"		2"	
Avg. I.D.	0.602		0.804		1.029		1.360		1.590		2.047	
Flow (gpm)	Vel.	psi loss	Vel.	psi loss	Vel.	psi loss	Vel.	psi loss	Vel.	psi loss	Vel.	psi loss
1	1.13	0.50	0.63	0.12	0.39	0.04	0.22	0.01	0.16	0.00		
2	2.25	1.82	1.26	0.44	0.77	0.13	0.44	0.03	0.32	0.02	0.19	0.00
3	3.38	3.85	1.89	0.94	1.16	0.28	0.66	0.07	0.48	0.03	0.29	0.01
4	4.50	6.55	2.52	1.60	1.54	0.48	0.88	0.12	0.65	0.06	0.39	0.02
5	5.56	9.91	3.16	2.42	1.93	0.73	1.10	0.19	0.81	0.09	0.49	0.03
6	6.75	13.89	3.79	3.40	2.31	1.02	1.32	0.26	0.97	0.12	0.58	0.04
7	7.88	18.48	4.42	4.52	2.70	1.36	1.54	0.35	1.13	0.16	0.68	0.05
8	9.01	23.66	5.05	5.79	3.08	1.74	1.76	0.45	1.29	0.21	0.78	0.06
9	10.13	29.43	5.68	7.20	3.47	2.17	1.99	0.56	1.45	0.26	0.88	0.08
10	11.26	35.77	6.31	8.75	3.85	2.63	2.21	0.68	1.61	0.32	0.97	0.09
12	13.51	50.14	7.57	12.27	4.62	3.68	2.65	0.95	1.94	0.44	1.17	0.13
14	15.76	66.71	8.84	16.32	5.39	4.91	3.09	1.26	2.26	0.59	1.36	0.17
16	18.01	85.42	10.10	20.90	6.17	6.29	3.35	1.62	2.58	0.76	1.56	0.22
18	20.26	106.24	11.36	25.99	6.94	7.82	3.97	2.01	2.90	0.94	1.75	0.28
20			12.62	31.59	7.71	9.51	4.41	2.45	3.23	1.14	1.95	0.33
22			13.89	37.69	8.48	11.35	4.85	2.92	3.55	1.37	2.14	0.40
24			15.15	44.28	9.25	13.33	5.29	3.43	3.87	1.60	2.34	0.47
26			16.41	51.36	10.02	15.46	5.74	3.98	4.20	1.86	2.53	0.54
28			17.67	58.91	10.79	17.73	6.18	4.56	4.52	2.13	2.73	0.62
30			18.94	66.94	11.56	20.15	6.62	5.19	4.84	2.42	2.92	0.71
32					12.33	22.71	7.06	5.58	5.16	2.73	3.12	0.80
34					13.10	25.41	7.50	6.54	5.49	3.06	3.31	0.89
36					13.87	28.24	7.94	7.27	5.81	3.40	3.51	0.99
38					14.64	31.22	8.38	8.04	6.13	3.76	3.70	1.10
40					15.41	34.33	8.82	8.84	6.46	4.13	3.89	1.21
42							9.26	9.67	6.78	4.52	4.09	1.32
44							9.71	10.54	7.10	4.93	4.28	1.44
46							10.15	11.54	7.42	5.35	4.48	1.57
48							10.59	12.39	7.75	5.79	4.67	1.69
50							11.03	13.36	8.07	6.25	4.87	1.83

Note: Loss is expressed as pounds per square inch per 100 feet of pipe. Shaded areas represent velocities over 5 feet per second and should only be used with caution.

The most important pipe sizing equation for gravity-fed systems examines the friction loss to determine the allowable pressure loss in the system. Once you determine the allowable pressure loss and velocity, consult table 6.3 to compare different pipe sizes to select the pipe size necessary for the application's gallons per minute. The tables are based on the Hazen-Williams Equation used later to determine pressure loss from the pipe. Table 6.3 gives the friction loss and velocity for SCH 40 pipe. Other tables are available online for other types of pipe from the Irrigation Association, or you can use the following formula:

$$Hf = [0.2083 \times (100/c)^{1.852} \times Q^{1.852} \div D^{4.866}]\ 0.433$$

where

Hf = friction loss per 100 feet; C = coefficient of retardation from the pipe material (140–150 for PVC); Q = flow volume (gpm); and d = inside diameter of pipe

Supplying Gravity-Flow Toilets

Toilets are a great indoor use of harvested rainwater. Using harvested rainwater for toilets requires a minimum pressure of 1 pound per square inch. With lower pressure the toilet takes longer to fill between flushes. Most cheap plastic float valves, such as the Flowmaster brand, require about 15 pounds per square inch to operate. For the toilet valve to work at 1 pound per square inch, I replaced the common float valve with a ball cock–style valve. These are the old-timey valves that have a large ball float opposite the valve. I recommend purchasing a high-quality, solid brass ball cock valve. Cheap plastic valves only last a few years, while the brass ball cock valves last at least a decade. So the added expense pays for itself.

Figure 6.11. A ball and cock valve works well to regulate toilet filling even with low water pressure.

To comply with Code, use purple piping for your rainwater harvesting distribution lines. I special-ordered the purple piping from a plumbing supplier. Using purple pipe ensures that any knowledgeable person working on the pipe system in the future will recognize that the piping conveys nonpotable water.

For toilet water, Code also requires using a debris excluder and a 100-micron filter. Downspout filters, first flush diverters, and coarse intake filters presented earlier provide the required debris exclusion. Install a filter used for drip irrigation systems to satisfy the requirement of a 100-micron filtration. To prevent the filter from reducing the pressure, which would require you to store the water at a higher point, use a minimum 1-inch filter. Most filters come with a mesh screen providing 150-micron filtration, and most irrigation suppliers will let you swap the 150-micron filter for a 100-micron filter. By combining the 1-inch filter with the 100-micron filtration screen, you will maintain nearly all pressure unless the filter becomes clogged. As shown in figure 6.7, I reduce maintenance by using a floating intake inside the pond to pull

Figure 6.12. A valve box outside our house contains a 100-micron irrigation filter connected to a 1-inch pipe siphoning water from the uphill pond. The filter ensures that no sediment enters the toilet or irrigation emitters. Purple pipes for plumbing comply with Green Plumbing codes. The hose is priming the siphon through the cleanout for the filter. After priming, the filter is returned to a downward position.

water from the cleanest area, keeping the filter clean for a long time.

Code also addresses water quality, requiring *Escherichia coli* (*E. coli*) of less than 100 colony-forming units (CFU) per 100 milliliters. *E. coli* indicates the presence of fecal contamination, and it is measured in colony-forming units, which gives an estimate of the number of viable bacteria in the sample. Contamination of rainwater could occur from a bird's defecating on the catchment area roof or in the pond. Services that test well water for private wells conduct the *E. coli* test. In South Carolina the Department of Health and

Environmental Control (DHEC) subsidizes the testing program to generate tests at a reduced cost.

Ponds and tanks have natural filtration processes using biofilms and solids settling that usually reduce the *E. coli* population to a safe level. The best way to ensure your water passes this test is by using first flush diverters and properly designing intake systems so they don't disturb the anaerobic sediment on the bottom of tanks and ponds.

Another Code water quality parameter for toilet flushing requires turbidity less than 10 nephelometric turbidity units (NTU). Measurements of turbidity are expressed in nephelometric turbidity units. The average person starts to see turbidity levels around 5 NTU and greater.[15] If you fill a glass jar with water and it looks perfectly clear, the water probably meets the 10 NTU requirements. You can obtain a more precise measurement, accurate down to 5 NTU, by using an easily built turbidity tube, a clear tube with a black and white disc at the bottom. When water is added to the tube, the amount of water required to obscure the colored disc from view is then compared to a chart to determine the NTU rating.[16]

MORE ABOUT OUR SYSTEM

Stephanie and I store our harvested rainwater in a storage pond that sits at an elevation 3 feet higher than the top of the toilet in our house (figure 6.2). After I installed a ball and cock valve in the toilet, we needed only 1 pound per square inch of pressure to operate the toilet. We understand that gravity-fed systems are limited, and we don't mind if the toilet takes a while to refill. We agree that 1 gallon per minute flow is reasonable.

I installed 200 feet of 1-inch piping from our pond to our toilet. The water also travels through a valve next to the toilet and 5 feet of ½-inch piping. To comply with Green Plumbing codes I installed a 1-inch filter on the line before its entry into the house. But before I started this project, I addressed the following questions:

- Do we have the needed 1 pound per square inch to run our toilet with a 1 gallon per minute flow through all our pipes and fittings?
- Will the velocity of flow exceed the recommended 5 feet per second?

First, I calculated the amount of pressure generated by storing the water 3 feet higher than the toilet, based on the fact that water generates pressure of 0.43 pound per square inch per foot of height above the point of use.

$$0.43 \text{ psi/foot} \times 3 \text{ feet} = 1.29 \text{ psi}$$

This confirmed that the storage height would provide a little more pressure than the minimum required to run the toilet, but I still needed to determine how much the plumbing reduced the pressure. The allowable pressure loss for the toilet is zero percent since the toilet valve won't work below 1 pound per square inch. I then looked at the pressure loss for the pipe and various components in the system. Two hundred feet of 1-inch piping runs directly to the toilet valve. Table 6.3 shows that for every 100 feet of 1-inch pipe with a flow of 1 gallon per minute we will lose 0.04 pound per square inch; therefore, total pressure loss from 200 feet of 1-inch pipe is 0.08 pound per square inch.

The flow also travels through a Spin Clean filter with the Code-required 100-micron filter. According to the pressure loss charts from the manufacturer for the filter, almost no pressure reduction is seen at 1 gallon per minute. However, the flow and pressure through the filter probably isn't high enough to activate the Spin Clean feature, and carefully cutting out the side of the filter with the Spin Clean mechanism may increase the pressure slightly if needed. The water also travels through 5 feet of ½-inch pipe and a ½-inch angle valve, which is equivalent to 7 feet of ½-inch pipe length in friction (determined by consulting manufacturer specifications for the valve). Therefore, a total of 12 equivalent feet of ½-inch pipe is in the system. Table 6.3 shows a pressure drop of 0.06 pound for 12 feet of ½-inch pipe. Next, I added up all the pressure losses to determine the total pressure loss.

200 feet of 1-inch pipe = 0.08 psi
1-inch filter = only critical when clogged
12 feet of ½-inch pipe = 0.06 psi

Total pressure loss from friction = 0.14 psi

Next I subtracted our pressure loss from friction from our static pressure to determine our dynamic pressure.

$$1.29 \text{ psi} - 0.14 \text{ psi} = 1.15 \text{ psi}$$

In theory, we have just enough pressure to operate the valve. In practice, the system has worked well with a refill time of the toilet clocking in at three minutes. One summer a drought left us without rain for forty days, and I had to switch the toilet water back over to city water for a week. A covered pond or tank would have retained water for a much longer time frame by preventing evaporation. The 100-micron irrigation filter requires cleaning twice a year, and we notice a pressure drop when the filter needs cleaning. We have a soil-bottom pond, so a tank or uncovered pond liner would probably maintain cleaner water. I have noticed a slight staining in the toilet bowl in the summer, and a biofilm exists in the toilet tank. In colder climates I would bury the plumbing deep enough to protect it from freezing to guarantee year-round use. We used PEX piping, which allows pipes to freeze and thaw without breaking. I switch the supply line over to the city water once things start to freeze up, then switch back over to pond water again in the spring.

Supplying a Drip Irrigation System

A gravity-fed irrigation system uses the same type of components as a standard irrigation system such as the one described in chapter 10. The difference is in how the water is supplied. Our storage pond sits at an elevation 10 feet higher than our garden. Water from storage travels through 200 feet of 1-inch piping, 100 feet of ¾-inch piping, and a filter before reaching the garden. Our garden has 160 feet of beds. Since we know the pressure is minimal, we use drip tape operating off 4 pounds per square inch to irrigate the garden. The drip tape has emitters spaced every foot that deliver 0.24 gallon per hour. Calculations are needed to size piping and determine if enough pressure is available.

First I determined the flow required at the garden. Since there are 160 feet of beds and each bed would have a single line of drip tape down the middle, we will need 160 feet of drip tape. The drip tape has emitters every foot delivering 0.24 gallon per hour.

$$160 \text{ feet} \times 0.24 \text{ gallon per foot per hour}$$
$$= 38.4 \text{ gallons per hour}$$

Let's convert this to gallons per minute by dividing by 60.

$$38.4 \text{ gallons per hour} \div 60 \text{ minutes}$$
$$= 0.64 \text{ gallon per minute}$$

Next, let's determine the water pressure at the garden. Since the garden lies 10 feet below the water storage, we use the following equation:

$$10 \text{ feet} \times 0.43 \text{ psi per foot} = 4.3 \text{ psi at the garden}$$

Now we add up all the friction loss from the pipes and filter in the system based on our flow rate of 0.64 (round up to 1) using table 6.3 and the manufacturer's literature for the filter.

200 feet of 1-inch pipe = 0.08 psi loss
100 feet of ¾-inch pipe = 0.12 psi loss
Filter = minimal but will change when clogged
Total = 0.2 psi

Let's subtract our friction loss from the pipe system from our static pressure gained from the elevated water storage to get the dynamic pressure.

4.3 psi from elevated water storage
− 0.2 from friction loss = 4.1 psi

We have just enough pressure to run the drip tape. Since there won't be a surplus of extra pressure available if the filter clogs, I have to keep a close eye on the distribution from the emitters and check the filter monthly to make sure everything is working properly. The 200 feet of 1-inch piping created as much friction as the 100 feet of ¾-inch piping, so running all 1-inch piping would have given us a little more pressure. If we used a ½-inch pipe instead of a ¾-inch pipe for the last 100-foot run, we wouldn't have enough pressure to run the drip tape properly.

Adding a Pump to the System

In my opinion, if a pump is necessary to increase the amount of water pressure in a harvested rainwater system, the design of the system failed. Since there are so many low-pressure options for using rainwater, you should only add a pump to a harvested rainwater system if you plan to use it for high-pressure applications unobtainable by good design, such as running a washing machine or a sprinkler system.

Irrigation pumps are usually different from pumps used to circulate aquaculture water, as presented in chapter 5. Aquaculture pumps move a high volume or flow of water at low pressure, or head. Irrigation pumps are low volume and high pressure. It's important to match the pump to the required application.

The life expectancy of a pump varies based on the type of pump, usage, and quality of water, but typically they last about ten years. In all pumps as pressure increases, flow will drop — this is called a pump curve. Manufacturers provide a graph depicting pump curve, which shows the relationship between flow and pressure.

To size a pump, you must consider the static pressure, friction loss from the distribution system, and the pressure of the required application to determine the pump's required maximum pressure, known as total dynamic head. I'll use an example here to explain the calculations.

A garden is located at an elevation that is 20 feet higher than the water storage. The garden contains 1,000 feet of drip emitter tubing requiring 25 pounds per square inch to operate and an 8-gallon per minute flow. A pump attached to 100 feet of 1-inch piping moves water from the storage to the irrigation system. How much total

Figure 6.13. A small pump moves water at pressure from a rainwater-harvesting tank (not visible). The box on the left has valves that control the flow of water from the tank to the pump. The gray fittings (called unions) allow the pipe to be disconnected from the pump for servicing or removal. The pump is plugged into a pressure switch, which engages the pump when the pressure drops so it's not constantly operating.

dynamic head must the pump provide to irrigate the garden?

Static head + friction loss + pressure required from application = total dynamic head

First, determine the static head. Since the water must pump up an elevation of 20 feet and each foot of head creates 0.43 pounds per square inch, multiply:

20 feet × 0.43 psi/foot = 8.6 psi

Next, determine the friction loss from the 100 feet of 1-inch piping delivering the water to the garden. Table 6.3 shows that 100 feet of 1-inch piping with a flow of 8 gallons per minute reduces the pressure 1.74 pounds per square inch from friction. The pressure needed to run the drip emitter tubing is 25 pounds per square inch.

8.6 psi (static head) + 1.74 psi (friction loss) + 25 psi (application pressure) = 35.34 psi (total dynamic head)

The pump needs to produce 36 pounds per square inch at a flow of 8 gallons per minute to supply the garden with its necessary water. We can now examine pump curves from various pumps to shop for the properly sized pump for our application.

Installing the Pump

Pumps perform better when pushing water rather than pulling water. Therefore, it's important to locate pumps below water storage so water flows by gravity from the storage into the pump. You can locate pumps above your water storage, but I recommend limiting the distance — usually no more than 7 feet of elevation depending on the pump. These situations may require a priming pot or self-priming pump to create the suction to pull water into the pump. Unless the pump is designed to run dry, the pump fails quickly without water. You may want to install a one-way check valve to prevent water from flowing out of the pump and back downhill into the storage when the pump is off.

Installing an irrigation pump is similar to installing an aquaculture pump, as described in chapter 5. You can purchase an in-line pump controller to detect when the tank is out of water. Or you can add a float switch, which floats in the tank and turns off the power supply to the pump when it detects that the water level in the tank is low.

Comparison of Pump Types

Some pumps need a pressure tank attached to the pump to prevent the pump from short cycling. A pressure tank is a steel tank pressurized with air and that contains a bladder. The bladder fills with water, and the air inside the tank maintains pressure on the water even when the pump is off. Once the bladder empties of water, the pressure drops slightly, and the pump turns back on. The bladder usually contains about 35 gallons of water. Since the water inside the bladder is available for use without the pump turning on, it prevents the pump from cycling on and off between short uses. The advantage of a pressure tank is that you can use a smaller pump to fill the bladder inside the tank because the volume of water in the tank is available if necessary. The disadvantage of a pressure tank is preventing the entire tank from freezing in winter. You may need to build a special pump house around the tank, and depending on your climate, you may need to heat the pump house.

On-demand pumps cycle on and off as needs dictate and don't require an attached pressure tank. Submersible pumps are placed inside tanks or ponds, simplifying installation. Some submersible pumps are attached to floats to pull water from higher in the water column of the pond or tank.

Designing a Gravity-Driven System

Having sufficient storage capacity is critical for a rainwater-harvesting system that provides a fundamental service such as water for flushing toilets. However, to estimate storage potential and also to calculate the required size of the rainwater-harvesting components, we first need to assess precipitation data for the specific site.

Assessing Rainfall Data

Planning a rainwater-harvesting system is much like preparing a family budget. We can't use what we don't have. So first we must determine our "rainfall income," since it directly dictates the amount of water we may harvest and use in our landscape.

Precipitation patterns vary drastically throughout the country and within individual states. I recommend visiting the website of the National Centers for Environmental Information (www.ncei.noaa.gov) or your state's climate office to find precipitation data for your area before starting a rainwater-harvesting project. Information is usually available on a county-by-county basis. Rainwater topics to consider are annual amount, monthly amount, intensity, frequency, and return period.

ANNUAL AMOUNT

We measure rainfall for a given site in inches of rainfall per year (snow is converted to equivalent inches of water). You can use annual rainfall data in combination with the area of the roof you'll use for catching water to determine the total amount of rainwater you can expect to harvest. However, in

cold northern areas, residents may have to shut off rainwater systems during winter months to prevent pipes from freezing. Therefore, rainfall during the winter may not be available for rainwater-harvesting systems in colder areas.

The catchment area of a building is measured by the horizontal plane or footprint of the building regardless of roof angles. A basic rule of thumb is that 1,000 square feet of roof area will harvest 623 gallons of water for every inch of rainfall. To estimate the amount of water you can harvest from your roof, use the following formula:

Catchment area footprint (square feet) × depth of rainfall (inches) × 0.623 = harvested water (gallons per year)

For example, the total building footprint of my house, porch, and garage measures 2,564 square feet. South Carolina's state climate office has determined that annual rainfall is 48 inches per year in the county where I live. In our mild winter climate we can harvest water year-round so winter months are included in the calculation.

2,564 (sq ft) × 48 in × 0.623 = 76,674 (gallons per year)

Realistically, some of the water will become entrapped in shingles and gutters and potentially leak from the system. I estimate that loss at 5 percent (0.05). Therefore, I multiply the total amount by 0.95 to account for expected losses.

MONTHLY AMOUNT

Since rainfall rates vary within the year, sites will usually have wet months and dry months. Ideally, the wet months correspond to the time when water is needed most — during the warm growing season when irrigation is in high demand.

Determining the monthly amount of water available gives a more exact amount of harvesting potential and helps guide storage volume calculations (presented later in this chapter). To determine rainwater-harvesting potential on a monthly basis, first determine the average amount of rainfall at the site for each month of the year — this data should be available from state climate offices. Next, use the formula from above for determining the quantity of harvested water per month. See table 6.4 for an example based on my area.

Remember to multiply each monthly total by 0.95 (the correction coefficient) to account for expected losses due to entrapment of water in shingles and gutters and potential leaks in the system.

INTENSITY, FREQUENCY, AND DURATION

Intensity of rainfall, measured in inches of rain per hour (iph), determines the pipe size necessary to convey water to the pond or tank. Rain intensity varies drastically throughout the United States and

the world. For design purposes it's important to know the duration and return period or expected frequency of rain events for a given area. I use duration and return periods to determine the probability of a particular magnitude of storm occurring within a year. In 1961 the Department of Commerce published Technical Paper No. 40, entitled *Rainfall Frequency Atlas of the United States,* which is available on the National Weather Service website. The atlas includes maps for the United States with durations of storms from thirty minutes to twenty-four hours and return periods of one to one hundred years. I use the return periods to determine the probability of a storm occurrence. For example, a storm with a ten-year return period has a 10 percent chance (1 in 10) of occurring in any one year.

An undersized rainwater system may cause water to overflow gutters and damage houses. The Uniform Plumbing Code (UPC) requires gutters and downspouts be sized to accommodate a sixty-minute-duration storm with a return period of

Table 6.4. Rainfall Total by Month

Month	A Monthly Rainfall Amount (inches)	B Catchment Footprint (sq ft)	C Coefficient to Convert to Gallons	D Correction Coefficient for Water Loss	A × B × C × D Total (gallons)
January	4.65	2564	0.623	0.95	7056
February	4.30	2564	0.623	0.95	6525
March	5.29	2564	0.623	0.95	8027
April	3.84	2564	0.623	0.95	5827
May	3.62	2564	0.623	0.95	5493
June	3.83	2564	0.623	0.95	5812
July	4.49	2564	0.623	0.95	6813
August	3.96	2564	0.623	0.95	6009
September	4.02	2564	0.623	0.95	6100
October	3.39	2564	0.623	0.95	5144
November	3.95	2564	0.623	0.95	5994
December	4.36	2564	0.623	0.95	6616
Annual	48	2564	0.623	0.95	72840

one hundred years. Currently NOAA is working to update precipitation frequency estimates in the Atlas 14 project.[17] If information for your state is unavailable from the Atlas 14 project, I recommend using the *Rainfall Frequency Atlas of the United States*.[18] I use the intensity and frequency of a one-hour, one-hundred-year storm event to determine the size of gutters, downspouts, and conveyance systems.

For some parts of the rainwater-harvesting system, we'll need to convert to gallons per minute. To determine the gallons per minute (gpm) from a catchment area during a one-hour, one-hundred-year rain event, first examine the rainfall frequency atlas maps. For example, my house is located in the upstate region of South Carolina, and the map indicates a one-hour, one-hundred-year rain event produces about 3.3 inches of rain over a duration of one hour. I use the following formula to convert to gallons per minute for the catchment area:

(Catchment area × depth of rainfall for 1-hour rain event × 0.623) ÷ 60 minutes = total rainfall (gallons per minute) for 1-hour rain event

To use the previous example of my house:

(2,564 [sq ft] × 3.3 [in] × 0.623) ÷ 60 min = 88 gpm

Therefore, I designed my rainwater-harvesting system to convey 3.3 inches per hour, or 88 gallons per minute, of water so that it can handle a rain event from a storm that has a frequency of one hundred years and a duration of one hour for my roof's catchment area of 2,564 square feet.

Sizing Gutters and Downspouts

Gutters and downspouts sized appropriately will convey water to pipe systems without backing up and causing damage to buildings and roofs. If rainwater harvesting is the goal, you will save a lot of money using gutters to move water to centralized harvesting locations. To maximize your rainwater-harvesting design, follow these steps to appropriately size gutters and downspouts:

Step 1. Determine rainfall intensity for your site for a one-hour, one-hundred-year storm, as described earlier in this chapter.

Step 2. Locate ideal downspout locations for rainwater harvesting on your site.

Step 3. Determine the catchment area for the roof above the potential downspout location by examining the square footage of the horizontal footprint.

Step 4. Using the data in tables 6.5 and 6.6, adjust gutter size, slope, and downspout size to achieve ideal rainwater-harvesting downspout locations. You may need to add downspouts to reduce the size of your catchment area.

Example: My house has a horizontal footprint of 2,564 square feet. It's a rectangular shape with a gable roof sloping lengthwise, splitting the drainage area in half. I want to position gutters along the length of the building and have two downspouts, one on each side of the house, to facilitate a central rainwater-harvesting system. The one-hour, one-hundred-year rainfall rate for my site is 3.3 inches per hour with a flow rate of 88 gallons per minute. The gutter contractor only has equipment to make a 5-inch gutter.

2,564 sq ft ÷ 2 = 1,282 sq ft horizontal footprint draining to each gutter

From table 6.5, I can determine that with a 5-inch-wide gutter and ¼-inch slope per foot, the gutters can handle the recommended flow rate.

Table 6.5. Sizing Gutters Based on Slope and Rainfall Rates

Gutter Slope	Diameter of Gutter (inches)	Maximum Horizontal Square Feet of Roof Catchment for Various Rainfall Rates Determined by 100-Year, 1-Hour Storm				
		2"/Hr	3"/Hr	4"/Hr	5"/Hr	6"/Hr
⅛"/ft Slope	3	480	320	240	192	160
	4	1020	681	510	408	340
	5	1760	1172	880	704	587
	6	2720	1815	1360	1085	905
	7	3900	2600	1950	1560	1300
	8	5600	3740	2800	2240	1870
	10	10200	6800	5100	4080	3400
¼"/ft Slope	3	680	454	340	272	226
	4	1440	960	720	576	480
	5	2500	1668	1250	1000	834
	6	3840	2560	1920	1536	1280
	7	5520	3680	2760	2205	1840
	8	7960	5310	3980	3180	2655
	10	14400	9600	7200	5750	4800
½"/ft Slope	3	960	640	480	384	320
	4	2040	1360	1020	816	680
	5	3540	2360	1770	1415	1180
	6	5540	3695	2770	2220	1850
	7	7800	5200	3900	3120	2600
	8	11200	7460	5600	4480	3730
	10	20000	13330	10000	8000	6660

Source: J. Mechell, B. Kniffen, B. Lesikar, D. Kingman, F. Jaber, R. Alexander, and B. Clayton. *Rainwater Harvesting: System Planning*. College Station: Texas AgriLife Extension Service, Draft version September 2009.

Table 6.6. Vertical Downspout and Drain Size Based on Flow and Rainfall Rates

Vertical Downspout or Drain (inches)	Flow (gpm)	Maximum Horizontal Square Feet of Roof Catchment for Various Rainfall Rates Determined by 100-Year, 1-Hour Storm					
		1"/Hr	2"/Hr	3"/Hr	4"/Hr	5"/Hr	6"/Hr
2	23	2176	1088	725	544	435	363
3	67	6440	3220	2147	1610	1288	1073
4	144	13840	6920	4613	3460	2768	2307
5	261	25120	12560	8373	6280	5024	4187
6	424	40800	20400	13600	10200	8160	6800
8	913	88000	44000	29333	22000	17600	14667

Source: J. Mechell, B. Kniffen, B. Lesikar, D. Kingman, F. Jaber, R. Alexander, and B. Clayton. *Rainwater Harvesting: System Planning.* College Station: Texas AgriLife Extension Service, Draft version September 2009.

Since the house will have two downspouts, one on each side, the volume of water is split in half.

88 ÷ 2 = 44 gallons draining to each downspout

From table 6.6, I can determine that the minimum size downspout or vertical drain for my house is 3 inches.

Conveyance and Maximizing Gravity

When I began experimenting with wet rainwater systems, I started small. I built systems simply to carry water over the edge of slightly raised ponds. However, a home improvement project forced me to push the limits of wet systems and helped me fully grasp the true potential of gravity. Our house sits several hundred feet from the main road at the bottom of our property. In spite of this inconvenient position, I still wanted to convey water via gravity from the roof of my house to the upper end of my property to store water at the highest point possible. My goal was to store water at a level that was higher than the toilets and the rest of my landscape but lower than the gutters on my roof. Since this point turned out to be several hundred feet away from the house, up a gentle hill, the project presented an engineering challenge.

I first started exploring dry conveyance systems. I pictured a trellis several hundred feet long with a pipe running along its top, gently sloping water from the gutters to the upper end of my property. Fortunately, a friend, Zev Friedman, mentioned the possibility of installing a wet system, saving me from an unsightly pipe running hundreds of feet through the entire length of my property. But how would I know that everything would work before committing to burying a bunch of pipe and building a complex system? Luckily, engineers have already pioneered this frontier via mathematics, removing risk from the equation.

As water travels through wet conveyance systems, it creates friction against the edge of pipes. This friction causes the water to slow down and back up, potentially overflowing the gutters. Let's say we have a horseshoe-shaped pipe with both sides of equal height. Without friction, water would enter one side of our pipe and simply exit the other side at the same rate of gravity. However, friction causes the water to back up slightly until the weight of the water overcomes the friction and pushes the water through to the other end of the pipe. We can measure the amount of water needed to overcome the friction in feet of head or height. As larger volumes of water move through a pipe, more friction is created, requiring more head to move the water through the pipe. If water moves through a smaller pipe, more friction is also created, as the same volume of water contacts more pipe surface. Pipe fittings also create friction, requiring more head. For example, a 1-inch-diameter tee fitting creates as much friction as 5 feet of straight 1-inch pipe (table 6.8).

By calculating the friction created by the volume of water moving through the wet conveyance system and the pipes used to move the water, we can determine an exact amount of head required to overcome the friction. The required head then dictates the height difference required between the gutter and water storage. The Hazen-Williams friction equation or table 6.7 can be used to determine the head required based on the flow of water and length of pipes in the system:[19]

Friction loss = $10.46L \left[(Q \div C)^{1.852} \div D^{4.871} \right]$

where

L = length of pipe (ft); Q = flow rate gpm;
C = friction coefficient (140–150 for PVC pipe);
and D = actual pipe inner diameter (in)

Table 6.7. Head Loss Due to Friction for Schedule 40 PVC Pipe

Flow (gpm)	Pipe Diameter (inches)					
	1	1¼	1½	2	3	4
1	0.09					
2	0.32	0.08				
3	0.67	0.18	0.08			
4	1.14	0.30	0.14			
5	1.73	0.46	0.21	0.06		
6	2.43	0.64	0.30	0.09		
7	3.23	0.85	0.40	0.12		
8	4.13	1.09	0.51	0.15		
9	5.14	1.35	0.64	0.19		
10	6.25	1.64	0.78	0.23		
11	7.45	1.96	0.92	0.27		
12	8.76	2.30	1.09	0.32		
13	10.16	2.67	1.26	0.37		
14	11.65	3.06	1.45	0.43	0.06	
15	13.24	3.48	1.64	0.49	0.07	
16		3.92	1.85	0.55	0.08	
17		4.39	2.07	0.61	0.09	
18		4.88	2.30	0.68	0.10	
19		5.39	2.55	0.75	0.11	

Example: Our house has a roof catchment with a footprint measuring 2,564 square feet. A one-hour, one-hundred-year rain event for our county would produce 3.3 inches per hour of rain, equating to 88 gallons per minute from our roof catchment area. We want to catch and transfer water from our roof through a single pipe into a pond at the highest point in the landscape, located 200 feet from our house. Using a water level, we determine the top of the pond sits 3.5 feet below the level of the gutters. Will friction in the piping system cause the water to back up and overflow the gutters?

Here's a list of the variables:

- All pipe: 4-inch Schedule 40 PVC
- 200-foot straight run
- Four 45-degree ells = 32 equivalent pipe feet
- Twenty couplings = 80 equivalent pipe feet
- Three straight-run tees = 12 equivalent pipe feet
- Total equivalent pipe length including fittings: 324 feet
- Water flow: 88 gallons per minute

From table 6.7 we see that 90 gallons per minute of water produces 0.52 feet of head pressure for every 100 feet of 4-inch pipe. Since we have 324 feet of pipe, we need to multiply the head pressure by 3.24 (324 ÷ 100).

Table 6.7 (*continued*)

Flow (gpm)	Pipe Diameter (inches)					
	1	1¼	1½	2	3	4
20		5.93	2.80	0.83	0.12	
25		8.96	4.93	1.25	0.18	
30			5.93	1.76	0.26	0.07
35			7.89	2.34	0.34	0.09
40				2.99	0.44	0.12
45				3.72	0.54	0.14
50				4.52	0.66	0.18
60				6.34	0.93	0.25
70				8.43	1.23	0.33
80				10.80	1.58	0.42
90				13.34	1.96	0.52
100				16.33	2.38	0.63
150					5.05	1.34
200					8.61	2.29
250						3.46
300						4.86
350						6.64
400						8.27

Source: J. Mechell, B. Kniffen, B. Lesikar, D. Kingman, F. Jaber, R. Alexander, and B. Clayton. *Rainwater Harvesting: System Planning*. College Station: Texas AgriLife Extension Service, Draft version September 2009.

Note: Head loss is expressed in feet of head per 100 feet of pipe.

Table 6.8. Equivalent Pipe Length of Fitting for Pressure Loss Equations for Schedule PVC

Diameter of Fitting (inches)	Equivalent Pipe Length (feet) for Fittings			
	90-Degree Ell	45-Degree Ell	90-Degree Tee	Coupling or Straight Run of Tee
1	3	1.8	5	0.9
1¼	4	2.4	6	1.2
1½	5	3	7	1.5
2	7	4	10	2
3	10	6	15	3
4	14	8	21	4

Source: J. Mechell, B. Kniffen, B. Lesikar, D. Kingman, F. Jaber, R. Alexander, and B. Clayton. *Rainwater Harvesting: System Planning*. College Station: Texas AgriLife Extension Service, Draft version September 2009.

0.52 feet of head pressure per 100 feet of pipe length × 3.24 = 1.69 feet of head pressure

In conclusion, since the pond sits 3.5 feet below the gutters and the head pressure created is only 1.69 feet, water will not back up and overflow the gutters during a one-hour, one-hundred-year rain event. To help reduce friction in conveyance systems, keep as many turns as possible high in the system before concentrating the flow of water into a single straight run.

Estimating Water Usage

To figure out how much water to store, you must first determine your demand, or how much rainwater you want to use, based on where you want to use it. Since the storage component of rainwater-harvesting systems represents the greatest cost, making the storage just the size needed and no larger will help keep costs low. Ponds also lose an enormous amount of water to evaporation and may need to be covered to keep the water in the pond and available for use. If a pond is used to store water, determining how much water is lost to evaporation will also help guide management and use decisions.

The EPA estimates the average American family uses more than 300 gallons of water per day. We use about 42 percent of this water indoors for purposes such as showering, washing dishes, and cooking. Using rainwater as your supply for these purposes is possible if you install the correct filtration, sanitation, and pumping systems. However, the cost for filtration exceeds a thousand dollars and has ongoing expenses to manage and replace filters or inject chlorine for treatment. Furthermore, most households can't harvest enough rainwater to supply all of their water needs unless they drastically reduce water usage or live in a very wet area.

Returning to the example of my system, in our humid wet climate on the East Coast, with 44 inches of annual rainfall, we can harvest an overall average of 210 gallons per day, if we have enough storage to capture all the falling water. That's still less than average consumption for a U.S. household, and it demonstrates why you'll probably need to make specific choices about what purposes you want harvested rainwater to fill. Common sense suggests harvested rainwater is best for nonpotable uses, such as irrigation, toilet flushing, and clothes washing. Toilet flushing and irrigation account for 50 percent of American water use and represent the easiest employment for harvested rainwater.

One way to figure out your monthly usage is to check your water bill or monitor your water meter. If you have a well rather than municipal water, you can install a water meter to do the same.

INDOOR VERSUS OUTDOOR DEMAND

Domestic water use is based on the number of people living in a house, daily water habits, and the number of guests and parties you have, as well as your location. Table 6.9 gives an estimate of the amount of water used indoors and outdoors in Denver, Colorado, and can be used to estimate daily water use for rainwater harvesting for domestic purposes.[20]

We base outdoor water use on the climate, type of plants grown, and type of irrigation system. A water-thirsty lawn located in a hot, dry climate and irrigated with a sprinkler uses a lot more water than drought-tolerant vegetation irrigated with an efficient drip irrigation system. In either case, when we install irrigation systems and meet our plants' water needs, our outdoor demand can easily exceed indoor demand. The rule of thumb for outdoor watering is 1 inch per week. When multiplied over an entire acre (43,560 square feet), 1 inch equals 27,138 gallons of water per week (0.623 gallons per square

Table 6.9. Typical Domestic Daily Water Use in Denver, Colorado

	Gallons per Capita	% of Daily Total
Potable Indoor Use		
Showers	11.6	7.0
Dishwashers	1.0	0.6
Baths	1.2	0.8
Faucets	10.9	6.6
Other Uses, Leaks	11.1	6.7
Subtotal	35.8	21.7
Nonpotable Indoor Use		
Clothes Washers	15.0	9.1
Toilets	18.5	11.2
Subtotal	33.5	20.3
Outdoor Uses		
	95.7	58.0

Source: C. Kloss. "Managing Wet Weather with Green Infrastructure Municipal Handbook Rainwater Harvesting Policies," EPA-833-F-08-010. December 2008.

foot). In comparison, indoor water use averages 300 gallons a day per household, or 2,100 gallons per week. Therefore, irrigating even a small amount of the landscape can easily use all the rainwater you collect. I use the following formula to estimate irrigation demand:

Square feet irrigated × 0.623 gallons per inch of water = irrigation needed per week

With drip irrigation some people estimate water needs based on the square footage of the entire planted area. Others estimate water needs based on the square footage in a limited area around the plant or along the wetted band created by a drip line. With perennials I prefer to base the square footage only on the area around the plants, hoping that as my plants grow they will find water in my swales and diversion channels.

Here's an example of how I figured out one water budget for an irrigation project. I planted seventy one-year-old blueberry plants that occupy 4 square feet per plant, totaling 280 square feet of planted area. Since I use a drip irrigation system, I conserve water by irrigating only the area around the plant. So I factor 4 square feet of irrigation per plant. But once the blueberries mature they will occupy 4,975 square feet of the landscape and require a lot more irrigation.

280 (sq ft) irrigated × 0.623 gallons per inch of water = 174 gallons of irrigation per week

While the 1-inch-per-week rule gives a general idea of water needs, the actual amount varies, depending on the time of year, type of plant, and location. For a more precise calculation, use potential evapotranspiration (PET). PET references the amount of water lost each month by a short grassy plant at specific locations. You can find your PET through your state climatology office. The PET usually represents a range, and you can adjust depending on your site's conditions. Windy, exposed sites lose more water than sheltered sites. Since PET is based on lawns, and lawns are water hogs, we multiply the PET by a factor called the crop coefficient. This reduces the PET to equate to particular plant uses. Every plant has a specific crop coefficient correlated to the type of plant and the age of the plant. Older plants use more water than younger plants. For example, one-year-old blueberries have a crop coefficient of 0.24 — they use a lot less water than lawn grass does.

HOW TO CALCULATE IRRIGATION WATER NEEDS

You can find a list of crop coefficients through your local Extension office, in the *National Engineering*

Handbook, or online. The coefficients are regionally specific based on climate conditions. With the following technique, called the Penman-Monteith reference evapotranspiration, you can determine precise amounts of water for specific plants at specific locations for every month.

Specific plant water requirement
= PET of site (in.) × K crop coefficient

For example, I planted seventy rabbiteye blueberries in a sheltered location. Each blueberry occupies 4 square feet per plant, for a total of 280 square feet. The crop coefficient for one-year-old blueberries is 0.24. I can set up a table to calculate the water requirements for these plants over a one-year period, as shown in table 6.10.

Once I know the specific plant water requirements, I multiply the water requirements by the square footage and conversion factor to obtain a

Table 6.10. Plant Water Requirements

Month	PET	K Crop Coefficient	Plant Water Requirement (inches)
January	1.8	0.24	0.4
February	2.2	0.24	0.5
March	3.8	0.24	0.9
April	5.0	0.24	1.2
May	5.4	0.24	1.3
June	5.4	0.24	1.3
July	5.4	0.24	1.3
August	5.4	0.24	1.3
September	4.6	0.24	1.1
October	3.0	0.24	0.7
November	1.8	0.24	0.5
December	1.4	0.24	0.3
Annual	45.2	0.24	9.6

Note: Potential evapotranspiration multiplied by the crop coefficient gives the water requirement for one-year-old blueberries.

Table 6.11. Monthly and Yearly Water Needs for Blueberries

Month	A — Plant Water Requirement (inches)	B — Planted Area (sq ft)	C — Conversion Factor 1"/sq ft	A × B × C — Gallons
January	0.4	280	0.62	69
February	0.5	280	0.62	87
March	0.9	280	0.62	156
April	1.2	280	0.62	208
May	1.3	280	0.62	225
June	1.3	280	0.62	225
July	1.3	280	0.62	225
August	1.3	280	0.62	225
September	1.1	280	0.62	190
October	0.7	280	0.62	121
November	0.5	280	0.62	87
December	0.3	280	0.62	52
Annual	9.6	280	0.62	1,667

Note: Rates are for seventy newly planted blueberries based on potential evapotranspiration rates for the upstate region of South Carolina.

Table 6.12. Total Water Demand

Month	Demand			Total Demand
	Toilets	Blueberries	Garden	
January	555	69	837	1461
February	555	87	1023	1665
March	555	156	1767	2478
April	555	208	2325	3088
May	555	225	2511	3291
June	555	225	2511	3291
July	555	225	2511	3291
August	555	225	2511	3291
September	555	190	2139	2884
October	555	121	1395	2071
November	555	87	837	1479
December	555	52	651	1258
Annual	6660	1870	21018	29548

Note: This summarizes the total demand from blueberries, the garden, and water for flushing toilets.

Table 6.13. Supply and Demand Balance for Rainwater Harvesting

Month	Total Supply from Rainfall (gallons)	Total Demand (gallons)	Balance (gallons)
January	7056	1461	5595
February	6525	1665	4860
March	8027	2478	5549
April	5827	3088	2739
May	5493	3291	2202
June	5812	3291	2521
July	6813	3291	3522
August	6009	3291	2718
September	6100	2884	3216
October	5144	2071	3073
November	5994	1479	4515
December	6616	1258	5358
Annual	75416	29548	45868

specific monthly amount of water for a specific size planting, as shown in table 6.11. The conversion factor is the reference rate of 1 inch of water for every square foot, or 0.62 gallons.

I repeat the process for plants in my 1,000-square-foot garden that have an average crop coefficient of 0.75 to determine the monthly gallons of water needed by the garden.

Finally, I combine all of my water demands into a single table (see table 6.12). From that, I can determine the four largest demand months of the year, then compare that with average rainfall rates (see table 6.13). Looking at the balance between supply and demand, I can see that rainfall is plentiful throughout the year and will likely meet my demands if I have enough storage capacity.

FIGURING MULTIYEAR STORAGE NEEDS

The next step examines the consequence of various storage capacities over a three-year period.

Table 6.14. Supply and Demand Comparison

Month	Total Supply from Rainfall (gallons)	Total Demand (gallons)	Year 1 (gallons)	Year 2* (gallons)	Year 3** (gallons)
January	7056	1461	5595	7000	7000
February	6525	1665	7000	7000	7000
March	8027	2478	7000	7000	7000
April	5827	3088	7000	7000	7000
May	5493	3291	7000	7000	7000
June	5812	3291	7000	7000	7000
July	6813	3291	7000	7000	7000
August	6009	3291	7000	7000	7000
September	6100	2884	7000	7000	7000
October	5144	2071	7000	7000	7000
November	5994	1479	7000	7000	7000
December	6616	1258	7000	7000	7000
Annual	75416	29548			

Note: This scenario assumes year-round rainfall and 7,000-gallon storage capacity.

*Starts with 7,000-gallon storage from excess of year 1

**Starts with 7,000-gallon storage from excess of year 2

Recommendations call for starting storage capacity calculations based on the sum of demand for the highest four months.[21] However, in my warm climate this may be excessive because the lack of freezing temperatures during the winter allows me to harvest rainfall year-round, ensuring storages are full heading into the high demand months of spring and summer. Also, the lack of a dry period and consistent year-round rainfall replenishes storages. I can set up a table such as table 6.14 to calculate the supply and demand balance for rainwater harvesting over three years based on information from table 6.13. Note that demand in this scenario is for a tank or covered pond and doesn't take into account water lost to evaporation.

Since most of the demand in this example is for outdoor irrigation, which is probably unnecessary during the winter, demand is realistically much less. If you live in a climate where you can't collect rainwater during the winter but have high demand from domestic use, you may run into a storage shortage in the spring, when demand for water will suddenly pick up, but you will have depleted the supply in your storage during the winter without being able to replenish it.

A similar situation would occur for climates with dry seasons, such as the western United States. Without the ability to harvest rainwater for part of the year, capacity would need to be increased to provide enough water for dry seasons. Mediterranean climates such as California's have a pronounced wet season during the winter when plants don't need irrigation. During the summer when demand is highest, rainfall is lacking, complicating the ability to store enough water for outdoor irrigation. In these dry summer climates, providing enough storage capacity to just flush toilets is a challenge. The average toilet uses 18.5 gallons per day. So over a six-month dry period, 3,330 gallons of storage are necessary just to flush toilets through the dry season.

When rainwater harvesting is lacking for part of the year, I recommend increasing storage capacity to provide a buffer for dry times.

Summing Up the Functions

1. Downspout filters filter debris from roof.
2. First flush diverters discard first flush of dirty water.
3. Pipes convey water to pond or tank.
4. Pipes, ponds, or tanks store water for flushing chicken coop manure into diversion channel (chapter 7).
5. Pipes, ponds, or tanks store water for flushing diversion channel nutrients into pond.
6. Ponds or tanks store water for flushing nutrients into the landscape.
7. Ponds or tanks store water for feeding swale.
8. Water from pond is siphoned to flush toilets.
9. Water from pond is siphoned to irrigate garden.
10. Wire mesh screen filters water.
11. Ponds provide water to chickens.
12. Ponds and tanks modify microclimates for plants, animals, and people.

CHAPTER 7

Chicken No Tractor
The Bio-Integrated Chicken Coop

When Stephanie and I bought our house in South Carolina, I immediately began calculating the cost of maintaining the acre and a half of lawn that came with it. Besides the fuel, the chemicals, and the upkeep of a lawn mower, maintaining a lawn requires a substantial amount of time. I calculated that I'd spend approximately 104 hours per year sustaining our new lawn. With a full-time job and a family, I view my time as a precious commodity. And after running the numbers, I realized I needed to find an alternate plan unless I wanted to develop a new hobby in lawn preservation.

I reviewed my options: live in a forest or find someone or something to mow my lawn for free. Although I enjoy a walk through the woods or a meandering hike in the mountains, there's something primal about wide, open spaces. As postulated in the prospect-refuge theory by British geographer Jay Appleton, humans are attracted to broad vistas — such as a large field — where we hover at the edge looking out with our backs safely covered. Our desire to create and maintain the open savanna probably derives from our earliest ancestors. In essence, the lawn is a part of our hunter-gatherer safety DNA and our subconscious satisfaction.

With an inner longing for lawnlike surroundings, I considered contracting out the task. But the true cost of lawn maintenance doesn't fall on the homeowner — our environment (i.e., our future generations) foots the proverbial bill. A study in Newcastle, Australia, determined that carbon dioxide emissions from lawn mowers represented 5.2 percent of total emissions, while hydrocarbon emissions garnered 11.6 percent of the total emitted.[1] Although catalytic converters can reduce the emissions 20 to 30 percent,[2] regulators and manufacturers have failed to make catalytic converters a reality for small engines. Based on EPA pollution limits, an operating lawnmower can pollute as much as six operating cars.

Caught in a crux between my primordial desires and my carbon footprint, I had an epiphany when a friend fortuitously posed a thoughtful question: "Why mow your lawn when chickens can do it for you?" Rotating chickens through the landscape to forage on vegetation results in a landscape with an open appearance — without the pollution of a mower. And so the birth of the bio-integrated coop began.

The bio-integrated chicken coop is the ultimate in low-maintenance chicken foraging systems. This coop lets chickens do the work by mowing,

Elements of the System

Why mow your lawn when chickens can do it for you? The bio-integrated chicken coop is an all-in-one mower, tiller, composter, and egg producer — not to mention a stage for back-yard chicken entertainment. In this chapter I'll explain how the bio-integrated coop:

▸ Turns chicken yards into gravity-fed mulch- and compost-producing machines.

▸ Uses chickens to prepare forage areas for seeding and annual beds for planting.

▸ Cleans the coop and grows feed for chickens, all with the turn of a valve. By connecting rainwater-harvesting systems with chicken coops, manure is flushed into specialized diversion channels. Plants grown in the fertile diversion channel are then fed back to the chickens as a feed.

▸ Automates chicken-watering and door-opening systems and locates coops for easy checks on automated systems.

▸ Builds a black soldier fly digester under coop for free, reducing the smell of manure and turning a waste into a chicken feed.

▸ Uses chickens to harvest minnows and pond organisms after ponds are drained for irrigation.

▸ Saves money by growing insects for feed.

tilling, making eggs, and composting. They eat vegetation, reducing the amount of time I spend mowing to eight hours per year. Chickens also have the ability to turn everything we don't eat into eggs, meat, and fertilizer. All the insects, seeds, plants, minnows, and food scraps we pass up or discard are a welcome treat for chickens.

The bio-integrated chicken coop provides all the benefits of farming chickens without the daunting work. Essentially, we turned our lawn into pasture for our chickens. Foraging systems that make use of portable electric fencing transform chickens into mowing and mulching machines. Coops are connected to rainwater-harvesting systems that supply water to flush manure into the landscape through diversion channels — cleaning, fertilizing, and irrigating all with the turn of a valve. Connecting the coop to a house or deck makes it easy to repurpose waste without the need to compost. Automated waterers and door openers in the coop eliminate the remedial daily burdens of caring for chickens.

Abandoning the lawn has afforded me an abundance of time to develop rainwater-harvesting systems, gardens, and fruit trees. Currently, we produce up to two hundred eggs per year from each hen and a bag of fertilizer every few weeks. We reduce our chicken feed purchases by providing diverse forage for the chickens. In essence, instead of spending time and money mowing a polluting lawn, our foraging chickens convert our lawn into food and money.

Free-Range Poultry Forage Systems

Why did the chicken cross the road? Because there was a large, enticing field to forage on the other side. When given the opportunity, chickens prefer to roam. Unless thick snow blankets the ground, chickens favor venturing out and only return to the coop for food or shelter. Chickens can get up to 30 percent of their dietary needs met from the landscape.

Figure 7.1. We abandoned the lawn around our home to create a chicken foraging system. Once a week we move the white electric fence to a new area. Our chickens feast on crops we've planted specifically for their needs.

A coop designed for foraging chickens maintains just enough roost space for chickens to sleep and retreat from foul weather. Laying boxes are provided for eggs, and water and feed are sheltered inside the coop. By taking advantage of chickens' preference to forage, the size of the chicken coop can be reduced drastically while maintaining chicken happiness and health.

Bigger isn't always better. A coop should be designed to house chickens, not humans. A secret to the success of the bio-integrated chicken coop is in its simple design: Humans access the coop from the exterior only. Why step in chicken manure if you don't have to? Everything from pouring feed to harvesting eggs is easy to accomplish via large

access doors positioned on each end of the coop. A chicken coop designed to house birds, not humans, is much more economical, creates a smaller footprint, and is easier to maintain.

A coop measuring 4 feet high, 4 feet wide, and 8 feet long (shown in figure 7.2) can house up to forty foraging chickens. To house that many chickens but also allow enough space for people to walk in and out of the coop, the dimensions would have to be tripled. And without access to outside forage, the same size coop would serve only eight (probably psychotic) chickens.

I've taken two approaches to foraging small flocks on the homestead: fencing chickens into confined areas or letting chickens run free and fencing them out of gardens, patios, and other sensitive places.

If chickens roam free, they move in packs throughout the woods and field edges. The birds will mainly scratch for high-protein insects and seeds, nibbling on the occasional fruit and plant. The birds seek out low bushes and safe places to hide from hawks that are constantly on the hunt for a tasty meal. Free-roaming chickens are somewhat useful in mature systems with larger plants and trees unsusceptible to damage. The large trees shade out the grasses and leave a layer of leaf mulch for the birds to rummage through.

With gardens and young plantings, free-roaming chickens make a mess of mulched areas, destroy raised beds, and eat up any seeds planted. Free-roaming chickens also poop and roost on porches and other inappropriate areas. In addition, a lack of fencing leaves chickens unprotected, and predators will find an easy meal. With free-roaming laying hens, eggs are scattered everywhere, making Easter egg hunts a daily event. In short, free-roaming chickens are great if your entire landscape consists of large mature plants, if you enjoy mowing your lawn once a week, and if you have a lot of free time to hunt for eggs and clean up poop.

By confining chickens with fencing or a movable coop, you force the chickens into a successional pattern that either benefits or destroys the landscape. When chickens first enter a new foraging area, they search out the high-protein insects that are too slow to escape. Once the insect population has been reduced, the chickens move to the energy-packed seeds hanging on the plants. After the seeds have been devoured, chickens eat plants. They start with the highly nutritious clovers and legumes, then move on to the young tender shoots of the grasses. With time the confined chickens will mow the grass down to a somewhat uniform

Figure 7.2. This compact chicken coop sleeps forty adult chickens and includes nine laying boxes and room for feed and water. The chickens forage outdoors during the daytime.

sward with tufts of tough grass such as fescue dotting the area.

Eventually the chickens start scratching and exposing bare soil and the roots of plants as they desperately dig for more insects and worms. Finally, if the chickens are left on the land too long, the landscape will become a barren, deserted moonscape dotted by the occasional dropping, devoid of all plant life.

We manage our chickens to control insects and weeds by holding the right number of chickens in the right size area for the right amount of time. When confined for a short period, chickens control insects. Confined for a little longer, they mow the grass and expose bare soil for planting seeds.

If confined for even longer, chickens remove all vegetation and provide a stiff dose of fertilizer.

Integrating the Coop

The central component in my system is a stationary coop at one end of an "alley" or "hallway." The alley, made of permanent metal fencing, is similar to a hallway in a home. It allows the chickens to access the "rooms"—pens enclosed by temporary lightweight poultry netting—that open off the "hallway." A long, narrow alley allows potential access to a large number of foraging pens. The alley, which is heavily mulched, also serves as a safe holding area during times of drought and cold, when I can't put the hens out on pasture to forage.

Keeping a chicken coop stationary has several advantages. If placed in ideal locations, stationary coops make egg collecting and maintenance easier. Stationary coops also have the benefit of creating a concentrated source of fertilizer available for collection and distribution to trees and gardens. However, unless manure is properly managed, odors may become an issue. I discuss manure-flushing systems to clean roosts easily and diminish odors later in this chapter.

With stationary coops the amount of fencing connected to the coop limits the amount of forage. Achieving up to an acre of forage area around a stationary coop is easy by combining permanent fencing and lightweight, movable electric fencing. To expand the forage area even more, I either enlarge the fencing or move the coop. Moving a coop offers a good option for larger flocks of birds in need of a large foraging range. However, moving a coop—even if it has wheels—is usually difficult. Unless movable coops are moved daily, manure still concentrates under the coop and necessitates proper distribution.

My system of an alley and foraging pens allows me to rotate chickens easily through a large area.

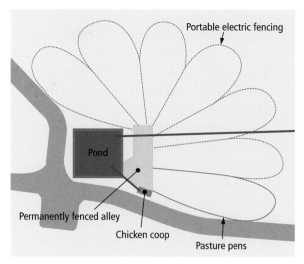

Figure 7.3. This schematic shows a foraging system in which the coop and alley are set up to allow chickens access to seven potential pasture pens and also to a pond.

The mix of permanent and temporary fencing limits the amount of total fencing required and thus prevents fencing from cluttering the landscape.

Designing an Alley

In my temperate climate, chickens seek out shaded areas in the summer months and sun during the winter months. I carefully located my alley in relation to buildings and trees to provide chickens with the basic need of shade in summer and sun during the winter. The alley should provide a good permanent microclimate adjacent to pasturing pens.

The alley is made from chicken wire with 1-inch holes small enough to contain baby birds. I secure the bottom of the chicken wire to the ground with sod staples to prevent chickens and predators from pushing under the fence. Sod staples alone might not deter all persistent predators, so you might want to install an electric fence or grow prickly pear cactus as presented later in this chapter. Folding the bottom 12 inches of the fence toward the middle of the pen and leaving the fence lying on top of the ground will discourage predators from digging under the fence. However, if electric pens are used in conjunction with the alley, folding the fence under is not necessary (see the "Protecting against Predators" on page 204). To keep the fence upright, I attach it to inexpensive metal T-posts hammered into the ground using a post driver. The chicken wire is difficult to stretch taut because of its thin nature, which gives the fence a shoddy appearance. Attaching the chicken wire to a thicker wire fence would give the netting a more refined look, but you'd have to cut holes in the thicker fence to create a passageway to the foraging pen.

A fence height of 4 feet contains heavy breeds. I usually have a few escapees in every flock. Clipping the flight feathers on a single wing before the hens learn to fly generally prevents potential breakouts. The only remedy for persistent runaways is a trip to the chopping block. Best to take care of business quickly before the rogue hen teaches the other birds bad habits.

I start the chicks inside the coop. For added warmth I seal the floor and all openings with cardboard and add a heat lamp. To prevent a manure mess I add a few inches of wood shavings on top of the cardboard. I allow the chicks to move into the alley as soon as they adjust to colder temperatures. After a few months I allow the birds to venture into a pen attached to the alley, once they are large enough to be contained by the electric poultry netting.

Access from the alley into the pen requires a gate. Building traditional gates is expensive and time consuming — not to mention, I would need a minimum of seven gates to access the multiple forage areas. Fortunately, my system only requires gates large enough for chickens to walk through, and chickens are small, so I solved this problem by building a small temporary tunnel that I

Figure 7.4. Clipping the flight feathers on a single wing will keep birds on the ground. This works best if done before birds get a taste of flight and should then be repeated after a molt.

relocate every time I move the electric netting. I have designed a makeshift chicken tunnel I like to call "bottomless 5-gallon bucket." Plastic 5-gallon buckets can be found at any local hardware or feed store. I simply saw off the bottom of the bucket, remove the sod staples securing the stationary chicken wire fence to the ground, and place the chicken tunnel under the fence. The flexible chicken wire is pliable and conveniently accommodates the bucket to create the perfect makeshift chicken gate.

My latest fencing endeavor involves a living fence. I planted prickly pear cactus (*Opuntia* sp.) around the perimeter of the alley fence. This particular prickly pear species contains large spines and grows to a height and width of 5 feet.

Figure 7.5. A 5-gallon bucket with the bottom cut off serves as a temporary hole in the alley fence, allowing chickens through to a foraging pen. The fence is secured to the ground with sod staples when the bucket is moved.

My hope is that the prickly pear will replace the metal poultry netting, as the metal deteriorates in three to five years. When we burn the prickly spines off the tender young pads in the spring, the cactus pads provide a delicious treat for us or the chickens. As a bonus, the fruits are refreshing, with a flavor similar to watermelons and strawberries. I easily propagate prickly pear cactus by breaking off its pads, letting them cure for a week in the open, then placing them bottom end down in the soil a few inches deep. A spacing of 3 feet between plants makes a hedge in three to five years if properly fertilized and irrigated.

For quite some time I've pondered how to get the chickens through the alley cacti fence and into the pens once the cacti are fully grown. I plan to build a lightweight bridge that I can place over the top of the cactus; then I'll move the bridge every time I move the electric fencing. I'm still a year away from testing out the technique, so the idea remains speculative.

Spiny prickly pears come with a major warning: The spines are extremely sharp, and puncture wounds are a serious threat. The chickens forage close to the cacti but appear wary of the sharp spines. We move slowly around the chickens to avoid startling them into the wrath of the spines. To make matters worse, the spines are likely inoculated with tetanus from the adjacent chicken manure. I made the mistake of chasing after an escaped chicken next to the cacti and received a major puncture wound. My tetanus shot wasn't up to date, and I quickly got a booster. I'll never chase an escaped chicken again when cacti are involved. Instead, bait escaped chickens back into the fenced area, or herd them through a gate, steering clear of the dangerous cacti.

STACKING FUNCTIONS IN AN ALLEY

There's nothing quite as satisfying as stuffing piles of autumn leaves into a shredder and watching uniformly chopped mulch stream out of the machine. Although not as instantaneous, given the right amount of space, time, and slope, chickens provide the same service, with the added benefit of fertilizer. In essence, the alley is like a slow-motion shredder that adds fertilizer as it grinds up material. Whole material enters the top, and shredded nutrient-rich material exits the bottom.

Since chickens use the alley area daily year-round, the soil there is stripped and barren. Rain would quickly wash away the soil if it were left bare. To prevent erosion I routinely add mulch material to the alley. Also, I design the alley to be as small as possible to limit the amount of bare soil. A width of 5 to 10 feet offers enough girth to allow the chickens to move comfortably within the alley. The number of pens necessary for foraging areas determines the length of the alley.

Ideally, the land slopes continuously at a shallow 0.5 to 2 percent grade between the highest point

Figure 7.6. *Left*, adding wood chips, straw, and weeds to the alley prevents bare soil from eroding when it rains. *Right*, when material is added to the top of the alley, the scratching action of the chickens, combined with gravity, moves material downhill.

Figure 7.7. Mulch naturally migrates downhill in a sloped alley and can be removed through a gate at the lower end. Whenever I need fertile fine mulch, I simply open the gate, and a pile lies ready at my feet.

and lowest point of the alley. The length of the alley should be positioned in a way that it cuts across the slope of the land at this grade. However, if a particular location is desired for the fence and the slope is too steep, heavy machinery could be used to terrace the alley into the desired slope. The shallow slope slows the flow of water over the length of the alley, keeping rain from washing soil away and pooling into puddles. We toss food scraps, weeds, straw, wood chips, and everything else that's edible and mulchlike over the fence at the upper end of the alley. The chickens' innate scratching behavior shreds the material as they dine on weed seeds and insects. Gravity then deposits the material at the lowest point of the alley in a heap of nutrient-rich material. I locate a people-size gate at this spot to make removal of the shredded material a cinch.

Managing the Foraging Pens

When I first disclose that I use electric fencing, some people assume it's a cruel technique, constantly shocking the chickens into submission. On the contrary, one shock usually instills a healthy fear of the fence and prevents further breakout attempts. Besides keeping the birds in, the fence must keep predators out. The electric shock discourages predators from breaching an otherwise flimsy fence.

Electric poultry netting is lightweight and easy to move. I loop the netting off the alley like petals unfolding from the black center of a daisy. Adjacent petals always share a common wall, as shown in figure 7.3. The illustration shows all seven petals, but in reality only one petal exists at a time. Each time I want to move the hens, I shut them in the alley, then move the fencing to the new pen area. Since the petals share a common wall, I only need to move two-thirds of the length of the fence each time as it leapfrogs around the alley. The alley fencing remains stationary at all times, but I rotate the electric poultry netting to offer the chickens new areas to forage.

It pays to follow the manufacturer's directions precisely when installing electronet. The key to ensuring a strong effective fence is the proper grounding system. I've visited several farms with electric fences that don't work because of improper grounding. The farmers only used a single grounding rod placed a few feet in the ground. Without an effective shock, the fences are worthless against predators. Any piece of grass or twig contacting the poorly grounded fence reduces the charge, making the fence even less effective.

Electric fences function by sending out a charge through the electric fence. When the fence is contacted, the current travels through the animal or person, then into the ground—the current only completes the circuit when the charge contacts the ground rods and returns to the opposite terminal on the charger. If ground rods are not appropriately placed, the circuit will never complete and create a shock. Recommendations call for placing three ground rods 6 feet deep and at least 10 feet apart. In addition, I place my ground rods in the bottom of a swale on contour. The extra moisture captured by the swale helps carry the charge and completes the circuit, making a very strong fence. Copper and galvanized wire or ground rods

should not be mixed to prevent electrolysis and corrosion from weakening the shocking power. (Always refer to the manufacturer's instructions for proper installation.)

Another common misconception with electric fencing is the need to loop the fence back onto itself. The electric shock and circuit completes when electricity travels between the fence and the ground rods. Therefore, only one end of the fence needs to connect to the charger for the system to work.

Designing the layout of the electric fencing system is simple. Starting at the charger, mounted in a protected area under my house, I use an insulated galvanized wire running along the top of the alley fence to bring the current to the electronet. To allow for connecting the galvanized wire to the electronet, I remove a 1-inch-long section of insulation from the galvanized wire every 10 feet (and next to a fence post).

Next, I install one end of the electronet adjacent to the alley fence post, making sure the electric net doesn't contact the metal fence post but is still close enough to prevent chickens from escaping. Next, I lay out the electric fence in a loop bringing the other end of the electronet back to the alley and install the opposing end of the electronet adjacent to the alley fence post. At this point the only upright electronet fence posts are the two ends attached to the alley fence, and the rest are simply laying on the ground in the proposed loop resembling a petal coming off the alley. Working from both ends attached to the alley fence, I install the rest of the electronet posts upright, tilting the posts slightly outward and rounding the corners instead of squaring them off to keep the fence taut. I finish the installation at the apex of the electronet loop with the final fence post pulling the rest taut.

Once I get to the end of the alley and the chickens have foraged through all the pens, I gather the electric fencing up in its entirety and move it to

Figure 7.8. Electric net fencing is tied to a fence post on the alley and attached to the insulated galvanized wire to charge the fence.

Figure 7.9. Electric poultry netting loops off the alley like the petal of a flower. A tree in the pen provides shade and shelter, encouraging the chickens to use the forage area.

the opposite end of the alley and start the process over. Most of the time I only need to move two-thirds of the fence at a time, saving labor and reducing the fence-moving process to about ten minutes. Another tip: If you can, move the fence the day after a rainy spell. That way it will be much easier to pull out and insert the posts, especially in clay soils.

Electric poultry netting is effective only for containing full-grown chickens. When I first used electric netting, I was surprised when my nearly two-month-old Rhode Island Red ran straight through the fence. At first I thought I had a Houdini hen on my hands, but I soon realized that chickens' feathers give them a much larger appearance than the actual size of their scrawny little bodies. As the seemingly large hen entered the foraging area, she poked her head through the small holes in the netting, and a stiff shock propelled her through in a magical display of dexterity. So unless you like

chasing chickens, it's best to wait several months before using the electronet.

Growing Forage

When I started growing my first flock of hens, I didn't realize the importance of forage. With the amount of money I spent on organic chicken feed and trips to the feed store (a 40-mile round trip) I could have bought free-range organic eggs for less at the local grocery store.

Although that was nearly twenty years ago, today's farmers generally agree that they reduce feed costs by 30 percent by growing forage for their chickens. Breaking free from commercial chicken feeds means growing as much food for chickens as possible in foraging pens. Insects provide a source of high protein for chickens as well as an exciting hunt. I place pieces of cardboard, logs, plywood, and piles of brush inside the pens and leave them in place until they decompose. Within weeks, insects bustle beneath the props. Using a rake or a hoe, I pull back a log to reveal the hiding insects, and the hens dive in, gobbling up the bountiful buffet. The chickens have learned to follow me around the pen as I conjure food out of hiding places. Sometimes I plan ahead and place piles of food waste into the next new foraging area to attract maggots. Rotating the chickens onto the food piles before maggots have a chance to morph into adults provides a safe and easy high-protein food source for the birds.

Piles of waste and debris in the paddocks work well, but the unsightly nature and smell relegate them to the outer unseen edges. Chickens obtain most of their food from plants. Long-lived perennial clovers and grasses lay the foundation for the forage. When allowed to mature, short-lived annuals such as millet and buckwheat provide highly nutritious seeds. Planted among perennials, these annuals help fill in the gaps when the perennials' growth slows.

In my climate clovers thrive during spring and fall, but other grasses persist through the heat of summer and the cold of winter. By far, chickens' favorite forage is clover. The basis of a good forage system is promoting the growth of this nutritious plant. The main perennial clover I grow is Durana white clover. Durana clover came from seed selected from vigorous clover patches in Georgia capable of withstanding drought, extreme heat, and soil pH as low as 5.4. An aggressive and persistent nitrogen fixer, Durana clover produces a dry weight of 4,000 pounds per acre per year of nutritious forage for chickens.

The chickens' preference for white clover is obvious as they search out the patches. The flowers attract bees and other beneficial insects, and the nitrogen-fixing capabilities of the plant fertilize all surrounding plants. Durana clover is adapted to all soils but pure sands and grows in most of the United States except the extreme South and desert states.

Drought and extreme cold knocks the clover back, but if the plant is allowed to flower and reseed, the clover will bounce back with a vengeance when conditions improve.

In the upstate region of South Carolina, white clover has a narrow planting window of mid-September. My chicks arrived in August, and since many months would pass before they were old enough to forage, I prepped the yard for their arrival by planting forage crops. However, before planting, I keyline ploughed our pasture, then added enough lime to bring the soil pH to 5.8 and fertilized with phosphorous and potassium to bring the major nutrients to satisfactory levels. I broadcast planted the clover with millet as a nurse crop. Since clover is slow to establish the first year, the millet provided adequate cover to protect the soil. When the millet surrendered to winter's temperatures, the established clover continued to grow.

I learned a critical lesson about the importance of proper soil testing when I didn't lime an area of my property. In an area designated for blueberries, an acidic soil-loving plant, I purposely left part of the land without lime. I still planted clover in this area, and I was astonished when the clover failed to grow because of the acidic soil condition. Cereal rye still performed well, but the experiment highlighted how critical pH adjustment is to plant growth.

Other less important perennial forages include Bermuda and native warm season grasses, which persist during our hot summers when the clover starts to fade. Dandelions, chicory, plantain, oxalis, and other perennial weeds round out the chickens' diet — each plant finds its niche when the weather permits.

Most of the protein in the forage comes from fresh new growth. Every plant has an ideal height at which to allow the chickens to commence grazing and an ideal height at which to stop them from

Figure 7.10. Grain rye ripens outside the alley as the chickens eye the awaiting forage.

Figure 7.11. I also grow grains outside the reach of the movable pens. When these crops mature, I cut them with a sickle bar mower.

Figure 7.12. The piles of grain and straw are gathered on a tarp and dragged to the chicken alley. The chickens feed on the grains and turn the straw into a nutrient-rich, seed-free mulch.

Figure 7.13. Chickens feast on grain rye harvested from the yard.

grazing, to have sufficient leaves for regrowth. Ideally, rotate chickens onto pasture that's 4 to 8 inches tall. If the chickens are frequently rotated onto fresh new growth, the pasture will provide a maximum amount of feed and protein. In contrast, if the chickens stay in the same pen for too long, they begin to deplete the leaves available to supply the plant with energy for regrowth. The plant must then use energy stored in the root system, and the whole plant is weakened as a result. A weakened pasture is susceptible to invasion by an army of weeds blown in by the winds or stored in the soil seed bank.

However, the weakened state of the overgrazed pasture is a great opportunity to seed an annual crop, enhancing the forage for the chickens' return. In this way the chickens are prepping the area for pasture improvement. The seeded annuals also help fill bare soil areas, preventing unwanted weeds from becoming established. I then rotate the chickens onto the forage while it's still young, or I wait, allowing the new forage to mature ripe seeds for the chickens to eat.

The type of annual forage planted depends on the time of year. When the weather cools in fall and all through the winter, I plant a cereal rye or wheat by broadcasting the seeds into the old pasture area. Cereal rye survives and germinates in cooler weather better than wheat.

The chickens savor the thick, plump, mature seed heads on cereal rye, but I have to cut the plants to give chickens easy access to the seeds. A normal rotary mower would have a difficult time in the tall grass and would shred the seeds, making them degrade faster. A better tool is a cutter bar or sickle bar mower. Similarly to hedge shears, the cutter bar cuts at the base, leaving the aboveground parts untouched. Once the rye is cut, the chickens can rummage through the plants foraging for seeds, or I rake the plants into piles of hay. Then I plop the hay on top of a tarp, drag it to the chickens' alley, and place it inside as mulch. Sometimes I stockpile hay with seeds inside our garage for later use.

During spring and summer when the weather warms, I sow a forage crop of millet, buckwheat, or sunflowers in the pasture where the chickens have just foraged. Browntop millet and buckwheat plants are short enough that the chickens can eat the seeds right off the plants. Buckwheat and sunflowers have the added benefit of producing beautiful displays of flowers, creating additional forage for bees while also attracting beneficial insects.

Summer plantings of annual forage are considerably more difficult to establish than fall and spring crops unless enough rain is available to support seed germination. I usually wait until an inch of rain is forecast to fall, then plant the summer

annual crops to ensure good establishment. I rely on the NOAA website precipitation forecast models to help plan seeding times. Planting seeds more than a few days ahead of a rain gives wild birds more time to eat the seeds. Tilling before seeding and raking in the seeds after broadcasting helps hide the seeds and keeps them moist, especially with larger seeds, such as buckwheat.

Fruiting trees and shrubs planted into the forage areas also feed the chickens when overripe fruit, flowers, and insects drop to the ground. However, be careful to let plants become well established before unleashing the flock to forage freely. I clearly recall coming home one day and finding my blueberry bushes had been massacred. Barely 6 inches tall, the baby bushes didn't stand a chance against those hungry birds. Only a few small leaves remained after the chickens got their fill. Since that fatal moment I have learned to protect young plants with a barrier of chicken wire looped around them. I move the chicken wire protectors every time I move the electronet to protect the next batch of young plants until they outgrow the height of the chickens. Other options include using small plastic cylinders called tree protection tubes or planting larger trees or shrubs. Anything beats the tragedy of chicken-mown blueberry bushes.

How Many Pasture Pens?

What is the right amount? For my humid, wet climate with regular rainfall, a stocking density of approximately 43 square feet per chicken works to create a uniform eating pattern. Given more space, the chickens may spot graze and only eat certain spots or certain plants in the part of the pasture where they've been penned. Ideally, the chickens are rotated every six to twelve days into a new area, which allows the foraged areas to regenerate. I move the chickens based on the height of the vegetation. I fence them into an area when the plant height is 6 to 8 inches. I move them when the height has decreased to 3 inches. Leaving the chickens in one area longer will eventually lead to bare spots, which can be an opportunity to sow seeds of additional forage crops. In this way the chickens help prepare the land for pasture renovation.

Though chickens readily eat young low-growing forage, they also scratch for insects around taller plants. In several of my pens a single tree provides shelter, protection, and a diversity of insects for the chickens' diet. Additionally, I locate two pens in wooded areas to give the pasture time to regrow.

The amount of time the pasture rests and the amount of time the chickens graze determines the number of pens needed for rotation. This simple formula helps determine the minimum number of pens for rotational grazing:[3]

$$\text{Days of rest} \div \text{days of grazing} + 1 = \text{number of pens}$$

In my climate, forage crops grow quickly during the spring and fall; therefore, pens might only need fifteen days of rest. However, in the heat of summer the pens may need thirty or more days of rest. For example, if thirty days of rest are needed and the chickens are moved every six days:

30 days of rest ÷ 6 days of grazing + 1 = 6 pens

Currently, I have a total of twelve pens, which gives me more flexibility with the types of forage I can grow. If I want to grow a forage crop to mature seed such as cereal rye, millet, or sorghum, the crop could take a minimum of 60 days and up to 120 days for seeds to ripen. Having more pens allows crops to be grown to a seed stage or tall enough to cut for hay that can then be used as mulch in a holding pen, as I explained earlier in this chapter.

For example, if the pen needs 60 days of rest and the chickens are moved every 6 days:

60 days of rest ÷ 6 days of grazing + 1 = 11 pens

Manure Management

Although I usually refrain from vulgarity, there's an old saying that sums up my approach to manure management quite well: all shit runs downhill.

Manure removal and dispersal represent a major portion of the chores in animal husbandry. Traditional techniques for small flocks include using shovels and loaders to remove manure. A modern technique involves moving the entire coop to distribute manure throughout the farm.

I've tried both techniques, and I definitely didn't like the shovel method. Movable coops are popular, but moving a coop, even on wheels, is usually more difficult than moving a small amount of manure. With movable coops, manure still concentrates under the coop and requires disbursement to distribute it properly. Instead, I decided to develop a new approach to managing manure.

Flushing with Rainwater

By connecting chicken coops to rainwater-harvesting systems, I take advantage of an often overlooked use for water—the ability to move nutrients. Stationary coops create a concentrated source of nutrients available for collecting and allows application of those nutrients in the form of manure to specific areas and plants in need of fertilizer. I apply the manure to cover crops in the garden and in small amounts to fruit trees and other plants to prevent burning. The only drawback is keeping unpleasant smells at bay. I like having a nutrient source at my disposal, but I don't want to clean the coop constantly to keep odors manageable. Using

Figure 7.14. A longer lasting PolyMax floor replaces a floor made of chicken wire. Manure falls through holes in the floor to the collection area below.

water to clean coops and disperse nutrients solves this problem. Fresh fertilizer is ready and waiting when needed but also disappears with the ease of flushing a toilet.

The comparison to flushing a toilet is apt! Store the water above the waste, lift a lever to release the water, and gravity does the rest. The trick is to place water storage as high as possible in the landscape, preferably just above the chicken coop area; then gravity is maximized to best distribute the manure downhill.

Of course, it's all in the design. My chicken coop has a floor designed to allow manure to fall through to the ground underneath, as shown in figure 7.14. I carefully carve a diversion channel under the coop and out across the landscape to nearby trees and shrubs. When I release harvested rainwater (stored in a pond or tank) to flush manure from under the coop, the manure-laden water flows precisely through the channels, carrying the manure and nutrients on the longest pathway possible to maximize the movement of

nutrients to plants. The more water that's available, the farther the nutrients and irrigation will seep into the diversion channel system. My system requires a minimum of 50 gallons of water to push the manure that accumulates beneath the 4 × 8-foot coop into a channel. Once the manure is out from under the coop, natural rainfall will continue to move nutrients downhill in the channels if there's insufficient harvested rainwater available to continue the flush. Close to the coop, I grow plants that are tolerant of high nutrients, such as Jerusalem artichokes, elderberry, and comfrey. Then I feed these potentially contaminated plants to the chickens. Further from the coop, I grow traditional fruit trees and aboveground crops for humans to eat.

I've seen many rain barrels, cisterns, and ponds sitting idly without a purpose. If situated at a slightly higher elevation than a chicken coop, these water-harvesting devices have a mission — push the poop from the coop. The simplest way to run the water to the coop is to connect the drainage pipe from a tank or pond to the coop. With good design the drainpipe from a tank or pond can open directly under the upper end of the chicken coop, making it easy to flush water into the diversion channel. The drain can also be part of the principal overflow system, allowing regular pond or tank overflow to flush through the coop. Connection to the chicken coop gives rainwater-harvesting devices a fixed function.

With wet rainwater-harvesting systems (see chapter 6 for more info), a substantial amount of water is stored in the pipes alone. Situating a drain at the lowest point of the pipes removes water to prevent freeze damage. My current rainwater harvesting system is composed of over 200 feet of 4-inch piping, which distributes water to a storage pond and stores more than 160 gallons of water inside the pipes alone. Flushing the coop

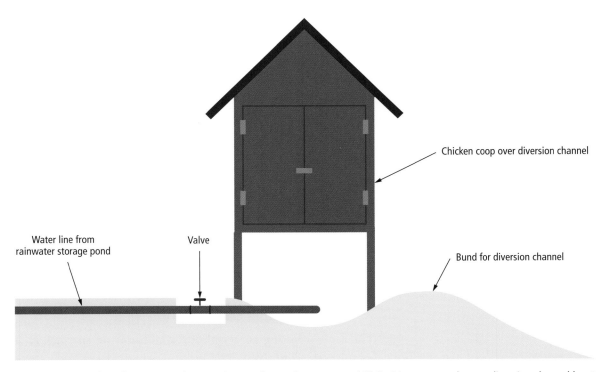

Chicken coop over diversion channel

Water line from
rainwater storage pond

Valve

Bund for diversion channel

Figure 7.15. A valve releases water from a rainwater-harvesting system uphill, flushing manure down a diversion channel located beneath the coop. A bund, or berm of soil, is sloped at 25 percent to form the diversion channel.

empties water stored in the pipes. The next rainfall refills the pipes and builds up the reserve in the pond or tank.

To connect the wet rainwater-harvesting system to the chicken coop, I first reduce the 4-inch pipe to a 2-inch pipe. Two-inch pipe is large enough to move the needed volume of water to the coop but small enough to cut the costs on valves and piping. The 2-inch pipe buried belowground slopes continuously downhill at least ¼ inch per foot and connects to a valve positioned outside the coop. I open the valve to release water from the pipe. The water flows beneath the chicken coop, which is positioned directly over the uphill end of the diversion channel, and flushes manure into the diversion channel and through the landscape.

Recently, I added another feature to my rainwater-harvesting system that employs the

Figure 7.16. Jerusalem artichokes grown in a diversion channel harvest nutrients and water flushed from the coop. The taller, dark green plants are in the section of the channel closer to the coop; thus, they receive the most nutrients. We feed the chokes to the chickens.

science of siphoning. By creating an adjustable air vent at the top of a wet rainwater-harvesting system, I can siphon water out of the pond to flush the coop. Now when I drain water out of the pond to harvest minnows, I'm also flushing manure from under the coop into the diversion channel.

Adding a Soldier Fly Digester

After an extremely wet summer I noticed a mass of soldier fly larvae growing in the diversion channel under the coop. This native fly proliferates wherever a readily available putrid waste source accumulates. The plentiful supply of manure constantly falling to the ground below the coop creates the perfect feedstock for rearing soldier fly larvae. This nonpest fly plays an important role in keeping the real pests, such as houseflies, out of the manure. (For more about soldier fly biology and habits, see chapter 9.) But the best benefit of these prized insects is the nutrition they provide for the chickens.

Before the soldier fly larvae pupate into adult flies, the chickens eat them as a high-protein snack. Thus, the soldier flies transform manure into protein-rich feed. The soldier flies also reduce the total volume of manure by 50 percent, reducing the need to remove wastes from the coop area.[4]

During dry years there's usually not enough moisture under the coop to allow larval development. During our wet summer, however, the diversion channel captures enough water from runoff over the landscape to moisten the manure for soldier fly development. Hence my experimental process began — to create the perfect environment for soldier fly production.

To grow larvae in manure under the coop, I adjust the moisture level of the manure. Too wet, and the larvae don't like it and may crawl off; too dry, and the population crashes or ceases to exist. Another integral part of larval production is excluding chickens from the larval rearing area. If chickens access the manure, they will eat the adult flies, thereby preventing procreation. The birds will also peck through the manure, eating any larvae that do hatch before they reach a large plump size. It's all a delicate balance.

Water the manure beneath the coop as if watering a garden — by hose, irrigation, or mister — and the larvae will flourish in all areas where they are native. I simply open the valve on my manure flushing system for a brief few seconds and wet the manure without washing it away. If water regularly flushes manure out of the coop, larvae will never have time to build up to meaningful populations. Saving a shovelful of larvae to reinoculate after flushing helps stabilize production.

As with other designs, I strive to work with nature rather than pointlessly swimming upstream. Larvae naturally climb uphill to escape the wet putrid manure before they pupate or go through their cocoon stage, so I locate the chicken yard or alley uphill from the chicken coop so the larvae crawl uphill into the chicken yard and into the hungry beaks of chickens. To prevent the chickens from feeding on the immature larvae but permit them to feast upon the escaping prepupa larvae, I install a barrier made from a metal roof to bridge the gap between the coop's floor and the ground. The metal nearly touches the soil, leaving a gap just big enough for larvae to crawl through.

On the other side of the metal skirt, the chickens await the escapees. Since the young larvae are protected by the metal skirt, there is a constant supply of larger larvae for the chickens during the growing season. The metal roof skirt surrounding the chicken coop also shades the larval growing area, supplying the darkness preferred by the soldier flies. Not all larvae will make it into the chicken yard, and some will inevitably escape, leading to a healthy population of adult flies for the

Figure 7.17. Soldier fly larvae feed on manure under the chicken coop. Larvae eventually follow their drive to climb uphill and out from under the metal foundation skirt into the chicken alley, where the chickens hungrily hang out.

next cycle. When I excavate nutrient-rich soil from the chicken alley, I take the soil that piles up next to the metal skirt on the coop, keeping a small gap under the coop to allow the larvae to escape easily.

If I really want to make my chickens happy, I scoop a shovelful of solder fly larvae and throw it into the chicken yard. It's like Christmas morning! They flock to the larvae pile, eagerly pecking away at their high-protein presents.

The Shovel Option

Although connecting a coop to a rainwater-harvesting system is a big time and labor saver, sometimes I need only a few scoops of fertilizer, and a shovel is the best tool to use. Manure from chickens and soldier fly larvae is a valuable resource to fertilize gardens and fruit trees. Stockpiling manure under the coop is like storing money in the bank. Whenever I need a readily available nutrient source, I make a withdrawal from the coop instead of purchasing fertilizer from a store. By designing the coop so that I can easily extract manure with

Figure 7.18. The roll-up foundation skirt on the chicken coop controls ventilation and gives easy access for shoveling out manure that falls through the floor of the coop and into the diversion channel below. The overhanging gable roof shelters tools.

a shovel, I save time and money and ensure my plants get a healthy dose of nutrients.

Designing a raised coop allows for accessing the manure under the coop, but the problem with raising the coop is airflow. With a large gap below the coop, convection pulls air up through the roost area. During the summer the airflow is beneficial; it keeps the hens cool. However, during the winter, the airflow from under the coop may be stressful to the birds and increase the amount of feed they consume. To solve this problem, I install a plastic roll-up foundation skirt on one side with the same technology used to build roll-up greenhouse sides (described in chapter 4 under "Roll-Up Sides"). In the summer I roll up the foundation skirt to cool the coop; in winter I roll it down to modify the coop microclimate as needed.

Connecting a Chicken Coop to a Deck

At our last July 4 party our guests gravitated toward our back deck. Not because of a refreshing pool or breathtaking fireworks, but because of our entertaining flock of Rhode Island Reds.

Instead of locating the chicken coop far from the house, I positioned it about 20 feet from my back deck. The chicken alley actually borders the posts and beams of the deck. From a comfortable seat at our outdoor table we can watch the hustle and bustle of the flock. My intention with this design was purely rooted in efficiency. The closer the coop is to the house, the less time I spend gathering eggs, refilling feed, and performing day-to-day maintenance. But the first day we watched our baby chicks explore the alley, I realized that by integrating the chickens' home close to ours I had created hours of free entertainment. There's something mesmerizing about watching chickens — possibly the same attraction that lures nature enthusiasts into woods and marshes to observe the behavior of birds, or maybe the same

joy we feel from watching a puppy playfully discover its new surroundings.

Feeding Chickens from the Deck

Our party guests grew even more excited when they learned they could clear their plates directly into the chicken alley. No longer spectators, they were interacting with the hens — and they loved it! From the edge of our deck our guests scraped the last few bits of food from their plates and watched as the chickens flocked over to devour the tasty morsels.

Connecting a chicken coop to a deck brings chickens into the heart of family outdoor life. Even when we dine inside, we still toss our food scraps to the chickens. They have learned to listen for our footsteps leading to the sliding glass door. When we step outside they are waiting like fish by a shore, hoping for a friendly passerby's handouts. We also collect food scraps in a small compost container by the kitchen sink. But instead of slaving away layering a compost pile, we simply toss our compost collection into the chicken alley. Our hens quickly turn our leftovers into fresh fertilizer and eggs, saving us lots of time, energy, and money.

Figure 7.19. Chickens turn food scraps into eggs and fertilizer. Connecting the chicken alley to the deck makes feeding the chickens easy and fun.

Ironically, the chickens' favorite food is chicken. In nature and in captivity chickens are quite cannibalistic, feeding on each other when the need arises. Chickens are omnivores capable of eating anything we can eat, and they crave the high protein in meats. On the other hand, meat and dairy are usually troublesome in compost systems, attracting unwanted rodents. As long as we give the food scraps to the chickens before they retire to the roost, the leftovers will be long gone before nocturnal rodents begin their nightly patrol. In essence, our chickens effectively control our rodent population.

WHAT NOT TO FEED

There are a few things chickens shouldn't eat. While I couldn't find any scientific literature defining exactly what is toxic to chickens, the well-known Murray McMurray Hatchery in Webster City, Iowa, recommends not feeding chickens the following:

- Raw potato peels: Peels, especially when they turn green, contain the alkaloid solanine, which is toxic.
- Garlic and onions: These are not toxic but may impart a flavor to eggs.
- Avocado skins and pits: Avocados contain persin, a fungicide toxic to chickens.
- Spoiled or rotten foods: These may contain toxins.
- Chocolate: This contains theobromine, which may be toxic to chickens.
- Raw meat: Feeding raw meat may lead to cannibalism.

Ways of Connecting

Chicken coops can connect to decks in several fashions. When housing only a few hens, the

Eliminating the Disposal

Our sink is equipped with a garbage disposal, usually a sought-after amenity in any kitchen; that is, until it's time to pay the bill from the septic pumpers. Recommendations call for pumping a septic system twice as much when a disposal is used, doubling maintenance costs. Our chickens are now our garbage disposals. But instead of grinding up the waste and piling up the bills, they are digesting the food and laying fresh eggs.

coop can connect directly to the deck. As the number of chickens increases, I would move the coop farther away to isolate the smell of manure. My friends have an innovative coop deck connection. By replacing the railing on their deck and connecting the chicken coop directly to the edge, the laying boxes open to the deck, making egg gathering a cinch. With my larger flock of thirty, I keep the hens about 20 feet away from our deck, joined by the alley fence. The coop is located downwind of predominant summer breezes so smells blow away from the house. Occasionally the wind shifts at night, blowing an unwanted odor toward the deck. Before spending time on the deck, I'll throw a little hay over the alley to buffer the smell.

Advantages

Connecting the coop to the house offers numerous advantages. The frequent trips required to

Figure 7.20. An alley extends under a deck, giving the chickens access to several more pens on the other side of the deck.

Figure 7.21. The electric poultry netting connects to an alley extending under the deck. The house becomes part of the fence system.

collect eggs and feed and maintain the flock are shortened, saving many hours of labor each year. In fact, every ten steps farther adds an extra 150 minutes a year to a task if several visits per day are made. If coops are placed close to houses and along main paths and roads, chicken care becomes an integrated part of everyday life. To see an overview of my property showing how I've integrated the coop into my landscape and added a second coop to serve my neighborhood — which I'll describe later in this chapter — check out the overall map of my home landscape in the appendix.

Protecting against Predators

Another perk to keeping the coop close is protection. Nature is divided between predator and prey. Chickens often fall into the latter category, and almost everything larger is a potential predator. Over the years I've lost countless chickens to lurking predators — hawks, raccoons, coyotes, even stray dogs. Stephanie quickly learned to not name the chickens because each loss would devastate her. But by building the coop close to the house, I have nearly eliminated the predator factor.

Chickens have their own built-in alarm systems. Much like the sensors on a car, if a presence triggers their sense of danger, they cause a ruckus. They start cackling and calling to alert the flock of a potential danger. Because of our chickens' proximity to our home, they alert us as well — and we come running.

One sunny Sunday afternoon Stephanie was reading on the front porch while the hens explored their new forage area in the front yard. A sudden squawking startled her, and within seconds she had thrown down her book and was halfway to the chickens. Meanwhile, I came running out the front door carrying the weapon closest to hand, my shoe. And our neighbor came barreling through our field on his golf cart, also stirred by the commotion of the flock. Although none of us bore a decent defense, the disturbance we created was sufficient to deter the hawk, and our fortunate hen scampered off, only a few feathers lost to the tumult.

Hawks and other birds of prey struggle when attempting to fly off holding a large chicken, forcing them to eat their prey on-site. I've foiled several attempts of hawks by spotting them and preventing the kill before they had a chance to finish the job. Without a successful kill the hawks seem uninterested in returning. Spotting and defending against

hawks are nearly impossible tasks if the coop is located far from the house.

Young chickens are most susceptible to predators, since they are lightweight and easy to carry. One morning a hawk snatched up one of our young hens, so I quickly locked the chickens inside the coop before rushing off to work. When I returned at the end of the day, the hawk was patiently perched on top of our fence staring at the coop door, awaiting the exit of the young hens. When I scared the hawk off, it politely flew to an adjacent tree to offer me a little more space. Luckily, Stephanie devised a successful plan. At her suggestion I placed fishing line in a spiderweb fashion over the top of the alley, connecting the line to the fence posts. The hawk did return, diving through the fishing line and snatching up a young hen. However, the hawk's near escape was spoiled by the fishing line. Navigating through the fishing line with a heavy hen gripped in his talons proved too difficult. So he dropped the hen back into the chicken alley and flew off into the trees. The same scenario repeated itself several more times before the hawk finally surrendered, and we haven't lost a chicken since.

The foraging pens bounded by electronet pose a different problem. Since the fencing is temporary, draping fishing line across the top is not practical. Unfortunately, this allows hawks easy access to foraging hens. However, having shrubs, trees, or a shelter in the foraging area offers the chickens an easy escape from danger. Trees and shrubs also motivate chickens to forage through the entire pen by giving them a shaded area of safety, or a home base, to venture from. If a pen has no trees and shrubs, the chickens will likely stay in the alley, hanging out by the coop and eating expensive feed while leaving the pen untouched.

Electric fencing is also an effective hawk deterrent. I've watched as a hawk swooped down to

Figure 7.22. Folding a 1-foot-wide expanse of the lower edge of the chicken wire fencing into the alley and stapling it to the ground prevents predators from digging under the fence.

snatch a hen, but the chicken quickly ran toward the electric fence. As the hawk grabbed the chicken, it also grazed the fence, which gave that predator a shock that sent it flying off, never to return.

Predators will dig under nonelectric fencing. I combine electric and nonelectric fencing to eliminate this issue. My theory is that a predator will walk the edge of a fence seeking a vulnerable area to breach. When electric and nonelectric fencing are linked, the predator walks the edge of the nonelectric fence and eventually reaches the electric fence. The shock scares them off permanently. I've noticed this behavior in dogs and assume that nighttime predators do the same. To improve the safety of nonelectric fencing, I fold a 1-foot strip of fencing toward the interior of the pen and bury or use sod staples to secure the fencing to the ground (see figure 7.22).

Door Systems

For many years I was a slave to my chickens. My daily plans hinged on the rise and fall of the coop door. Regardless of how late in the evening my head hit the pillow, I still had to be awake early the following morning to let the chickens out. Countless evening plans were interrupted, canceled, or delayed because of my nightly duty. Now that I've switched to an automated door opener, I'll never again need to rush home to lock up the chickens or fretfully worry that I forgot to let them out.

Since predators are usually likely to strike at night, closing a coop door after the chickens retreat to the roost at dusk ensures a restful night's sleep — for both the hens and for me. Coop doors are especially useful in protecting chickens from predators without the added safety of an electric fence. However, opening and closing a coop door requires a lot of time and energy. Also, if the chickens aren't released at daybreak, they lose valuable forage time, which increases feed costs.

The sliding gate is simple to construct. A square piece of plywood or metal, slightly larger than the door opening, is sandwiched against the wall. Use a string or rope to pull the square piece of plywood up, exposing the opening in the wall. Commercially available automatic coop door openers are able to raise and lower the string. With automation a small motor rotates a pulley that winds the string up and down. The motor is then plugged into a receptacle timer to control its operation.

When electricity is not available, sliding gates also make great manual doors. A system of pulleys can carry the rope from the sliding gate to the ledge of a deck or beside a back door, allowing it to be opened and closed from afar. Why make an extra trip to the coop to open the door when a rope can make the trip for you?

Automatic Waterers

Manually filling watering devices for poultry consumes an enormous amount of time. Before

I became liberated by the automatic waterer, I spent over thirty hours a year providing water for my chickens.

I took a class on small flock management from a renowned author on the subject. He highly discouraged using automatic watering devices with poultry because he thought it was necessary to check on the chickens to ensure they had water. Manually filling the water was his daily ritual. My teacher had a valid point because when chickens run out of water for even a few hours they become stressed, and egg production will decline for quite some time. However, a good design will allow you to check on automatic waterers easily, ensuring everything is functioning properly.

I like to connect the chickens' alley to a pond. Chickens and foraging animals should never have access to the entire edge of a pond because they will quickly denude vegetation, causing erosion and water quality issues. However, giving chickens access to a small strip on the edge of the pond allows them to access water when necessary — and gives me peace of mind that a backup plan is in place. If the chicken area is uphill from the pond, the pond will stay fertile and green. If the chicken area is downhill from the pond, the pond will stay clear. As a bonus, chickens eat tadpoles and duckweed (a small floating plant) directly from the pond, reducing feed costs. Orienting the connection so the predominant winds blow the duckweed into the chicken area keeps a continuous supply of feed at hand.

With my current poultry setup, connecting the primary fence to a pond was not possible. So to ensure I could monitor the chickens' water supply easily, I positioned the automatic waterer within sight of my kitchen window for easy observation. The waterer sits immediately inside the coop door. A mirror hangs on the wall behind the waterer, shedding extra light on it. From my kitchen I can

Figure 7.23. No, the mirror isn't there for the chickens to check their feathers. Light reflecting off the mirror shines on the waterer, allowing us to see from a distance (through the coop door) whether the waterer has water in it. A drain hole below prevents water from pooling.

easily see the view through the open coop door and tell whether the bowl has water in it. I've considered using automatic water nipples, another common and inexpensive technique to provide water, but they lack the visual assurance that the system is working, since you can't see any water at the nipple.

It's important to protect automatic water lines from freezing in colder areas. I bury the pipe to the required depths and insulate as much as possible when it's aboveground. Chickens will eat foam pipe insulation, commonly found at hardware stores, so the pipe insulation must be protected with a rigid pipe cover to prevent chickens from eating the foam. Alternatively, electric heat tape, a product commonly used to keep pipes from freezing, can be attached to pipes and plugged in when needed. The combination of insulation, heat tape, and a heated base for the waterer prevents freezing and extends automated water for chickens through the winter in colder climates. For additional insurance I use PEX pipe instead of rigid pipe for the supply line to the waterer. When the water in PEX pipe freezes, the pipe expands instead of cracking like rigid pipe, and the system will return to normal once the water thaws.

Placing an additional automatic waterer in the foraging pen area adds another layer of protection. In the summer I connect garden hoses to automatic watering bowls in the pen. The presence of the additional waterer also makes the chickens more likely to stay in the foraging pen, since they don't have to return to the coop for water. When I move the pen, I simply reposition the portable waterer. In the winter I don't use an additional waterer since it's difficult to protect from freezing.

All waterers should be cleaned at least weekly. I keep a toilet-scrubbing brush and a spray bottle filled with a 10 percent solution of bleach next to a nearby hose bib to make cleaning easy.

Supplemental Feed

Foraging chickens find an average of 30 percent of their food intake from pasture but supplementing with a commercial feed is still necessary. I place a feeder inside the coop and keep it stocked at all times. The feeder is raised above the floor using a small rope tied to the ceiling. As the chickens grow I adjust the height to match their height, which prevents the chickens from scratching, scattering, and wasting the feed.

While the waterer is placed directly inside the coop opening, the feeder is placed farther inside.

With a small coop opening measuring about 1 foot high and wide and the feeder out of sight, wild birds and other critters have a difficult time stealing the expensive feed. I've tried using a sheltered range feeder in the chicken alley but found that I was feeding every bird in the neighborhood. In addition, placing the feeder inside the coop makes it a little more difficult for the chickens to get to the feed, since they spend most of their time outdoors. My hope is that the extra layer of difficulty encourages them to forage.

As you decide on coop location, consider how you will store feed and refill the feeder. If the coop is close to parking or a road, the heavy 50-pound bags of feed won't have to be carried far. I store the feed in a 32-gallon galvanized metal trash can

Figure 7.24. As the chickens grow, the height of the feeder inside the coop is raised using a rope on a pulley to prevent chickens from wasting feed.

next to the coop for easy access. The trash can is raised above the ground on concrete blocks to prevent it from degrading from ground contact. Apply a squirt of silicone where the handle penetrates the lid of the trash can to ensure it will stay completely watertight.

Egg Management

The last topic to cover is the collection and cleaning of eggs. Much like harvesting vegetables, collecting eggs is often considered the prized reward from all the hard work — the daily return from an ongoing investment. However, even egg collection can seem daunting if a proper design is not in place. Since egg management can consume the most time in caring for poultry, I carefully consider every aspect.

I provide one laying box for every three to four hens. Each laying box should measure no larger than 12 inches tall by 12 inches wide to prevent two hens from fighting over a single box. Large boxes also promote roosting inside the laying box. Wherever chickens roost, manure accumulates, so preventing roosting in laying boxes reduces manure cleanup. A sharply sloped roof prevents chickens from roosting above the laying boxes and thus also minimizes manure accumulation inside the boxes.

I've seen many operations in which hens lay eggs everywhere except in the laying boxes.

Figure 7.25. Laying boxes located in the darkest part of the coop with access to the outside through cabinet-style doors. Pine straw nesting material prevents chickens from breaking eggs and keeps the eggs clean.

Chickens prefer to lay in the deepest, darkest area of the coop. If laying boxes are placed next to busy waterers, feeders, or entrances, the birds won't use them. The problem with locating the boxes in the deepest area is access. Who wants to walk through piles of chicken poop to the darkest area of the coop to collect eggs? I solved this problem by building cabinet-style doors on my coop that open

Cleaning the Eggs

I'll begin with a confession: This is my least favorite part of farming chickens. But as with all chores, no matter how messy, they must be done. So break out the rubber gloves, grab a brush, and start scrubbing! Here are the steps to properly cleaning eggs:

Step 1. Gather eggs in an egg basket.
Step 2. Rinse the eggs in warm water (90° to 100°F). Cold water makes the contents shrink, creating a vacuum inside the egg potentially sucking in harmful bacteria. I place the basket in the kitchen sink and use a removable faucet head to spray the eggs with warm water.
Step 3. Scrub the eggs with a brush to remove any stains or debris, and place the eggs into a clean egg basket. You can also scrub the eggs under warm running water.
Step 4. Dip eggs in a warm water sanitizing solution containing approximately 200 parts per million food-grade bleach (1 tablespoon per gallon).
Step 5. Rinse the eggs with warm water.
Step 6. Let the eggs air dry in the egg basket, then place them in cartons for refrigeration.

Figure 7.26. Egg baskets and sanitizing solution used to clean eggs.

Figure 7.27. Larger operations use compressed air and bubbles to clean eggs submerged in a bucket.

from the back wall of the laying boxes; this makes egg collection as easy as retrieving a dish from the kitchen cupboard!

There's another aspect to consider in egg management: protecting the eggs prior to collection. Before I figured out how to protect the eggs successfully, I lost up to a third of production from hens pecking the eggs or covering them with manure. By placing nesting material in the laying boxes to protect the eggs, I reclaimed my one-third loss. But maintaining the nesting material in the laying boxes was an uphill battle. I made the mistake of using wheat straw — straw frustrates the hens, and they scratch it out of the nesting box. A wise neighbor pointed out the benefits of using pine straw. The hens quickly form the pine straw into a round nest and rarely damage the eggs after laying them. Egg removal is even easier with rollaway laying boxes. If you slope the floor or purchase special rollaway inserts, eggs will roll into a protected compartment. Once they are in the safety of the compartment, damage and manure contact is eliminated. Both of these options are extremely helpful in transforming egg collection from a messy, dreaded chore into an exciting, anticipated pleasure.

Integrating a Coop into Your Neighborhood

If you have more than a dozen hens, dealing with the surplus of eggs is relentless. Collecting and cleaning twelve eggs every few days is enjoyable. However, collecting and cleaning twenty-four eggs every day becomes a part-time job. Eliminating the need to collect eggs solves the final hurdle in chicken maintenance, reducing labor and time to insignificant levels.

Many hands make light work. This concept applies to most tasks, even to egg collection. The idea of automatic egg collection occurred to me after Thanksgiving dinner. Pots and pans cluttered

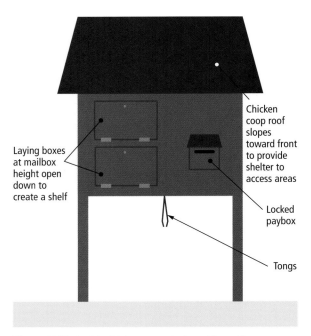

Figure 7.28. In this chicken coop designed for selling eggs to neighbors, doors to the laying boxes are at mailbox height and fold down for egg gathering. An adjacent pay box secures payment.

Chicken coop roof slopes toward front to provide shelter to access areas

Laying boxes at mailbox height open down to create a shelf

Locked paybox

Tongs

the sink, and plates were stacked high on the counter. At first, cleanup seemed nearly impossible, but when the whole family chipped in, the chore happened effortlessly. The automatic egg collection system works on the same principle of shared labor. Neighbors share in the harvesting and cleaning of the eggs by purchasing eggs directly from the coop with a pay box.

Egg collecting becomes an experience for the neighbors, suiting their individual needs. For the mom in a rush, the coop is designed for drive-through-style harvesting: roll down the window, lift up the door, and collect the eggs. For the retired neighbor who has more time, an area to sit and watch the hens can be provided. For the little ones all sorts of nooks and crannies lie waiting, with tasty morsels of minnows or worms to be collected and thrown to the hens.

To connect the coop to your neighbors, locate the coop close to the entrance of the property. County and city ordinances, neighbors, and home-owner associations will likely have a say in the exact location unless you try to pay off as many people as possible with discounted or free eggs. I placed ours adjacent to the driveway 50 feet from the road. A convenient turnaround allows cars to get back to the street easily after pulling up next to the coop.

Position the laying boxes at a comfortable height for people to access through a car window. Mailbox height provides a good guideline: It's generally recommended to be 42 inches from the ground to the base of the box. The door for the laying boxes should extend no more than 12 inches when open to prevent the open door from hitting cars parked next to the coop. If hinges are attached on the bottom and the door opens down like the door on a mailbox, the door becomes a table on which to manage egg collecting. Chains on both sides of the door hold the door in a level position. Neighbors must bring their own egg cartons, but kitchen tongs hanging on a hook are at hand for those who prefer not to touch the eggs. A wall-mounted drop box safe sits on the side of the coop to accept payment, securing money behind a tamperproof metal door.[5]

Connecting the coop to the neighborhood takes the bio-integrated chicken coop to a whole new level. Similar to a park or a small market, the coop is a meeting place where neighbors run into each other and share in a bit of conversation or simply exchange greetings. In the day-to-day bustle of our busy lives, we barely know our neighbors. Our independent lifestyles rarely rely on interaction with the folks next door. However, by introducing concepts such as the bio-integrated chicken coop into our own front yards, we are slowly helping to build bio-integrated neighborhoods.

Summing Up the Functions

1. Chickens produce eggs and meat.
2. Chicken manure provides fertilizer made available to plants via flushing through a diversion channel to downhill areas.
3. Chicken manure also provides fertilizer for an adjacent kitchen garden and fruit trees, moved via shovel and wheelbarrow.
4. Foraging chickens mow grass.
5. Chickens' scratching prepares ground for planting forage crops.
6. Chickens' scratching also prepares annual beds for planting.
7. As they feed, chickens control insect pests and weed seeds.
8. Design of the coop allows easy feeding of kitchen wastes to chickens.
9. Coop provides feed and automated water for chickens.
10. Coop provides shelter and protection from predators for chickens.
11. Automated door opens in the morning and closes at night.
12. Soldier fly larvae grow in chicken manure under the coop and later become feed for chickens.

CHAPTER 8

Harnessing Heat

Bio-Integrated Compost Applications

My father took recycling seriously, and composting was a major part of the operation. I still remember lugging the bucket of food scraps out to the compost pile in the backyard and carefully covering the waste with lawn clippings. The bucket was nearly as big as I was, so I would spin the bucket around in circles, then drag it when I felt dizzy. Our family's food scraps never produced enough compost for our hungry garden, and my father was always on a search for larger sources of waste, stockpiling the material in places that must have made my mother full of joy.

I remember the day I came home from school and saw large piles of pulped fruit waste from the local Odwalla juice factory in the driveway. My father seemed part magician, part mad scientist as he inoculated the waste with strange concoctions, fortifying the compost with astrological components. The waste would always transform into a beautiful, black, rich amendment for the garden that to this day still feels miraculous.

From an early age I understood the magic in transforming waste into a beautiful product, and as a scientific farmer I've sought to maximize its potential. In this book I write about waste management in the context of two separate farm processes. In the wintertime we make hot compost from wastes, and we divert heat from the piles to provide free energy to heat our greenhouses. That's what this chapter is about. During the summer, when I don't need heat in the greenhouses, I feed waste to soldier fly larvae. The larvae turn the waste into high-value products of feed, fuel, and fertilizer; I describe this project in chapter 9.

Compost Heating: Past and Present

Capturing heat from compost piles is nothing new. Since at least the 1940s, bales of decomposing straw have been used to provide heat and carbon dioxide for plants inside greenhouses. During the 1970s French inventor Jean Pain pioneered compost heating systems with his 20-foot-wide, 10-foot-tall, round wood chip compost piles. Pain used coils of polyethylene pipe wrapped inside the circular compost pile. When water passed through the pipes at 1.1 gallons per minute, the water heated to 140°F. Built by hand and chipped on-site, the piles contained a fine composition of wood chips pulverized by Pain's specially designed chippers. The massive piles of wood chips continued to heat for eighteen months with no turning.[1]

Elements of the System

Traditional composting requires a lot of work, but the only benefit it produces is a rich soil amendment. A bio-integrated composting system can also produce heat for your greenhouse or home. At Clemson University we use bio-integrated compost piles to convert farm and cafeteria waste into heat for our greenhouses and ponds. As long as you have access to a large stream of waste, you can apply these same techniques at your farm or homestead. In this chapter you'll learn how to:

▸ Cheaply and easily find materials needed to build a hot compost pile
▸ Easily build and turn compost piles and increase heating time frame with basic farm equipment
▸ Build and maintain the pile in contact with the wall of the greenhouse so that heat passes through the wall, yielding 4°F of extra cold protection for the plants inside
▸ Cheaply combine hydronic heating systems with compost piles
▸ Generate more heat than hot water heaters with less expense
▸ Sift and use finished compost

Another interesting system developed at the New Alchemy Institute on Cape Cod in the 1980s used compost piles located inside insulated chambers within a greenhouse. The compost piles were accessed through removable panels on the outside of the greenhouse. During the coldest time of the year the compost chambers were manually loaded with new compost material every four to five days to provide a continuous heat supply. Air blown through the compost piles and into a grow bed filtered toxic ammonia and introduced beneficial carbon dioxide into the greenhouse to promote plant growth. The compost heating system kept the greenhouse at an amazing 23° to 35°F warmer than outside nighttime minimums.[2]

Currently, a company called Agrilab Technologies sells commercial compost heating systems. These systems place compost piles on top of insulated slabs. A fan draws heated vapor from the compost pile through vents embedded in the concrete slab and into heat exchangers that transfer heat into water. You can use the heated water anywhere domestic or commercial hot water is needed or use it in radiant floor heating systems for buildings. Since this system does not involve any pipes inside the compost pile, as Jean Pain's system did, it is easy to use a tractor to build and turn the pile. A system built to compost cow manure produces up to 120,000 Btus per hour for radiant floor heating.[3]

While all of these systems are great heat generators and extractors, they lack the two basic characteristics required by busy farmers: cheap and easy. Jean Pain admitted that he discontinued his system because of the time involved: His large piles had to be carefully built and dismantled by hand. The New Alchemy system required moving two pickup truck loads of material by hand every week. The Agrilab system, with its expensive heat exchangers and slabs, is cost prohibitive. I needed a system I could build, turn, and aerate with the front-end loader on our small tractor, a system with cheap and easily accessible materials that would give me a substantial payback for my investment in time and money.

I developed several different systems to meet these needs; each system produces between 4,000

Figure 8.1. Wood chips are an essential part of the large compost piles we build, and we use a manure spreader to mix ingredients. We construct the piles along the wall of a greenhouse so we can take advantage of compost heat to warm the greenhouse interior.

and 15,000 Btus per hour. I calculated the payback based on a price of $2.39 per gallon of propane, and we get about $25 in heat for every hour spent building the system. Once the rich pile of nutrients and soil-building products the system produces are factored in, the value per hour is much greater. Materials in the cheaper systems cost less than $300 — paying for themselves in a few months.

We use five different techniques to extract heat from compost piles:

1. Direct transfer of compost heat through the wall of the greenhouse.
2. Pipes embedded inside a concrete slab under the compost pile.
3. Pipes embedded inside the compost pile. The front-end loader on the tractor installs and extracts the pipes.
4. Pipes embedded inside the compost using a rotary plow to install the pipes.
5. Biofiltration and swirl separators embedded inside compost piles. The filtration systems connect to pond pumps to heat ponds inside greenhouses.

The Science of Composting

How does a compost pile transform waste into rich soil and free heat? Composting utilizes microorganisms — bacteria, fungi, and actinomycetes — to decompose organic wastes. When these organisms combine into a large mass, the compost pile acts like a single organism, breathing, feeding, and drinking much like any other animal. By understanding what's happening on a micro level, the compost steward is prepared to revive a sick or dying pile just like a doctor understands the inner workings of the human body to prevent and treat illness.

The first step in caring for and taking full advantage of a compost pile requires an understanding of the needs and products of the compost organisms.

NEEDS	PRODUCTS
Carbon-rich organic wastes	Heat
Nitrogen-rich organic wastes	Carbon dioxide
Oxygen	Plant nutrients
Water	Humus and organic compounds

Of all the microorganisms in a compost pile, aerobic bacteria are the most important for hot composting; a single gram of soil may contain over a million bacteria. The bacteria use the carbon in organic wastes for energy, and the nitrogen in organic wastes to build proteins in their cells for growth and reproduction. As the bacteria consume the carbon, heat is released.

Contrary to popular belief, you do not need to add bacteria to compost piles. If you supply all the correct ingredients to the pile, the bacteria naturally flourish, and composting innately occurs. In fact, with all dead organic material, composting and decomposition happen regardless of what we do; we would have to fight hard to prevent it. The Egyptians dehydrated and pumped chemicals through the bodies of their dead kings to fight off the power of decomposition so the dead could live forever.

The two main ingredients for a compost pile are carbon and nitrogen. While decomposition and composting happen effortlessly, having the correct ratio of these ingredients enables a compost pile to heat quickly. Building a compost pile is a lot like making a pot of chili: There are a thousand different recipes, but the main ingredients are meat, beans, and spices. The cook usually looks through the fridge first, taking inventory of what's on hand, then goes to the store as a last resort to buy the remaining ingredients. Likewise, a farmer searches through the farm and surrounding community to pull the ingredients of the compost pile together based on what's readily available.

Woody brown material such as wood chips, leaves, and straw contain a large amount of carbon. Green material such as food scraps, grass clippings, manure, and urine contain a large amount of nitrogen. The ratio of carbon to nitrogen defines every type of compost ingredient (see table 8.1). A pile with an initial carbon-to-nitrogen ratio between 30:1 and 70:1 favors bacterial decomposition and heat generation. High-carbon material mixed with high-nitrogen material creates the perfect ratio for bacterial decomposition. To determine the final ratio of a mix, add up the carbon and nitrogen parts, then divide by the total number of parts. For example, if one part wood chips with a carbon-to-nitrogen ratio of 100:1 were added to one part vegetable scraps with a ratio of 15:1, the equation would look like this:

$$(100 + 15) \div 2 = 57.5$$

Therefore, the final carbon-to-nitrogen ratio would be 57.5:1, which is favorable for bacterial decomposition and heat generation.

The amount of heat released by a compost pile depends on many factors but usually follows a general pattern that starts out with little heat, then peaks at a high temperature, and slowly declines before reaching ambient temperatures again. If

Table 8.1. Carbon-to-Nitrogen Ratios of Common Compost Ingredients

Materials High in Carbon	C:N
Autumn Leaves	30–80:1
Straw	40–100:1
Bark	100–130:1
Wood Chips or Sawdust	100–500:1
Mixed Paper	150–200:1
Newspaper or Corrugated Cardboard	560:1

Materials High in Nitrogen	C:N
Manure	5–25:1
Vegetable Scraps	15–20:1
Grass Clippings	15–25:1
Coffee Grounds	20:1

Source: N. Dickson, R. Thomas, and R. Kozlowski. *Composting to Reduce the Waste Stream.* Ithaca, NY: Northeast Regional Agricultural Engineering Service, 1991.

you're planning on implementing a compost heating system, it's important to keep in mind that heat will not be available for several days after the pile is constructed, and the amount of heat will vary depending on the biological activity in the compost pile.

Gathering Compost Ingredients

The first step in building a hot compost pile is locating a large source of carbon and nitrogen to make the pile. Why is large important? Because large compost piles generate more heat. Cool composting occurs in small piles slowly built, or piles built with improper materials. As an Extension agent I heard constant complaints from homeowners about compost piles that wouldn't heat. I had to sympathize; I've never seen a small compost pile

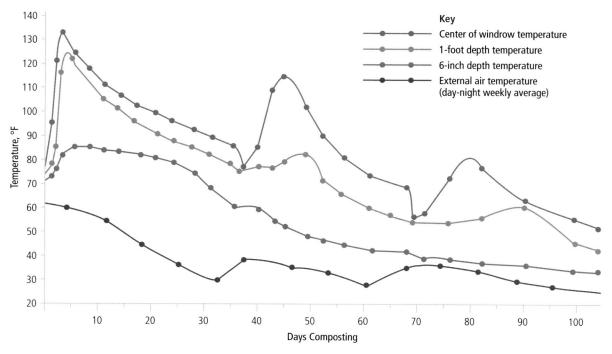

Figure 8.2. Compost heat in relation to location in compost and time. Before each temperature spike the windrow-style compost pile was turned.[4]

made from kitchen scraps, leaves, or lawn clippings heat up, either. The trick to hot composting is building a pile large enough to generate the needed heat. According to *The Compost Manufacturers Manual,*[5] the ideal size for proper fermentation or heat generation in composting is a pile 10 feet wide at the base, 5 feet wide at the top, and at least 9 feet long. While it's possible to build a pile this size by hand, I'm far too lazy to attempt the feat.

The ideal shape for a tractor-made compost pile is a windrow — a long, narrow, snakelike mound. The front-end loader of a tractor conveniently accesses the middle and sides of a windrow pile. A windrow-style pile also fits nicely against the side of a greenhouse if you want to create a greenhouse connection. In windrow-style compost piles the internal temperature increases as you move toward the center, with maximum temperatures deeper than 1 foot.

How big should the pile be? For the system we've built at Clemson University, I strive for 35 cubic yards, or an area at least 4 feet tall, 6 feet wide, and 40 feet long. Ideally, use 45 cubic yards to build the pile at least 8 feet wide. A full-size pickup truck with the bed mounded high with material holds about 3 cubic yards. While 35 cubic yards may seem like a lot of material, it's easily accessible with some investigative perseverance.

Carbon Waste Sources

Buying a heater brings instant gratification. You go to the store, bring home a heater, hook it up, and voilà! Heating with compost, on the other hand, is like planting a garden: Preparation, planning, and patience lead to a bountiful harvest. Composting involves finding waste resources, building the pile, and tending to the needs of the organisms generating the heat. Like a garden, the success of the compost heater depends on the skills of the gardener.

Wood chips are a consistent and free carbon waste source. Usually frowned upon by compost aficionados because of their slow decomposition rate, wood chips provide the perfect option when you're looking for longer-lasting heat. One man's waste is another man's treasure. Tree service contractors, utility crews, and municipal waste services are constantly producing wood chips. They usually have to pay to dispose of the chips, so they'll happily give them away for free. The typical dump truck can carry about 7 cubic yards of wood chips. After only a year of working to develop relationships with tree workers, I found myself often turning down truckloads of chips because of lack of storage space.

Getting enough wood chips delivered all at once may prove difficult. Plan well ahead of time, and stockpile the wood chips close to the compost pile location. Most of the time required to build the compost pile is spent moving material around, so the shorter the distance material has to be moved, the better. If the wood chips age a bit as you stockpile enough material, it only improves the final product.

If wood chips aren't available, find animal manures with bedding from confined animal operations where manure accumulates. Horse stables and poultry farms are good places to start, and they may see it as a favor and give it to you for free. The bedding mixed with manure and urine usually makes the perfect carbon-to-nitrogen ratio to generate heat.

For those using manure-based composts on vegetable farms, the United States Department of Agriculture offers guidelines for creating safe compost. Safe composting of manure in windrows requires maintaining the pile between 131° and 170°F for at least fifteen days and turning at least five times during that period. The heat reduces the harmful bacteria in the manure, which makes the compost safe to use around vegetables up to the day of harvest. Alternatively, incorporating the manure into the soil 120 days before vegetables are harvested allows enough time for the soil to reduce harmful bacteria in the manure. It's easier to not follow the turning and heating guidelines and treat the manure-based compost like fresh manure by tilling it into the ground well before harvest and letting the soil sterilize any noncomposted manure.

Nitrogen Waste Sources

Nitrogen is the second major ingredient of hot compost piles. Finding sources of nitrogen could be easy and cheap or expensive and difficult, depending on what's available on the farm. Nitrogen is a component of the amino acids and proteins that make up the cells of the bacteria that consume the carbon in the compost. The more

nitrogen available, the more bacterial bodies to consume the waste. Since bacteria generate heat as they consume carbonaceous materials, more heat is generated when the right combination of carbon and nitrogen are available to maximize bacterial growth. With manure-based compost piles, urine and manure may provide adequate nitrogen, but you may need to add more carbon to balance the pile. Wood-chip-based compost piles usually require additional nitrogen to reach high temperatures quickly.

At the Clemson Student Organic Farm (SOF) we use food waste for our nitrogen source and mix the food waste with wood chips at an approximate rate of two parts wood chips to one part food waste. The mixture creates good pore space for aeration and drainage (which are important for maintaining a supply of oxygen in the pile), leading to high

Figure 8.3. David Thornton delivering food waste collected in bins ready to load directly into the manure spreader for mixing with wood chips.

temperatures. A steady stream of food waste is collected in the school cafeterias in 35-gallon rolling trash bins. The bins hold about 125 pounds of food waste that's shredded, pulped, or whole. The lids on the bins lock, preventing animal intrusions and spills. Collected daily, the bins are rolled onto a truck with a lift gate and transported to the farm for immediate use. We divert about one month's worth of waste generated from two of the cafeterias on campus to build three 40-cubic-yard piles.

A CSA operation could implement a composting program by asking its customers to bring food wastes to the farm when they come to pick up their produce share. However, composting off-farm food waste may require a permit from county and state regulators.

Mixing Materials

When I first started building compost heating systems I simply placed the wood chips and food waste in layers, and once the pile was large enough, I left it alone. Several weeks later the pile would finally start generating heat. During the lapse of time, the pile would develop odors from anaerobic conditions; it would reach a maximum temperature close to 135°F, but it never stayed hot for long.

Since then I've learned the importance of mixing the material. Mixing aerates the food waste with the wood chips, and both essential food groups, nitrogen and carbon, have adequate contact. If you mix rather than layer, compost piles reach temperatures of 150° to 160°F within a week and usually maintain high temperatures of over 140°F for five weeks.

There are two easy mixing techniques. At Clemson University's composting facility, staff use a concrete slab as a mixing pad. Using a front-end loader, they spread the wood chips and food waste onto a concrete slab and mix the material on the slab. Then they scoop it up and place it on the compost

Figure 8.4. A mixing pad at the Clemson University compost facility used to mix wood chips and food waste for composting. The white pipe is connected to a blower for static aeration of the compost pile.

pile or push it into an adjacent compost pile. Mixing pads simplify the unloading process since the truck can dump the food waste right onto the pad.

At the farm we use a power-take-off-driven manure spreader to mix the material. A manure spreader is like a trailer with chains and blades at the base. The blades pull the manure toward the back of the trailer, where beater bars mix, aerate, and fling the material out the back. We layer wood chips inside the manure spreader with food waste and park the trailer with the back facing the spot where we want to build the compost pile. The power take-off drive on the tractor then spits and mixes the material into a pile. This excellent mixing process leads to high temperatures and little odor.

It takes about three minutes per 35-gallon bin full of waste to process and mix material for the compost pile. Overall, it takes six hours to build a 40-cubic-yard compost pile containing food waste from approximately 120 bins, along with about four dump truck loads of wood chips.

The manure spreader also allows us to build compost piles efficiently at a distance from the

Figure 8.5. Food waste from rolling bins is easily dumped into the front-end loader by hooking the handle of the bin on the top of the bucket, then lifting the bottom of the bin up using the handle on the bin as a fulcrum point. Once in the bucket, the food waste is easily moved into the manure spreader or spread on a pile.

ingredients. For example, we have a small space the perfect width of a compost pile sandwiched between two greenhouses. The area sits too far from our compost ingredients to bring compost to the location scoop by scoop with the front-end loader. However, the manure spreader holds eight buckets worth of compost material. We add the food waste and wood chips to the manure spreader and drive it to the destination to make a remote pile quickly.

Finally, the manure spreader makes applying the finished compost a breeze. Once the compost is ready, we can quickly load it back into the manure spreader and spread it into the fields, making room for more heat-generating compost piles.

Most farms and homesteads don't have access to an unlimited supply of freely delivered food waste and must find other avenues of adding nitrogen to wood chips to maximize heat production. The easiest way to add nitrogen is by sprinkling a water-soluble form on top of the pile of wood chips and watering it in. For example, a high-nitrogen material such as

Figure 8.6. A manure spreader does a great job of mixing food waste and wood chips for composting. The well-mixed material reaches high temperatures and generates few odors.

Figure 8.7. The manure spreader also delivers compost ingredients to remote locations and tight spaces, making compost piles possible in otherwise inaccessible areas. I set up plywood strips to block some of the flinging action of the spreader to prevent material from hitting the greenhouse.

blood meal, which contains 13 percent nitrogen, creates a hot compost pile when added to wood chips. Fifty pounds of blood meal would treat about 14 cubic yards of wood chips, quickly bringing temperatures to high levels. The blood meal-and-wood chip–based compost is safe to use fresh without the hassle of following the compost guidelines required of manure-based composts. Another source of nitrogen is old feed or bags of damaged fertilizer. I once used expired fish feed and some cottonseed meal that had gotten wet to boost nitrogen levels. I was so glad I hadn't thrown it away.

Another unconventional but natural option is the use of human urine. Human urine averages 17 percent nitrogen, is completely sterile, and may be chemical-free, depending on your intake of food and medication. When compared to the price of blood meal, the average person produces four dollars' worth of nitrogen a day in the form of urine. If you save your urine for a few weeks, you could capture enough nitrogen to treat the same 14 cubic yards of wood chips that a 50-pound bag of blood meal would treat. I don't use urine at the Clemson University farm because of our organic status, but I do use it at home.

Wood chips alone with a water-soluble form of nitrogen may not contain the right mix of small and large particles for good heat generation. Wood chip material varies drastically based on how many leaves and small branches are in the mix, the type of chipper, and the age of the teeth in the chipper. Jean Pain used a specialized chipper to shred chips into finer pieces, giving him a good consistency of small to large pieces. Mixing sawdust or another type of fine material may help the pile. Experiment with available material to see what works best — does it heat and for how long? Keep records of your mixing rates and techniques as you go, and your compost will eventually yield satisfying results.

Compost Pile Dynamics

Everything requires maintenance — your house, your car, and, yes, even your compost pile. It's not enough to just build a compost pile; it's important to give it a little TLC. Fortunately, the list is not as long as the to-do list for your house, but you'll need to keep the following matters in mind: oxygen and water.

Oxygen and Water

All living things need water and oxygen. Usually, those needs are obvious. If my dog lacks oxygen for a few minutes, he will die. When I forget to fill his water bowl, he places his head in the bowl and looks at me with sad eyes.

The need for oxygen and water is more intuitive in a compost pile, like the watering needs of a garden. When it's hot, sunny, and windy, the astute gardener knows his plants will need regular water. The gardener probes the soil and feels it to check the moisture content. When a compost pile is hot, the pile sucks moisture from the ground beneath it.[6] A lot of water evaporates from the pile, too, and the pile requires regular irrigation. You can determine whether a compost pile has the proper amount of moisture by digging into the pile and feeling the material. It should feel moist, like a wrung-out sponge, but not dripping wet.

In terms of oxygen, too much water can be a problem. If garden soil stays saturated with water, the gardener builds raised beds to bring more air into the soil. Likewise, if your compost pile is too wet, you'll need to add loose material such as wood chips or straw or turn the pile to correct the problem. A foul odor indicates excessive moisture in the pile. The optimum moisture content for microbial growth is between 50 and 60 percent.[7] If compost-building material is dry, water the pile as you build it to ensure it reaches the proper moisture content. Also water the pile if the rains stop or increased temperatures evaporate the water.

AERATION

Natural ventilation aerates windrow compost piles. The heat of the compost rises through the center, pulling oxygen into the pile from the edges. Natural ventilation is similar to the natural draft that occurs in fireplaces. As a fire initially starts in a fireplace, smoke may enter the room. As the fire heats the chimney, the heat inside the chimney pulls air and smoke up through the chimney and out of the house in a natural draft. Similarly, as a compost pile heats, the rising hot vapors pull oxygen into the center of the pile.

When a compost pile is built adjacent to a greenhouse, the greenhouse wall impedes natural ventilation of the compost. To remedy the problem, place a 4-inch-diameter perforated pipe along the bottom edge of the greenhouse wall before you build the pile, similar to the CO_2 extraction pipe shown in figure 8.22. The pipe will provide for airflow into the compost pile on the greenhouse side. Since the pipe lies at the deepest part of the compost pile, it's easy to avoid damaging it with the tractor when you work with the compost.

Because we build our piles on a slightly sloped site, any water that enters the pipe flows downhill, providing drainage for roof runoff that migrates through the compost pile.

Another way to introduce oxygen is by turning the pile, as described below. While turning the pile aerates the mix, the main benefit comes from loosening the compost material and improving natural ventilation.[8]

CONTROLLING MOISTURE

We cover our compost pile with a removable roof to control moisture. By lengthening the roll-up side on the greenhouse to extend over the compost pile, the roll-up side becomes a roll-up roof. A separate piece of greenhouse plastic is then added between the top of the roll-up side and the base of the greenhouse to prevent compost and compost gases from entering. The roof provides shade and rain protection for the compost and the soldier fly digesters inside the compost pile. We heap up compost materials until the pile is just a bit shorter than the point at which the compost roof connects to the greenhouse. This leaves a small open space between the compost roof and the top of the compost pile, extending the entire length of the pile. Hot gases exit through this space at the upper end of the compost pile, indicating that the slope of the greenhouse assists convective removal of compost gases.

When we first unroll the roof over the pile, the pile temperature increases. However, because the roof impedes the flow of oxygen through the pile, the temperature quickly falls after the initial boost. However, the roof offers other benefits to compensate. During extended rainfall, covering the pile prevents oversaturation. And when the pile is covered, decomposition occurs right up to the edges, reducing the need to turn the pile. And if odors become an issue, we just roll the roof over the compost pile during farm events to keep the smell at bay. Then we roll it back up after the event to promote natural ventilation (Figure 8.13 shows the rolled-up roof). I'd like to try using woven polypropylene landscape fabric as a roofing material. It may provide the right balance of oxygen exchange, heat retention, and complete decomposition needed for a compost pile.

SPECIAL MOISTURE CONSIDERATIONS

When you use leaves and grass clippings as compost ingredients, layers of moisture-resistant mats can form inside the compost pile, preventing water from penetrating. The leaves lie on top of each other like shingles on a roof, flattened by the weight of the compost. To solve this problem, simply mix the leaves with wood chips or other bulky ingredients to break up the mats, or don't use leaves at all.

Another consideration with moisture and air is the compost's ability to conduct and store heat. A moist compost pile efficiently conducts or moves heat, which is important in a system where pipes inside the pile are extracting heat. A compost pile with 50 to 60 percent moisture conducts heat about 30 percent better than a compost pile with a moisture content of 20 percent. However, research indicates that compost piles conduct heat poorly over an entire range of moisture contents and tend to be self-insulating.[9] Also, as the compost ages, material decomposes and shrinks, which reduces the air between compost particles. Since air acts as an insulator, an older compost pile with less air conducts heat better than a young, fresh pile. Compost thirty-five days old conducts heat twice as well as compost seven days old.[10]

While a moist pile may conduct heat poorly, it does have the advantage of storing significantly more heat or having a higher heat capacity than a dry pile. A compost pile with 50 to 60 percent moisture will store about 50 percent more heat than a compost pile with 20 percent moisture.

I would like to research the effects of placing drip irrigation tubing on top of compost piles situated above heat extraction pipes and determine a possible irrigation schedule to maximize heat extraction without oversaturating the compost pile and causing heat collapse.

Turning the Pile

After four to five weeks the heat of the compost pile may diminish. Stimulate more heat activity by turning the pile to mix the material and improve natural ventilation. The turning process

Figure 8.8. A 40-cubic-yard compost pile adjacent to a greenhouse is turned using the loader on a tractor.

also allows the material on the edge of the pile to completely decompose, leading to a better finished product. During turning you may add more food waste as fuel to digest the high-carbon woodchips. Be prepared for temperatures to briefly decrease after turning the pile, but the temperatures will rise again. If my compost needs more moisture, I add it while I'm turning the piles.

Adding water-soluble nitrogen to the pile also stimulates more heat if turning isn't possible. You should harvest at least two months of heat from a pile of slowly decomposing wood chips. Adding nitrogen and turning the pile achieves higher temperatures but may ultimately shorten the overall heating time frame.

Connecting to Greenhouses

We located our compost pile against the side of our high-tunnel greenhouse. The greenhouse serves as a support for one side of the pile, and the compost pile generates heat that transfers through the greenhouse wall to the interior. Consequently, this greenhouse offers 4°F more protection against cold outside air than a similar greenhouse without the compost connection. The compost pile not only heats the greenhouse but also prevents heat from escaping through the thin wall of the greenhouse, even after the compost heat itself has dissipated. The north wall of a greenhouse is a logical place for a compost pile because the compost can provide heat and insulation but the pile doesn't block sunlight during winter.

It might seem ideal to locate a compost pile inside a greenhouse to capture all the generated heat. However, compost piles off-gas ammonia, carbon dioxide, and other toxic gases. If a small greenhouse contains a large compost pile, the amount of gases emitted from the pile would reach toxic levels for plants and humans. By locating the pile outside the greenhouse we capture less heat, but we easily maintain the pile with a tractor, and

the greenhouse wall separates the greenhouse environment from any gases released from the compost pile.

I've tried several techniques to connect compost piles to greenhouses, and what works best is reinforcing the greenhouse wall with plywood or plastic 55-gallon drums filled with water. I first tried doubling the plastic wall covering, which worked fairly well, but we found that the tractor bucket damaged the greenhouse bows and stretched the plastic when we used it to remove compost from the pile.

Thus, to create additional support we added plywood to the side of the greenhouse, and then we had better results. We covered the outside face of the plywood, too, to protect it from moisture in the compost pile. With the plywood there, we can lightly push against the greenhouse with the

Figure 8.9. Sidewall of greenhouse from inside with compost pile pressed up against it and a crop of beets ready to harvest. Plywood was left off some sections to make room for soldier fly digesters (see chapter 9).

Figure 8.10. This concrete slab slopes away from the greenhouse to drain compost leachate, which flows into a retention basin. The white bins are soldier fly digesters (see chapter 9). An outer layer of plastic in the greenhouse wall protects plywood from moisture when compost materials are piled against the greenhouse wall.

tractor to extract the compost. A cinder block wall strong enough to handle the full force of the tractor would work even better.

After the initial experiments, we poured a concrete slab foundation along the greenhouse wall to make turning and extracting the compost easier. I carefully graded the area to slope away from the greenhouse so waste oozing out of the pile wouldn't contaminate growing crops. I also created a diversion channel that would direct runoff from the compost area into a retention basin to capture the water and nutrients for repurposing in the system.

We placed the slope of the foundation at 1.5 percent to the southwest to match the slope of the greenhouses. The improved drainage may help

Figure 8.11. This small retention basin captures runoff from the compost pile. The nutrient-rich water can be pumped back onto the pile as needed to fertilize and irrigate the compost materials.

prevent the formation of anaerobic zones that tend to develop in the base of windrow compost piles.

Connecting to Hydronic Systems

If you have an existing hydronic heating system or plan to build one, a compost pile connected to the system provides free heat. As explained in chapter 4, hydronic systems heat water, then distribute the heated water to multiple locations using a closed loop pipe system. The piping can be run through the concrete slab underneath the pile or through the pile itself. Locate the compost pile adjacent to the return line bringing water back to a commericial hot water heater (if used). The closer the compost pile is to the return line, the less heat will be lost in transit and less pipe and insulation will be needed to transfer the heat. However, consider access for equipment and storage of composting materials in the final placement of the compost pile.

Extracting Heat through the Slab

Compost temperatures are usually about 20°F cooler at the slab-compost interface than at the center of the compost pile. However, there's still enough heat in the slab to allow it to serve as a heat exchanger. Running pipes through the slab to collect heat allows me to turn the compost pile easily without having to deal with pipes inside the compost pile. Although more expensive than all the other techniques I use, the concrete slab is very low maintenance once in place.

Here's how we set up this type of system at the SOF: After creating the sloping site for the slab (see chapter 4 for directions on creating a sloping site), we dug a trench as a footer around the slab perimeter and positioned forms for the concrete along the edges. The upper lip of the forms corresponded to the slope needed for the slab. Next, we put in place a layer of extruded polystyrene insulation, which served to prevent heat from escaping from the slab into the soil underneath. We then placed gravel and concrete-reinforcing wire in the concrete slab area. Using cable ties, we secured ¾-inch PEX pipe to the concrete-reinforcing wire. About 700 feet of piping, with pipes spaced approximately 6 inches apart, fit in the 10 × 40-foot slab area. The PEX pipe is attached to the return line of the hydronic heating system, allowing water to pass from pipes inside the greenhouse through the piping in the slab before heading back to the water heater. Thus, heat extracted from the slab is helping to rewarm the water, so that the water heater doesn't have to do as much work. You may want to refer to the plan view of the greenhouses and hydronic heating system in the appendix for a visual overview.

David Thornton, organic and biofuels project coordinator for Clemson University, designed the PEX system for the slab and divided it into five separate heat extraction zones, as shown in figure 8.12. Each zone has an input and output valve connecting to the hydronic return line. With this arrangement we can direct water into just one zone or into multiple zones, so water passes only through the portion of the slab covered by compost. This allows us to construct the pile in stages rather than having to keep the whole slab covered whenever we want to extract heat from it.

With all the zones running and a 40-cubic-yard compost pile sitting on top of the slab producing a temperature of 156°F, I calculated that the system generates 15,600 Btus per hour with water running through the pipes at 2.4 gallons per minute. However, with maximum heat extraction the compost pile temperature starts to drop after several days, reducing the ability to extract maximum heat. A slower rate of water flow through the pipes or pulsing the system to allow the

Figure 8.12. Valves connected to a hydronic return line feed pipes embedded inside a concrete slab. Each valve serves a separate zone inside the slab, with a total of five zones available. An identical valve box with five valves is connected to the output from each zone.

compost pile to recover helps prevent drops in compost temperature.

Direct Heat Extraction

In Jean Pain's system he added about 900 feet of piping to a 40-cubic-yard compost pile, extracting around 10,000 Btus per hour of heat. Water is circulated at a rate between 1 and 2 gallons per minute.[11] Jean Pain's water pipe system is effective and inexpensive. But a tractor trying to turn a compost pile full of piping would rip the pipes to pieces. The amount of piping inside the pile creates an obstacle every time you want to turn the pile. In addition, building and deconstructing the pile becomes an enormous chore because of the tangle of extensive pipe. Nine hundred feet of piping is also difficult to compile into a bundle for storage when the compost pile is not in use. To overcome these issues I've created a piping design I can remove easily from a pile and reinstall in an existing pile.

When Pain built his compost piles, the piping available for use at the time was polyethylene. Polyethylene pipe can withstand the high temperatures of compost; it's flexible and bends easily but lacks strength and inconveniently kinks. Cross-linked polyethylene pipe, better known as PEX, revolutionized plumbing. PEX pipe is flexible yet tough, and unlike regular polyethylene pipe, it remains rigid as temperatures increase, resisting kinking when bent. The flexible yet rigid nature of PEX pipe permits us to use the tractor to pull the tubing out of the compost pile without damaging the tubing. We can then turn the compost to aerate and boost temperatures and add new compost material if necessary. By designing the compost piles for accessible pipe extraction, we overcame a major hurdle in compost heating: the ability to work the compost piles with tractors.

The center of the compost pile generates the hottest temperatures, so we place heat extraction pipes in the center to extract the most heat with the least amount of piping.

The slab heat extraction system on the bottom of the compost pile generated 22 Btus per foot of pipe. The Jean Pain system generated around 11 Btus per foot of pipe. I have experimented with running a single loop of pipe through the pile and a double loop of pipe. A single 1-inch PEX pipe loop running down the middle of the compost pile generates 63 Btus per foot of pipe. Thus, at a length of 66 feet running through the center of a 40-cubic-yard compost pile, the single loop generated 4,193 Btus. When I split the pipe into two loops down the center, 143 feet of piping generated 8,280 Btus.

With a flow rate of 2.4 gallons per minute, the single loop didn't cause compost temperatures to drop. However, a double loop through the middle did cause a compost temperature drop when we extracted maximum heat. As we added more piping, we should have reduced water flow to prevent compost temperatures from decreasing. I'd like to conduct future research to determine the ideal

Figure 8.13. Components of the compost heating system: Insulated pipe in center extracts heat from the compost for use in the hydronic heating system. Brown pipe irrigates the pile with emitters every 12 inches. Green boxes protect valves connecting the hydronic heating system with the pipes in the middle of the pile and inside the concrete slab.

pipe length and flow rate for maximizing heat extraction using the least amount of pipe material.

I can extract and replace one or two loops of piping through the center of the compost pile using a simple technique I developed with our small tractor. Once we've extracted the piping, we turn the pile to generate more heat or remove the finished compost and build a new pile, using the tractor to do all the work. Since the compost pile shrinks with age, I've also removed the pipe, then added a new layer of mixed material on top with pipes inserted into the new material.

INSTALLING PIPES IN A PILE

Before building a compost pile for direct hydronic heating, make sure you've completed the following steps:

Step 1. Stockpile enough carbon-based material nearby so you can complete the compost pile in a short time frame.

Step 2. Locate enough nitrogen-based material to heat the carbon-based material sufficiently.

Step 3. Grade the area to shed water away from the greenhouse and fields to prevent leachate from the compost pile from contaminating growing areas.

Step 4. Purchase enough 1-inch PEX pipe to run the entire length of the compost pile at least twice, with two elbows to make the 180° turn at the end of the compost pile.

First, we build the whole compost windrow needed for heat extraction. Then we dig a trench in the top of the windrow with the bucket of the

Plowing for Pipes

In an effort to generate more heat and compost, I started building compost piles in places that are inaccessible with the tractor. I soon found that digging the trench in the middle of these compost piles by hand to install pipes was tedious, so I started looking for an alternative. I discovered that a rotary plow attached to the walk-behind tractor made the job easy. A rotary plow is a rotating corkscrew dragged through the soil by the tractor. The plow tosses soil to one side as it moves forward. Usually, the tool is used to till soil deeply, inverting the layers and making raised beds with the side cast action. However, the tool also works well in digging a deep trench in the middle of the compost pile. First, I build a shallow slope on one end of the compost pile that acts as a ramp so the walk-behind tractor can climb to the top of the pile. Next, I make two passes, one down and one back, throwing soil in the opposite direction away from the middle to dig the trench. Once the pipes are in place, I cover them using a manure hook to pull the compost material over the pipes. A manure hook is like a hoe except it has tines that penetrate easily into the pile of compost.

Extracting the pipe from the pile presents more of a challenge. I use a longer chain or rope since I can't pull up the tractor next to the pile. Alternatively, two to three people can pull the pipes out of the pile similar to playing a game of tug-of-war with the compost pile. Once I remove the pipes I can place a layer of new material in a strip a few feet wide down the center of the pile. Then I run the plow through the new material to incorporate it.

Because of the danger of running a walk-behind tractor on top of a steeply sloped pile, I don't recommend this technique for the novice operator. However, with the right skill and a large enough compost pile, the technique works well.

Figure 8.14. Making a trench in a compost pile using a rotary plow.

Figure 8.15. Ropes tied to the loop in the buried PEX pipes make it easy to find and extract the pipes.

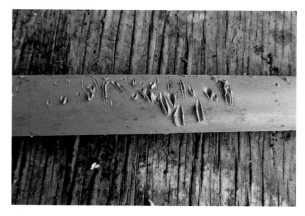

Figure 8.16. Be careful when inserting a compost thermometer into a compost pile with PEX pipe buried in it. These gouges were made by the pointed tip of a thermometer.

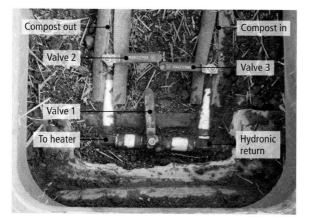

Figure 8.17. Valve 1 is inserted into the return line of the hydronic heating system. Next, a tee and valves 2 and 3 are placed before and after valve 1. By turning valve 1 off and valve 2 and 3 on, water will divert into the compost pile pipe loop.

Figure 8.18. Pipe is split into two loops and installed in the center of a compost pile.

tractor by placing the bucket in the top of the windrow and moving it back and forth. We lay PEX piping in the trench in a single or double loop. I try to maintain 6 to 12 inches of space between the pipes to facilitate maximum contact with the hot compost material. Ideally, the piping lies in the center of the compost pile, where it will be surrounded by 18 inches of compost once the trench is filled in. When the pipe is new it's rather inflexible and coiled tightly. Placing cinder blocks on top of the piping helps hold it in place until compost fills the trench.

At the other end of the compost pile farthest from the connection with the hydronic heating return line, I place two elbows in the PEX pipe, 12 inches apart. The elbows enable the pipe to turn sharply at the end of the loop, and their presence also aids in extracting the pipe. I also tie a polypropylene or polyethylene rope through the loops and leave the rope hanging out of the pile. The rope helps me find the pipes when I'm ready to extract.

I use the front-end loader to push compost material over the pipes. Be sure to remove all the cinder blocks as the trench is filled.

I use three valves to tie the compost heat extraction pipes into the return line on the hydronic heating system, as shown in figure 8.17. The idea is to stop the return flow and divert it through the pipes that are buried in the compost pile. With the three-valve arrangement, the compost pile is easily shut off when not in use and turned on when needed. (See chapter 4 for more information regarding hydronic heat plumbing.)

REMOVING PIPES FROM A PILE

I've had difficulty removing the pipes if more than 1 foot of compost covers them. To prevent any problems I first scrape off the top layer of the compost pile with the front-end loader. Next, I turn off the valves that allow water to circulate through the compost pile, in case a leak develops during pipe extraction (I haven't had a leak yet, but this is a good precaution). Then I attach the hooks on a tow strap to the PEX elbow fittings at the loop end. I attach the tow strap to the tractor, then gently lift the tractor bucket. The lifting action causes the tow strap to tighten. Next, I back the tractor up and pull the pipes out of the compost pile.

More about Pipes

It's important to know the temperature of the water entering and exiting the pipes embedded in the concrete slab or directly in the compost pile. When the entering water's temperature is greater than the exiting water's temperature, it's time to stop diverting water through the compost loop and manipulate the pile to improve temperatures or start over again. To calculate temperatures, insert a thermometer with a male threaded fitting into a female threaded tee in the water line. A probe inside the pipe records the temperature by lying in the pathway of the water.

I use data loggers with temperature sensors to keep track of temperatures over a longer period of time. The sensor is placed inside a thermowell, which is a hollow metal finger that can be inserted into a tee in the pipe, as shown in figure 8.20. I fill

Figure 8.19. *Left*, a tow strap is attached to PEX elbows at the end of a pipe loop through a compost pile; the other end of the strap hooks onto the bucket of a tractor. *Right*, the tractor then pulls the pipes out of the pile.

Figure 8.20. The stainless steel metal fitting called a thermowell is like a hollow finger that is inserted into the pipe through a tee fitting. The small white box is a data logger for recording temperature.

the thermowell with glycerin and place the sensor inside the thermowell. I then seal the top of the thermowell with a glob of silicone caulk.

To retain heat and prevent insect problems, insulate all hot water distribution pipes, as mentioned in chapter 4.

Take care when adding PEX pipe to a system that includes iron pipe. PEX pipe is permeable to a small amount of oxygen, and iron pipe corrodes with oxygen present. If the two types of pipes are used together, corrosion will occur in the iron pipes and pumps, eventually clogging the system and reducing the flow rate of the water. To solve this problem, use PEX pipe coated with an oxygen barrier, and add an anticorrosion chemical to the

hydronic system to prevent corrosion. Finally, adding antifreeze to the system prevents pipes from freezing when not in use but is expensive if a leak occurs. I would only use antifreeze in a zone 6 climate or colder and recommend a propylene glycol–based antifreeze to reduce toxicity. You can always leave the system on to protect pipes from freezing on cold nights.

Bio-Integrated Heat Exchanger

Now that we've discussed how to use compost to heat water for a heating system, let's discuss how to use compost to heat water for storage in a pond. For decades gardeners have diligently fed their compost piles to create rich, fertile soil. By combining these hot compost piles with biofiltration systems for ponds, gardeners can also heat, filter, and store water for later use.

A biofiltration system consists of

- A pump that pushes pond water into the biofiltration system
- A swirl separator, which filters solids out of the water
- An ebb-and-flow "biofiltration barrel" in which the water sits while it is warmed by heat from the compost pile and is filtered by bacteria
- A bell siphon to draw water out of the biofiltration barrel and back into the pond

We make the swirl separator out of a sealed 55-gallon drum, which we position next to the greenhouse in the compost pile area. The biofiltration barrel is also a 55-gallon drum, but it is filled with aggregate. For more details about biofiltration systems like this, refer to "About Aquaponics" in chapter 5. When hot compost surrounds these drums, it "cooks" them, warming the water and aggregate inside. Once the barrel with gravel fills, the bell siphon automatically draws the water out of the barrel and back into the pond as heated, filtered water. An additional pipe inserted into the top of the biofiltration drum acts like a snorkel, supplying air for bacteria growth and proper functioning of the bell siphon. Since the ideal temperature for bacterial biofiltration is 86°F, the compost stimulates better filtration through heat. The constant influx of cool pond water prevents the aggregate from becoming too hot. And the aggregate stores heat between flushing cycles, allowing water to heat to higher temperatures during the next flood cycle.

For the bell siphon to function properly, the barrel should be located close to the pond. In addition, the drainpipe preferably should be a straight line from tank to pond, with a maximum of one or two elbows. If the drainpipe is too long or has too many curves, it prevents water from quickly exiting the tank, and the bell siphon won't function. The drain line back to the pond should be sloped at least 1 percent.

When I tested our system, the swirl separator captured 1,390 Btus per hour, and the ebb-and-flow biofiltration tank captured 1,160 Btus per hour with a compost temperature of 158°F and a flow rate of 2 gallons per minute.

Using these types of barrel heating systems to form the wall between the greenhouse and the

Figure 8.21. The tops of these two barrels are open to expose the heat exchanger components inside. Pond water enters the swirl separator on the right; the white PVC line is to drain settled solids. The barrel on the left is an ebb-and-flow biofiltration system. The small gray wire is one of three temperature sensors. Connected to the pond pump, the heat exchanger filters and heats water for the greenhouse pond.

compost pile provides a rigid structure that supports the compost pile well. The solid wall created by the barrels also makes it easier to extract the compost with the tractor. A series of swirl separators could be added in line with each other, with a drainpipe located on the inside of the greenhouse for easy access to drain solids out of the barrels.

Careful design maximizes the benefits of the heated water. If the heated water first enters a small propagation pond, you can make use of the heat for transplant production. The propagation pond then spills over into the greenhouse pond with every ebb and flow of the drum. The insulated greenhouse pond then stores and releases the heat into the greenhouse. For more information on heated propagation ponds and greenhouse ponds, see chapters 5 and 12.

Extracting Carbon Dioxide from Compost Piles

The importance of carbon dioxide for plant growth didn't occur to me for the first twenty years I spent growing plants, even though I'd read about results indicating 30 percent more plant growth and fruit production in an enhanced CO_2 atmosphere. I had to see it to believe it. Luckily, I had the chance — due to a happy accident.

Since our compost pile was leaning against the greenhouse, inevitably small holes were torn in the greenhouse plastic. The holes only measured a total of a few square feet, but those openings allowed an enormous amount of carbon dioxide to pass from the compost pile into the greenhouse. After working inside the closed greenhouse on a cool winter day, I felt unusually lightheaded. Investigations with a carbon dioxide sensor indicated the concentration inside the greenhouse was 2,400 parts per million. In comparison, atmospheric

carbon dioxide is currently at 398 parts per million. Luckily, the carbon dioxide levels inside the greenhouse were still within OSHA's health and safety standards of 5,000 parts per million total weighted average.

The accidental experiment occurred when I planted the high carbon dioxide greenhouse and an adjacent greenhouse at the same time with the same plants. The growth difference was phenomenal, and I finally saw what 30 percent looks like: plants on steroids. The concentration was actually too high, and I started seeing some toxicity symptoms — brown spots on the leaves — along with the improved growth. Fortunately, plants only need a small increase in the amount of carbon dioxide to see dramatic effects. Levels of carbon dioxide are commonly enriched to 1,000 to 1,500 parts per million in commercial greenhouses to improve plant growth.[12] Plants only use the carbon dioxide during daylight photosynthesis, and use increases with light intensity. Carbon dioxide use also increases with temperature, allowing daytime temperature

Photosynthetic Chemistry

Photosynthesis is a process in which plants use light energy to convert water and carbon dioxide into chemical energy for use by the plant. An examination of the photosynthetic reaction shows the importance of carbon dioxide:

Carbon dioxide + water + light energy
→ carbohydrates + oxygen

to be raised 5° to 9°F above normal stress points for plants.[13] Since plants can tolerate higher temperatures with enriched carbon dioxide environments, greenhouse vents can remain closed, storing more heat in soil and water for nighttime use.

Once I finally realized the importance of carbon dioxide, I began experimenting with ways to capture and use free carbon dioxide from compost. Compost gas contains carbon dioxide—a beneficial component for plants at levels up to 1,500 parts per million during daylight, but detrimental at levels above 2,000 parts per million. Compost gas also contains ammonia and a host of sulfide-containing compounds that may cause toxicity to plants and humans and odor problems within the greenhouse. Compost gases vary depending on the composition of the compost and may change as the compost pile ages.

It's important to filter out toxic compounds and meter the proper amount of carbon dioxide into the greenhouse to benefit plants. Compost gas may contain hydrogen sulfide produced during anaerobic decomposition. This deadly gas smells like rotten eggs. Consider compost gas hazardous until you test the air for hydrogen sulfide, carbon dioxide, carbon monoxide, and methane using a multimeter gas detection device. Additionally, I recommend venting the greenhouse by opening doors and roll-up sides and turning off the compost gas before entering. Post warning signs on the greenhouse doors indicating the potential danger.

Now that you know the risks, here are the advantages: Filtered compost gas may increase greenhouse carbon dioxide levels. Traditionally, fossil fuels such as propane are burned to generate carbon dioxide for greenhouses. The practice is common in the north, where greenhouses remain closed most of the winter. Closed greenhouses prevent carbon dioxide and money from floating out the vents. In the southeastern United States we rarely use carbon dioxide because of our need to vent greenhouses during the winter. However, if a cheap or free source of carbon dioxide is available, the prospect makes sense.

Supplying CO$_2$ to a Single Greenhouse

When you connect a compost pile to a single greenhouse, carbon dioxide inevitably finds its way into the greenhouse. Even if there are no holes in the greenhouse wall, carbon dioxide may reach levels high enough to improve plant growth. If you need more gas, cut a hole in the plastic between the greenhouse and the compost pile. I recommend starting with a hole 1 inch in diameter for every 500 square feet of greenhouse floor area.

I've found a way to filter the compost pile gases as they enter the greenhouse. I place a pocket of finished compost mixed with peat moss, at least 1 foot thick, between the greenhouse and the compost pile. To do this, mix together some compost and peat moss. Then, at the hole in the greenhouse wall, pull the compost away from the wall and insert a plug of the peat-compost mix into the hole. Wire or netting placed over the hole before the plug is added prevents the finished compost from falling into the greenhouse. Once the plug is in place, push the compost back up against the side of the greenhouse. The plug serves as a living and mechanical filter for the compost gas as it flows out of the pile and into the greenhouse.

I use a carbon dioxide meter to gauge the carbon dioxide level inside the greenhouse and adjust as needed. If I want to increase the carbon dioxide level, I place a cover on top of the compost pile to force more gas into the greenhouse. Otherwise, I enlarge the hole or make more holes to let more carbon dioxide enter. If carbon dioxide concentrations are too high, tape plastic over the holes or insert a flue pipe with an adjustable damper through the hole and close the damper.

Supplying CO₂ to Multiple Greenhouses

Since compost piles generate a massive amount of carbon dioxide, a single compost pile can supply carbon dioxide for multiple greenhouses or to multiple locations within a large greenhouse. This type of system requires a considerable amount of mathematical

Figure 8.22. This perforated PVC pipe along the base of a greenhouse wall connects to underground pipes that branch off to multiple greenhouses. After the compost pile is built, gases rich in carbon dioxide will flow into the perforated pipe and be pulled through the distribution system.

Figure 8.23. This small radon fan pulls compost gas out of the compost pile and pushes it through a manifold system to multiple greenhouses.

calculation to ensure that the system will deliver the amount of CO_2 desired to produce good results. But without the calculations to guide you, you may engineer a system that just doesn't work.

Let's start by describing the components of the system.

The first component is a perforated pipe along the base of the wall of the greenhouse. Polyethylene pipe lasts longer than PVC under the extreme heat of an active compost pile. The pipe is attached to a fan to pull gas, rich in carbon dioxide, out of the compost pile. A system of insulated piping then delivers the gas to additional greenhouses. At each endpoint inside a greenhouse, a perforated pipe buried under the soil releases the gas. The soil filters ammonia and toxic compounds from the gas, releasing the filtered carbon dioxide into the greenhouse.

Figure 8.24. In each greenhouse a section of perforated pipe delivers the carbon dioxide. This pipe will be covered with 2 feet of finished compost with peat moss mixed in to serve as a filter for the compost gas as it moves up out of the pipe.

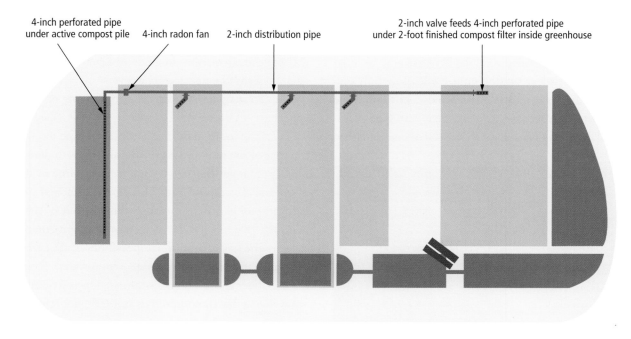

4-inch perforated pipe under active compost pile 4-inch radon fan 2-inch distribution pipe 2-inch valve feeds 4-inch perforated pipe under 2-foot finished compost filter inside greenhouse

Figure 8.25. This diagram shows the compost gas distribution system to multiple greenhouses. The greenhouse on the left is directly connected to the compost pile and doesn't need carbon dioxide distributed through a pipe.

The math involved is important to determine how much carbon dioxide is needed to raise the CO_2 level in the greenhouses to the desired level; how much CO_2 may be lost because of leaks to the outside; how much compost gas will be pulled out of the pile and how much carbon dioxide is present in that gas; the role that friction plays in the system; and how powerful a fan needs to be to move the desired amount of gas through the system.

CO_2 CALCULATIONS

First, we must determine how much carbon dioxide plants use per hour. Then we need to account for the fact that greenhouses are not airtight. So if we add CO_2 to the atmosphere in the greenhouse, some of it will be lost by leakage to the outdoors. If we factor together the total amount of carbon dioxide our plants need and the amount that will be lost from air exchange, we'll know the total amount of carbon dioxide we must supply for optimum plant growth. We'll also have to calculate how much carbon dioxide is in compost gas, and from there we can figure out how much compost gas we need to draw to meet the need.

On average, plants use 0.002 to 0.004 cubic feet of carbon dioxide per hour per square foot of greenhouse floor for optimal growth. I use the following formula to determine total plant usage of carbon dioxide in a greenhouse:[14]

$$\text{Total } CO_2 \text{ use} = \text{greenhouse floor area} \times \text{plant usage rate}$$

Farm example: Our greenhouse measures 14 feet wide by 45 feet long, and our plants use an estimated 0.003 cubic feet (the middle of the optimum range) of carbon dioxide per hour per square foot.

Total CO_2 use = (14 ft × 45 ft) × 0.003 (cu ft/hr/sq ft)
Total CO_2 use = 1.89 (cu ft/hr)

Since we have five greenhouses this size, we multiply this by five. This shows that 9.45 cubic feet of carbon dioxide per hour are required to meet the needs of all the plants in all of our greenhouses.

Another factor we have to account for is that most greenhouses are not completely airtight. Thus, if we enrich the atmosphere inside a greenhouse with extra CO_2, some of that CO_2 will be lost as outside air infiltrates, replacing the CO_2-rich air. This amount of air exchange varies and depends on the wind protection around the greenhouse and the construction materials, as well as the venting of the greenhouse. On average, in a closed double-layer polyethylene film greenhouse, half (0.5) of the greenhouse air is being exchanged every hour from leaks.[15]

We can account for infiltration loss using the following formula, where 0.000001 is a conversion factor to change parts per million (ppm) to cubic feet (cu ft) and the assumed concentration of atmospheric CO_2 is 300 parts per million:[16]

Infiltration loss = greenhouse volume ×
air changes per hour × 0.000001
× (desired CO_2 level − atmospheric CO_2)

To continue with the example of our farm, the greenhouse measures 14 feet wide by 45 feet long with an average height of 7 feet and our desired CO_2 concentration is 1,300ppm.

Infiltration loss = 14 ft × 45 ft × 7 ft ×
0.5 changes/hr × 0.000001 × (1300 − 300)
Infiltration loss = 2.2 cu ft CO_2/hr

Thus, if we have raised the CO_2 level in a greenhouse to 1300 parts per million, the infiltration loss

calculation shows that we will lose more CO_2 to leaks than the total amount our plants could absorb. So we must account for this loss in our total amount of gas distributed. To determine the total amount of carbon dioxide we need per greenhouse, we add plant usage of carbon dioxide to the amount lost.

(Plant usage) 1.89 cu ft CO_2/hr
+ (infiltration loss) 2.2 cu ft CO_2/hr
= (total needed) 4.09 cu ft CO_2/hr

Since we have five greenhouses, we need to multiply this figure by five to total 20.45 cubic feet of carbon dioxide per hour. In order to increase the CO_2 level in the five greenhouses to 1300 parts per million, we need a supply of 20.45 cubic feet of carbon dioxide per hour.

Compost gas isn't pure carbon dioxide, so we must take into account the percentage of carbon dioxide in the compost gas. Research shows that the amount of carbon dioxide in the compost gas is based on the flow rate of air through the compost. This is represented as the actual cubic feet minute per dry tons per day, or the ACFM/DTPD. When gas is sucked out at a rate of 100 ACFM/DTPD, the concentration of carbon dioxide is 4 percent, or 40,000 parts per million.[17] When gas is sucked out at a higher rate of 400 ACFM/DTPD, the concentration of carbon dioxide drops to 2 percent, or 20,000 parts per million. I now understand why breathing concentrated compost gas is not a good idea.

Farm example: Because our system extracts only a small amount of gas from the compost pile, we assume the higher concentration of 4 percent and use the following formula:

Total CO_2 needed ÷ CO_2 concentration
of compost gas = total compost gas needed
20.45 cu ft/hr ÷ 0.04 = 511 cu ft/hr

The final stage of the CO_2 calculations is to convert cubic feet per hour to cubic feet per minute. This step is necessary because fans are sized based on the cubic feet of air moved in one minute (cfm).

511 cu ft per hour ÷ 60 minutes per hour
= 8.52 cubic feet per minute (cfm)

Thus, 511 cubic feet of compost gas per hour equates to 8.52 cubic feet of compost gas per minute.

GAS DISTRIBUTION CALCULATIONS

Now that we know how much gas we need to support all our greenhouses, let's do the calculations to determine the size fan needed to move the gas from the compost pile through the manifold system and filters into the greenhouses. To properly size our fan, we must determine the pressure drop as air is pulled through the compost pile and pushed through the filter. The presence of the compost material slows down airflow, creating a defined pressure drop measured in inch water column (inch WC) or static pressure. To complicate matters, the porosity of the compost changes over time. A new compost pile contains loose material, so air moves freely throughout the pile. Over time, settling and decomposition reduce the amount of air space inside the compost, making it more difficult for the fan to pull out compost gas.

The velocity or speed of the air moving through the compost also has an effect since faster moving air encounters more friction. Also, the size of the compost pile affects the total length the air must travel through the constricted material. A larger pile (or a thicker filter) increases the amount of pressure needed to pull air through. Use the following formula and table 8.2 to determine the pressure drop as air moves through compost and filter material:[18]

Pressure drop = K × V × L

where

K = coefficient for permeability of material;
V = velocity of flow (feet per second); and
L = length of airflow through the material (feet)

Farm example: Our compost piles consist of a mixture of two parts wood chips and one part pulped food waste. This mixture probably has a similar porosity as the two-to-one wood chip to sludge mixture in table 8.2. We determined that the velocity of airflow needed was 8.52 cubic feet per minute from previous calculations. Our compost pile was 4 feet thick when first built, but decreased to 3 feet over time. Our filter consists of 2 feet of screened compost.

Since we determined our airflow in cubic feet per minute, we need to convert to feet per second by dividing by 60.

Cubic feet of compost needed
8.52 cfm ÷ 60 = 0.142 fps
Pressure drop compost = 1.245 × 0.142 fps × 4 feet
Pressure drop compost = 0.7 inch WC
Pressure drop filter = 1.421 × 0.142 fps × 2 foot filter
Pressure drop filter = 0.4 inch WC

Table 8.2. Pressure Drop Coefficients for Compost Materials

Substrate	Coefficient (K)
Wood Chip:Sludge 2:1	1.245
Wood Chip:Sludge 1:1	2.482
Wood Chips, New	0.539
Screened Compost	1.421

Source: R. Haug. *The Practical Handbook of Compost Engineering.* Boca Raton, Fla.: Lewis, 1993.

Adding both pressure drops together tells us the total pressure drop created by the compost pile and filter, or 1.1 inches of water column. As mentioned above, the porosity of compost decreases as it ages. However, if new material isn't added on top of the pile, the length of flow is reduced as the pile shrinks. The two factors should balance out, allowing us to assume that the total pressure drop stays similar over time.

Next, we need to factor the pressure exerted on the system by the pipes. As gas moves through pipes, friction between the wall of the pipe and the gas exerts pressure. As the size of the pipe decreases, more friction is created, increasing the amount of pressure needed to propel air through the pipe. Here's one way to picture this concept. Imagine blowing air through a cardboard paper towel roll (without any paper towel on the roll). Then imagine trying to blow air through a small straw. It takes more force to blow through the straw — that's the friction factor!

The surface texture of the pipe also affects the friction loss. For example, a rough corrugated pipe creates more friction than a smooth pipe because the ridges slow the flow of the air. The amount of fittings and turns in a pipe also create friction, adding to total pressure loss on the airflow. The friction in the fittings is usually expressed in equivalent feet of pipe. For example, a 90-degree elbow on a 2-inch pipe is equivalent to 5.7 feet of straight 2-inch pipe.

Determine the total pressure loss by measuring the total length of pipe and adding the extra length created by the fittings. Consult online charts from websites such as www.engineeringtoolbox.com for friction loss due to fittings and total pipe length. The friction loss (pressure drop) is represented in inches of water column and added to the pressure drop produced by the compost pile and filter.

Farm example: I determined we needed 85 feet of 2-inch PVC pipe to move compost gas from the pile to the greenhouse. I also determined we'd need to use a tee and an elbow fitting for each outlet. After consulting the online chart, I adjusted the length of pipe to approximately 100 feet based on friction from the fittings. Friction loss for 100 feet of 2-inch PVC pipe based on an airflow rate of 10 cubic feet per minute is 0.3 inches of water column. If we needed to reduce the friction loss in the system, we could use a larger pipe. Finally, we add the pressure drop produced by the pipes to the pressure drop produced by the compost and filter system to calculate a final pressure drop of 1.4 inches of water column.

FAN SELECTION

Compost gas contains corrosive ammonia and sulfur compounds that corrode metal. Plastic fans resistant to corrosion are available, although the smallest ones I could find moved more air than necessary and cost over $1,100. I ended up using a small fan designed for removing radon from underneath houses at a cost of $130. The fans pull a small amount of air out of gravel and run continuously. However, these fans are not resistant to corrosion. The fans should last for many years, with a fan bearing being the first thing to fail.[19]

Fans are rated by the cubic feet of air moved per minute, or cfm, and the amount of pressure they can push or pull rated in inches of water column. The more pressure placed on the fan, the less air the fan will move. For example, a fan that moves 500 cubic feet per minute at 0 inches of water column may only move 35 cubic feet per minute at 2 inches of water column. To conserve energy and money, size your fans to handle the capacity (cubic feet per minute) and pressure with the lowest energy use.

Our computations showed that we needed a fan capable of moving 8.5 cubic feet per minute of compost gas through a pressure of 1.4 inches of water column. Our calculations also helped us determine that we only need a small amount of gas for the amount of greenhouse space served because of the high concentration of carbon dioxide in the gas. So we can use smaller fans and distribution networks to move the gas, creating enormous savings in installation and ongoing energy consumption.

CONDENSATION MANAGEMENT

As hot compost gas cools throughout the distribution network, condensation inevitably forms inside the pipe. Condensation pooling inside the pipe or fan shortens the fan's life and blocks airflow. Prevent condensation by installing the fan in a vertical position. Also, place the least amount of piping possible above the fan. If there is no piping above the fan, condensation cannot form and rain down on the fan. All piping that attaches to the fan should slope at least 1 percent toward a low point. At the low point install a sewage cleanout fitting with a removable cap. To drain condensation as soon as it forms, drill a ⅛-inch hole into the lower side of the removable cap.

Additionally, less condensation will form inside the warm pipes if the distribution piping is insulated. We placed our pipe inside an insulated trench shared with the hydronic tubing (see chapter 4 and figure 4.21).

Compost versus Fossil Fuels

How do the economics of compost carbon dioxide injection systems compare to conventional systems? Conventional systems either use compressed carbon dioxide distributed through small, perforated pipes or generate carbon dioxide from burning fossil fuels such as propane and kerosene inside the greenhouse. The upfront costs of compost systems are comparable to conventional systems. The compost system requires a fan and a distribution network. Conventional systems require burners, regulators, pressure gauges, and distribution networks.

The real savings with compost carbon dioxide systems is the freedom from the ongoing expense of fuel or compressed carbon dioxide. Compost piles give us carbon dioxide for free. At our farm we would need 272 gallons of propane to provide our required carbon dioxide for eight hours per day for 180 days. At the time of writing the price is $2.39 per gallon. This equates to an annual expense of $651 in fuel costs. Even if we have to replace our fan every year or purchase an expensive corrosion resistant fan, the compost carbon dioxide system is substantially cheaper.

BUDGET FOR COMPOST CARBON DIOXIDE INJECTION SYSTEM

Installation Materials

$375 Pipes, fittings, insulation

$130 Fan ($1,100+ for long-lasting, corrosion-resistant fan)

$15 Timer

$520 Total cost, installation materials

Annual Costs

$14 Electricity to run 80 watt fan for 180 hours at $0.12/kwh

$65 Replacement fan (fan costs $130, but is replaced every other year)

$79 Total yearly expense

Using Finished Compost

We value finished compost as a precious resource on our farm. The best finished compost is well aged, is dark in color, and lacks heat. We apply compost based on the farm's needs. In our fields and orchards, compost suppresses disease while adding organic matter and nutrients to the soil. In our plant nursery operation, the compost becomes the potting soil for our larger containers. For smaller pots and flats, we sift and mix the compost with other ingredients, cutting our potting soil costs in half.

When compost is applied in the field or orchard, nutrients from the waste are returned to the soil, and biologically active components of the compost combat plant diseases. On a small scale, simply shovel the compost onto your beds using a wheelbarrow. On larger scales, quickly distribute the compost with a manure spreader in the fields. Apply compost about ½ inch thick, and either leave it on the surface or till it into the ground.

Compost Potting Soil Mixes

Containers and pots restrict plants' roots, limiting the plants' access to water and oxygen. Potting soils overcome these restrictions by holding large amounts of water and containing aeration components in the soil. Our potting soil mix for containers 1 gallon size and larger consists of 100 percent finished unsifted compost. The organic matter in the compost holds moisture much like expensive peat moss–based potting soil mixes. Undecomposed pieces of wood chips in the compost help loosen the material, improving drainage and aeration, similarly to perlite and bark in commercial blends.

Creating a potting soil mix for seeding flats and other small pots requires a little more engineering and care than does soil for larger containers. I've never purchased a commercial blend that performs as well as my homemade blends. So making our own potting soil mix is an essential cost saver.

Figure 8.26. This simple compost sifter is designed to fit over two wheelbarrows or the loader on the tractor. Chelsi Crawford is working the sifter.

To create a potting soil mix for small containers, first sift the compost to remove large chunks of wood. There are two basic types of compost sifters: vibrating screens and trommel screens. I built our vibrating screen sifter using ½-inch hardware cloth attached to a simple frame. I designed the frame to fit over the top of two wheelbarrows or our front-end loader, giving us multiple options if equipment is in use. When two people grasp the handles on the ends of the sifter and shake the frame back and forth, the finer compost falls through the hardware cloth. (A single person can accomplish this task by suspending the sifter from a roof.)

In a trommel screen sifter, the screen is shaped into a circular drum. The drum is commonly placed at a slight angle with the upper end open to receive the unsifted material. As the screen drum rotates, material falls through the screen and into a collection area below. I've seen small, motorized units online I'd like to build for the farm.

If you need to sift a lot of compost, you may want to rent a large motorized sifter or hire someone to sift the compost for you. The large sifters come in both vibrating screen and trommel screen styles.

After sifting the compost, we carefully mix it with peat, perlite, and a fertilizer. Peat provides the fibers and organic matter that help bind the mix together. All peats have a low pH, which you can adjust by adding lime.

We add perlite to provide aeration to the mix and prevent compaction. Perlite is a mined volcanic rock that is heated, causing the rock to pop

Figure 8.27. A large vibrating screen sifter.

Our Potting Soil Recipe

Here's our secret recipe for potting soil. In this recipe, a single 5-gallon bucket is one unit. Our compost tumbler mixes 80 gallons at a time.

1 unit peat moss
⅛ cup dolomitic limestone powder for every
    1 unit peat moss
2 ounces 8-5-5 fertilizer for every 1 unit
    total ingredients
1 unit perlite
2 units sifted compost
5 gallons water for every 40 gallons dry
    potting mix (or until you can squeeze a
    few drops of water out)

Remember to mix together the first three ingredients, then add perlite and sifted compost, and mix again. Do one final mix after adding the water.

like popcorn and expand in size. During expansion tiny bubbles form inside the rock, rendering a lightweight product. Perlite's pores will fill with either air or water, both of which are necessary for containerized plants.

I superaerate the mix by adding perlite to take the guesswork out of irrigation. When seedlings of different ages and pots of different sizes are mixed together, watering needs vary drastically over small areas. By superaerating the potting soil mix, I can irrigate the plants that need water the most without worrying about overwatering the plants that need

less. So the ideal potting soil for small pots not only holds moisture but also provides ample oxygen.

I use an organic commercial fertilizer blend derived from blood meal, bone meal, feather meal, meat meal, and sulfate of potash with a guaranteed analysis of 8-5-5. Before adding it to our potting soil mix, we grind it into a fine powder. A few weeks after the seeds have germinated, we use a large flour sifter to sprinkle the same fertilizer over the flats to add more nutrients. The flour sifter allows precise applications of fertilizers without having to rely on stinky liquid fish emulsions used in most organic operations.

We mix the potting soil ingredients in a large compost tumbler. The tumbler holds four wheelbarrows full of material and provides a superior mixing action. Before we started using the tumbler, I would mix material in wheelbarrows, on tarps, and on patios with a shovel. The job was laborious, and yet it left material inadequately mixed. The compost tumbler is the perfect potting soil mixer for our small farm. As production expands, I want to experiment with using the manure spreader to mix compost ingredients.

First, we mix together the peat, lime, and fertilizer. Then we add the compost and perlite and mix the materials again. Finally, we add water and mix it once more. Once wet, the lime and fertilizers activate, making the mix better with age as nutrients become available and the pH changes.

This mix works well for filling seed flats and other containers, but I don't use it to *cover* seeds. With small seeds such as lettuce and some flowers, I choose not to cover the seeds at all. Covering the seeds reduces germination, as the small seeds are smothered by the soil.

With larger seeds and seeds of the brassica family, covering the seeds renders a strong upright plant. Recommendations call for covering the seeds with soil one and a half times the width of the seed.

Figure 8.28. A compost tumbler that mixes potting soil ingredients sits outside the greenhouse where we seed transplants and store potting soil ingredients. A wheelbarrow is placed under the tumbler and the cylinder is rotated until potting soil falls out. The tumbler holds four wheelbarrows full of material.

However, workers at the farm commonly make the mistake of covering seeds with too much soil. To prevent this, I cover the seeds with vermiculite, a mined mineral that expands when heated, like perlite. Unlike perlite, it has the ability to absorb large amounts of water, helping to prevent seeds from drying out. The seeds also push through the vermiculite even when buried too deep. If the seed flats are overwatered, the sterile nature of vermiculite prevents oxygen-blocking algae from forming on the surface of the potting soil.

Future Uses

With such a cheap source of carbon dioxide readily available in compost piles, I'm searching for more ways to use it. I envision connecting piles to drip irrigation systems in fields, pulling compost gas out, filtering it, then pumping it through drip systems when not in use for irrigation. Using compost to amend the soil and air would also make full use of this valuable resource.

SUMMING UP THE FUNCTIONS

1. Compost pile consumes cafeteria and farm waste.
2. Compost turns wastes into nutrient-rich soil.
3. Placed against a greenhouse wall, the compost pile provides direct heat for the greenhouse.
4. Heat generated by compost windrow heats multiple greenhouses by adding supplemental heat to the hydronic system.
5. Compost pile heats ponds through heat exchangers.
6. Compost pile produces organic matter and nutrients to enrich fields and serve as potting soil ingredients.
7. Soldier fly digesters are inserted inside the compost pile. The pile heats the digesters.
8. Compost pile produces carbon dioxide to improve plant growth.

CHAPTER 9

Feed, Fuel, Fertilizer

Bio-Integrated Fly Farming

During the last decade the farm and garden world has been abuzz about the black soldier fly. Native to North America, *Hermetia illucens* currently has a worldwide distribution in tropical and warm temperate regions. This small insect plays a large role in the decomposition of plants, animals, and feces by quickly composting waste and generating a high-protein feed for chickens and more.

Unlike common houseflies, adult black soldier flies aren't found in kitchens, buildings, or picnic areas. Therefore, they aren't vectors of disease or filth. In fact, black soldier fly larvae suppress housefly larvae by 95 to 100 percent by outcompeting them in the manure. Meanwhile, they reduce manure mass by about 50 percent over four hours when properly fed.[1] The adult fly measures ¾ inch long and looks similar to a wasp. You usually see the large flies hovering above compost piles or manure, depositing eggs nearby. The adults live for five to fifteen days, mating and reproducing. Their lack of functioning mouths indicates they live only to reproduce.

I first encountered black soldier fly larvae feeding on the fresh waste near the top of my home

Figure 9.1. *Left*, an adult soldier fly mimics the look of a wasp. *Right*, larvae make a great feed for chickens, prawns, pigs, and other animals.

Elements of the System

Quicker than any compost pile, soldier fly larvae digest food waste and manure, reducing it to miniscule amounts. Not only do they turn waste into nutrient-rich compost, they also self-harvest into containers for use as a high-protein feed for chickens and prawns. This chapter shows you how to:

► Start a soldier fly digester
► Maximize production by feeding larvae
► Harvest, store, and process larvae
► Extend the growing season for larvae

Figure 9.2. This commercial ProtaPod soldier fly digester with a curved projection on the side allows the soldier fly larvae to crawl up, then fall into the hole at the top.

compost pile. The large larvae, about an inch long, had distinct segments all along their writhing bodies. The maggots were moving in a mass in a rotten watermelon placed on the compost. As they feed, they rid the waste of harmful bacteria and convert it into larval biomass. Unlike earthworms, black soldier fly larvae tolerate a wide variety of temperatures and moisture levels in waste. The larvae also consume waste faster than earthworms. And best of all, you can raise the larvae in special containers called digesters, from which the larvae will self-harvest into a bucket for collection as a feed for animals.

At the Student Organic Farm (SOF), we use soldier fly digesters equipped with ramps to take advantage of the larvae's natural tendency to vacate the waste before pupating. After they've fed voraciously on the waste, the larvae pull themselves up the ramp with a specialized mouthpart and fall

into the bucket. Every one to five days, we collect the larvae and feed them to chickens, fish, prawns, or anything else in need of a high-quality protein and fat.

My first attempts at raising soldier fly larvae involved homemade digesters constructed from recycled worm bins. I have also worked with David Thornton, organic and biofuels project coordinator for Clemson University, to attach larger bins to greenhouses for season extension. Though the larvae consumed massive amounts of waste, our harvest was lacking. We lost many larvae to cracks and crevices, and some larvae never left the bin because of poorly designed ramps. Dr. Craig Sheppard at the University of Georgia developed a design for the most successful large larval rearing digesters. Constructed of concrete, the digesters are equipped with ramps on two sides sloped at angles between 35 and 40 degrees. A gutter is located at the end of each ramp. The larvae crawl off the ramp and into the gutter, then fall into a bucket. Commercial plastic digesters are now available and work well.[2] They come in two styles, a large round digester with a diameter of 4 feet (ProtaPod) and a smaller rectangular digester (BioPod). Both have ramps ascending either side, directing larvae into a hole and bucket for collection.

Starting a Soldier Fly Digester

Like any other project, the hardest part of growing soldier fly larvae is just getting started. First, find a good location for your digester, an area protected from the sun and rain. Placing a lid directly on top of a digester prevents airflow and may cause temperatures to reach lethal levels. Situating a digester under an overhanging roof allows better ventilation and temperature control. Because of this, I like to locate digesters under a large roof overhang on the north side of a house or building.

Robert Olivier, founder and CEO of Prota Culture LLC, a company specializing in the bioconversion of waste through insect farming, recommends applying a layer of gravel on the bottom of digesters for drainage and aeration. He then covers the gravel with landscape fabric or a coir mat made from coconut husk. Larvae will eventually degrade fabric material, but the gravel will remain.

Several weeks after the last frost in your area, place a few pounds of waste material inside the digester to attract adults for egg laying. Native ranges for black soldier flies are climate zones 7 and higher, primarily in the southeast of the United States. In these areas adults usually find the waste material and lay eggs, so larvae appear naturally. However, if you live in a dry or cold climate, you may have to import larvae by mail order to get your brood going.[3] Consult your local Extension office to determine if black soldier flies are present in your area. Check with local jurisdictions before importing a nonnative fly.

Once the female adult finds the waste, she lays a clutch of about five hundred eggs in a crevice near the waste. The eggs hatch in about four days in temperatures over 80°F. You should see larval activity within a week or two after placing food waste in your digester. Before larval activity strengthens, the waste material may smell. That's why I recommend using only 2 to 4 pounds of waste during the larval seeding stage. Odors rarely occur once larval activity is dense, unless you overfeed the larvae. Foods such as cooked grains or moist chicken feed tend to be less smelly than other types of waste.

Moisture is another consideration when seeding the digester. Once the digester is active, you will add waste material daily, and the addition of new material maintains high moisture levels. But since you're not adding new material daily during the seeding process, the waste material may become too dry. You may need to add moisture, similarly to watering a garden. Covering the waste material with a piece of shade cloth or muslin helps retain moisture and provide habitat for egg laying.

Houseflies may take up residence in the waste during the first few weeks until the soldier fly larvae eventually exclude them. Since housefly larvae have a lighter color, move faster, and don't reach the same size as soldier fly larvae, you can identify them easily in the waste material. Research shows that flooding basins with 2.5 centimeters of water before adding chicken manure from caged hens situated over the basin gives nearly 100 percent control of houseflies.[4] Soldier fly larvae tolerated the moist conditions, feeding on manure around the edges of the basin and eventually into the center. Since commercial digesters have drainage holes in the bottom, you would need a separate basin or tub to attract the soldier flies with flooded feed for this technique to work.

Figure 9.3. Three soldier fly digesters sit next to an in-vessel composter under a carport at Clemson's Cherry Crossing Research Facility. The lids have spaces between the slats to allow airflow.

Robert Olivier also recommends using an attractant spray to speed up the colonization process. In his freezer he stores a small jar of the juice that exudes from his digester. When he's starting new digesters, he defrosts the jar of juice; wraps a rubber band around a small stack of cardboard, about the size of a deck of playing cards; sprays the juice on the stack; and places the scented cardboard next to a small amount of waste material in his digesters. The adults lay eggs in the crevices of the cardboard, and he moves the cardboard stacks loaded with established eggs into new digesters to seed them for quick establishment.

Figure 9.4. Larvae feeding on bottom of digester.

Lessons Learned the Hard Way

Avoid letting uncontrolled rain or water enter your digester. Our digester flooded when its lid broke, and we had a mass exodus of larvae into an adjacent greenhouse. Birds, rats, and skunks terrorized the greenhouse beds over the next few months as they scratched into the soil, digging up pupating larvae. That's also when we learned the valuable lesson of using overhanging roofs to shelter the digester rather than an attached lid.

The following spring, adult soldier flies filled the greenhouse early in the season, emerging from pupae that had overwintered in the greenhouse soil. I now place larvae in the greenhouse at the end of the year to give the adults an earlier start. The larvae in the greenhouse start a few weeks earlier than at an adjacent site 1 mile away. I simply sprinkle ½ gallon of larvae on the ground in a dry location and cover it with a layer of mulch or compost. I top it with a small square of hardware cloth secured by a cinderblock, to prevent predators from eating the larvae.

If your digester is completely escapeproof, it's probably a good idea to continually release some of the collected larvae into a protected area to pupate into adults for maximum egg laying. Sandy Lin, a graduate student at Clemson University, is researching "techniques to rear black soldier flies to supplement natural populations in waste composting systems." Her research compared finished compost, vermiculite, and wood chips as a pupation substrate and found wood chips increased successful adult emergence.

Figure 9.5. This shade cloth layered over plastic protects soldier fly digesters and a compost windrow underneath from rain and the heat of the sun.

Feeding Larvae

Similarly to livestock or vegetable crops, soldier fly larvae require a constant supply of food and water. As a dedicated soldier fly farmer, I start feeding my soldier fly larvae as soon as I see colonized larvae writhing in the waste. I try to feed them daily or at least once every three days by placing manure or food in the digester. The larvae constantly amaze me by how quickly they work. I'll pile a large shovelful of waste into the middle of the digester and within hours they tear it apart, spreading it evenly over the surface for better access.

To prevent material from becoming anaerobic or attracting houseflies, I provide small amounts of feed more frequently rather than piling up larger amounts. Craig Sheppard at the University of Georgia concluded that feeding larvae daily with amounts they consume within four to six hours results in the best larvae growth. Over-feeding leads to odors because bacteria have an opportunity to colonize the waste. Stirring and aerating the material with a shovel also helps prevent odors and allows you to add more feed at one time. I feed 2 to 3 pounds of food waste per square foot of digester per day.

The ideal temperature range for larvae is between 84° and 104°F. Lower temperatures cause a drastic reduction in feeding and movement, while higher temperatures could kill larvae or cause them to prematurely migrate. Karl Warkomski of ProtaCulture places ice cubes in his digesters when quick cooling is needed. Locating digesters in shaded areas protected from direct sunlight reduces the risk of overheating.

If you feed your larvae frequently with moist food in a shady area, they probably won't need additional water. If not, you may need to water your larvae like you would water a garden — spray the larvae feed with a hose or automated misters until moist. Larvae tend to aerate and dry out material with time. Sheppard's research indicates that a moisture content for chicken manure between 55 and 80 percent works well. The material should look and feel (if you dare) wet, but not wet enough for water to drip out when squeezed.

Larvae may crawl up the vertical sides of digesters and escape if the sides are moist. A strip of Velcro placed along the upper edge creates an impassible barrier for the larvae, directing them to only crawl up the harvesting ramps. Commercial digesters are designed with a curved lip at the top to prevent larval escape.

Figure 9.6. The larvae have fallen into this 5-gallon bucket and are now crawling around on the sides. The clear vinyl tube directs the larvae through a hole in a lid placed on top of the bucket to prevent the larvae from escaping.

Harvesting and Storing Larvae

By far, the best part of farming black soldier fly larvae is the efficiency of the harvest. If I could design systems for my vegetables and eggs to self-harvest, I'd have a lot more time on my hands.

The larvae undergo six developmental stages before they enter the prepupa stage—which is when they self-harvest. During the previous stages the larvae are white; however, they turn a darker color in the prepupa phase. They also stop feeding and develop a hooked mouthpart to aid in climbing and digging.[5] As the prepupae crawl up the harvesting ramps, the waste they were feeding on rubs off. They reach the top of the ramp cleaner before falling into the collection bucket.

"You are what you eat" also applies to black soldier fly larvae. The type of food they consume determines the dry matter conversion rate of feed into larval biomass. High-fat and -protein diets result in a conversion rate between 16 and 19 percent.[6] Larvae fed chicken manure produced a dry conversion rate between 7 and 8 percent.[7]

The pupa or cocoon stage lasts about one week. Therefore, the prepupae should be collected at least weekly to prevent adult flies from emerging. We collect and feed larvae every weekday and skip feeding and collecting on the weekend. To compensate for the weekend lapse, we give the larvae a little extra feed on Fridays.

After harvest we compile the larvae into 5-gallon buckets. Either we use the fresh larvae shortly after harvest or we process them for later feedings. If we're planning on processing, we place the bucket in a chest freezer to kill the larvae. We either keep them stored in the freezer or remove them from the freezer and spread them on screens in a greenhouse. The larvae defrost and dehydrate, and we pour the dried larvae back into 5-gallon buckets for storage in a cool, dry place.

David Thornton has developed a technique using his oil press to extract oil from black soldier fly larvae. After dehydrating the larvae he combines them with dried sunflower seed.

"The fiber in the sunflower seeds creates the pressure needed to extract oil from the black soldier fly. Because sunflowers contain less protein, they also reduce the protein of black soldier flies from 40 to 17 percent, making the meal a better feed for animals like chickens," said Thornton. He found that a mix of 30 percent black soldier fly to 70 percent sunflower seed works best. The oil from the sunflower–soldier fly mix is then added to oils harvested from the cafeteria to make biodiesel for the Clemson University facility fleet.

Harvesting Compost from a Digester

The leftover waste or compost in the soldier fly digester contains nutrients and organic matter, making it a perfect amendment to potting soil, gardens, or compost piles (see table 9.1). When we analyzed the fertilizer value of the compost at the Clemson University Agricultural Service Laboratory, the major nutrients of nitrogen, phosphorus, and potassium are in a percent ratio of 1.64-0.35-0.49. As mentioned earlier, larvae reduce the overall bulk of manure added to a digester by at least 50 percent.[8] With food waste I've observed much greater reductions. If your digester is a few feet deep, it should only need emptying once a year. If your digester walls are 1 foot tall or less and you feed the larvae a low-conversion-rate material such as manure, you may need to remove the accumulating compost in the digester more than once a year. I'm constantly amazed at how much waste goes into the digester and how little compost comes out.

Black soldier fly larvae also reduce the amount of nutrients and harmful *E. coli* in the feed material. If not used, nutrients can become pollutants. From 40 to 55 percent reduction in nutrients is

Table 9.1. Nutrient Analysis for Black Soldier Fly Compost from Food Waste

	Wet Basis	Dry Basis	Lbs/ton Wet Basis
Ammonium Nitrogen	0.02%	0.06%	0.40
Total Nitrogen	0.58%	1.64%	11.68
Carbon	13.10%	36.82%	262.06
Carbon:Nitrogen (C:N) Ratio	22:44	22:44	
Phosphorus as P_2O_5	0.13%	0.35%	2.51
Potassium as K_2O	0.18%	0.49%	3.52
Calcium	0.35%	0.97%	6.93
Magnesium	0.05%	0.14%	1.02
Sulfur	0.04%	0.12%	0.88
Zinc	18 ppm	50 ppm	0.04
Copper	11 ppm	31 ppm	0.02
Manganese	81 ppm	228 ppm	0.16
Iron	2563 ppm	7202 ppm	5.13
Sodium	502 ppm	1411 ppm	1.00
Aluminum	2574 ppm	7232 ppm	5.15
Organic Matter	23.63%	66.40%	472.58
Electrical Conductivity (soluble salts)	0.87 mmhos/cm		
pH	7.1		

Source: Agricultural Service Laboratory, Clemson University. Courtesy of David Thornton.

typical when animal manures are used as a feedstock.[9] Larvae reduced but didn't eliminate *E. coli* and salmonella in chicken manure.[10]

If you remove the waste during the larval production season, the colonization process must start over again, reducing total larval yields and increasing housefly pests.[11] In temperate climates larval activity starts to slow down in fall and eventually stops completely as temperature and day length decrease. For upstate South Carolina, activity slows during September and stops in October. The seasonal cycle of the soldier flies creates an opportunity to empty the digester of accumulated waste.

I treat the waste from the digester as I do manure from animals. The nutrient content is similar to chicken manure. I either add the waste to fields before I plant fall cover crops or use the waste to fertilize compost piles to generate heat.

Extending the Active Season

The cooler temperatures in fall trigger the larvae to build up greater fat reserves. The large larvae dig into the soil in a protected area to pupate through the winter. When temperatures return to 80°F, adults emerge, and mating commences. In our area this temperature rebound happens in late winter or spring.

Since most waste management strategies require year-round processing, interest in black soldier fly season extension abounds. At Clemson University, for example, the large volume of food waste is

Figure 9.7. My first attempt at connecting a soldier fly digester to a greenhouse. The digester has a screen to provide shade under the greenhouse plastic. Once the plastic rolls down, pipes lock it in place. Larvae crawling up the ramp would roll back into the basin because the angled boards limit harvest. A gutter on the end works better.

produced during fall and winter, when students are on campus. But this is when the black soldier fly is dormant. To convert the abundant waste into a high-protein feed, we need to extend the larval growing season.

Two major hurdles to season extension exist: maintaining eggs to replenish harvested larvae and generating cheap heat to maintain larval activity. At Clemson University we're experimenting with solutions to both of these problems — raising flies in cages so eggs can be harvested and keeping digesters warm with heat from compost piles.

Within the next few years we hope to have a working model for a low-cost technique to extend the growing season for black soldier fly larvae.

Harvesting Eggs

In nature, soldier fly adults exhibit what's called a "lekking" behavior during mating. Males rest on the leaves of trees adjacent to larval growing areas. If another male approaches the resting male, the two grapple in an aerial battle, with the dominant one returning to the leaf to rest. If a female approaches the resting male, the two grasp each

Figure 9.8. Graduate student Sandy Lin is misting these mating cages inside a Clemson University greenhouse.

created by the corrugation in the cardboard should measure 2 × 3 millimeters. We place the egg trap 1 inch above moist attractant media such as chicken feed inside a plastic tub or tray and raise the tray 16 inches above the floor of the cage by placing it on top of two cinder blocks.

Research shows that soldier flies only deposit their egg clutches when temperatures rise over 79°F. Furthermore, 80 percent of egg clutches were deposited when the humidity rose over 60 percent.[17] You may need humidifiers and heaters to achieve the desired temperature and humidity in greenhouses and rooms used to rear eggs.

Drs. Craig Sheppard and Donald Booth determined that each egg weighs 0.0276 milligram. Weighing cardboard egg traps before and after oviposition gives an estimate of the number of eggs harvested. Special care must be taken to ensure that the cardboard hasn't absorbed moisture that would affect the weight. Alternatively, you can get a less accurate estimate by counting the clutches — each egg clutch averages five hundred eggs. Soldier fly larvae grow well at densities of 2.5 larvae per square centimeter (16 per 2 square inches or 2,322 per 2 square feet).[18]

Once the soldier flies oviposit their eggs in cardboard egg traps, hatching occurs in three and a half days at 86°F.[19] Dr. Sheppard found that eggs placed directly on food were less likely to hatch, possibly due to fungal attack. Sheppard places the eggs in a plastic cup until they hatch, then places the young larvae on fresh food. Approximately 5,000 three-day-old larvae fill a volume of 0.17 to 0.27 ounces.[20]

Lighting is not important for oviposition, but it does affect mating.[21] In the absence of proper light, mating will not occur, and the soldier flies will lay sterile eggs. Bright sunlight, provided by placing mating cages inside greenhouses, stimulates the most mating.

other in an aerial encounter, and copulation occurs in flight.[12] Females then seek out waste material for depositing eggs, or oviposition, nearby in dry crevices. Oviposition occurs mostly between 2:00 p.m. and 3:00 p.m.[13]

Successfully rearing eggs during the dormant season requires temperature-controlled cages, suitable for mating and egg-laying behaviors. Purchasing or making cages out of screen material works fine. The cages are set up in heated greenhouses and must be large enough to allow the lekking behavior to occur. Cages sized approximately 5 × 5 × 10 feet and 6 × 4 × 5 feet have proven successful.[14] Commercial cages have zippered and sleeved openings for easy access to the inside, as shown in Figure 9.8[15]

The prepupae are placed inside cages and allowed to pupate into adults. Mating begins two days after emergence, and oviposition occurs at day four.[16] We tape the egg traps, which are composed of three layers of double-faced corrugated cardboard and measure 1 × 2 inches. The holes

Recently, researchers have experimented with artificial lighting for mating. Since mating and oviposition require high temperatures, mating inside an insulated building would reduce heating costs compared to mating in cages inside a greenhouse. Greenhouses lose an enormous amount of heat through glass or plastic walls, thus requiring more energy. For example, it takes one-tenth the amount of energy to heat an insulated building with R-13 walls as it does the same size greenhouse, even if it's insulated with a double layer of plastic.

Figure 9.9. Cardboard egg traps next to a wet attractant feed.

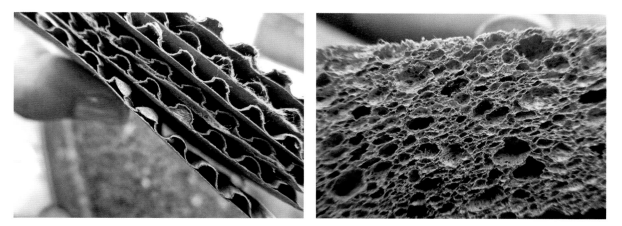

Figure 9.10. The soldier fly eggs in this cardboard trap (*left*) and sponge (*right*) are starting to hatch.

Several types of lighting failed to produce mating. However, a 500-watt quartz-iodine lamp (halogen) with a light spectrum between 350 and 2500 nanometres produced mating at a rate 61 percent of the sunlight control.[22] Comparing light from different wavelengths, the research suggests that wavelengths between 450 and 700 nanometres were influencing mating behavior.[23] Halogen lamps usually have a glass cover filtering harmful UV light; it's not clear whether the protective covering was removed during the experiments.

Conjugal Cooler

At the SOF we have a small walk-in cooler measuring 8 feet × 10 feet. No longer used for refrigeration, the walk-in cooler is basically a large insulated box. We've used it as a curing shed for sweet potatoes and a germination room for seed flats, but more recently we started using it as an egg-rearing chamber for black soldier flies. The 500-watt halogen lamp that's recommended to promote mating not only provides the needed light but also produces heat to help maintain the high temperatures. We also supplement heating when necessary with a small electric radiator heater equipped with a built-in thermostat.

Soldier fly rearing is completely compatible with germinating seed flats inside the cooler. I use automated misters to water the seedlings when we use the cooler for germination in the spring. Stacking functions, the misters also create the humidity needed for adult mating. I set the timer to go off three times a day for one minute each time, which keeps the cooler humidity high. A hole at the lowpoint in the cooler floor drains excess water.

If eggs are needed during the warmer part of the year, we equip the cooler with a solar chimney and bathroom fan to draw cool outside air through the cooler. The solar chimney is a 5-inch galvanized pipe used to vent gases from a gas heater. I insert the pipe through the side of the cooler near the top, next to the halogen light. The pipe is painted black so when the sun hits it the hot air created inside the pipe pulls cool air by convection through a floor vent placed on the bottom of the cooler. A bathroom fan connected to the solar chimney also assists airflow when needed. During warm weather I run the light at night for fourteen hours, from 9:00 p.m. to 11:00 a.m., the coolest part of the day. The fan runs while the light is in operation; then the solar chimney vents the cooler during the daytime. I place window screen over the fan housing to prevent flies from escaping. I also remove the glass cover from the light to expose flies to the full spectrum of light and place aluminum window screen over the light housing to prevent flies from flying into the light and dying.

I place two trays inside the cooler. One of the trays is filled with finished compost or wood chips; I keep the compost moist and place the prepupa larvae on the compost. They quickly crawl in and start pupating two weeks later. To attract the adults to an egg trap, I place a tray with an inch of chicken feed on the bottom on top of cinder blocks stacked two high. I fill the tray with enough water to submerge the chicken feed 1 inch. If the chicken feed is only moist, it quickly starts to mold. Submerged under water, it ferments, which attracts the females, but doesn't mold. I place a screen or cloth on top of the tray to prevent the flies from drowning. I then place cardboard egg traps made from a few layers of cardboard taped together on top of the screen. I spray the cardboard with watered-down exudate seeping from the bottom of active digesters.

If the screen covering the attractant is made from normal window screen, the flies lay eggs through the holes in the screen material. Once the larvae hatch in the attractant, I transfer them to digesters and add a new batch of fresh moist chicken feed attractant to the tub. Alternatively, a fine mesh screen or cloth prevents flies from laying eggs through the screen and compels egg-laying inside egg traps. Encouraging egg laying in traps has been inconsistent for us.

Since most farms have walk-in coolers and most coolers sit idle during the winter, off-season soldier fly rearing is an easy option. Retrofitting our cooler cost less than a hundred dollars—a real bargain compared to heating a greenhouse and equipping it with mating cages.

Heating Digesters with Compost

This is a work in progress. We know that combining food waste or manure with wood chips in large mounds creates hot compost. The hot compost piles represent a free source of heat for greenhouses and ponds, as described in chapter 8. Recently, we've experimented with inserting digesters inside compost piles to keep black soldier fly larvae warm and active throughout the winter. Though it's an inexpensive technique, as well as new and unproven, it warrants more exploration.

Properly built compost piles reach temperatures around 150°F for approximately five weeks. So it's easily achievable to keep compost temperatures around 120°F for periods spanning the entire winter. Combining hot compost piles with soldier fly digesters during winter seems like a logical fit for cold climates; however, management is complicated. When placed inside compost piles, digesters indeed reached ideal temperatures for

Figure 9.11. I'm using the tractor to place hot compost around two ProtaPod soldier fly digesters. The digesters are connected to the greenhouse, with harvesting buckets inside the greenhouse.

larval activity. However, our first and only attempt cooked the larvae when we covered the compost pile with a tarp. The tarp trapped excessive heat, causing the larvae to leave in search of a cooler place. Feed material inside the digester also dried out quickly with the extra heat from the compost pile, and irrigation was needed.

Other potential hurdles included pollutants from compost gas. Black soldier flies' tolerance of carbon dioxide and other gases in compost piles is unknown. Using a carbon dioxide meter, I measured concentrations at the surface of media inside active digesters under an open shed and found readings ranging from 1,500 to 5,500 parts per million. Carbon dioxide readings under a compost tarp may reach 300,000 parts per million.

Our latest efforts aim to control temperature, moisture, and compost gases. To maintain moisture we inserted automated misters attached to poly tubing used for drip irrigation into the digester. We're still working on a system for temperature and compost gas control.

Inserting a commercial digester like the Prota-Pod into the top or side of a windrow-style compost pile may be a workable method. A high roof that prevents rainwater from entering the digester but allows excess heat and carbon dioxide to escape would be important for a successful system.

SUMMING UP THE FUNCTIONS

1. Soldier fly larvae consume cafeteria waste.
2. Soldier fly larvae consume farm waste.
3. Soldier fly larvae convert waste into a high-protein feed.
4. Soldier fly larvae convert waste into nutrient-rich compost.
5. Dehydrated larvae are pressed to extract oil for biodiesel production.
6. Compost pile heats soldier fly digesters.
7. Walk-in cooler grows eggs for soldier fly digesters.
8. Greenhouse extends adult fly growing season.

CHAPTER 10

Taking It to the Field
The Bio-Integrated Field and Garden

A little forethought pays off in the long run in designing the layout of your fields and gardens. Without proper drainage, fields turn into ponds, roads turn into rivers, plants drown from sitting in puddles, and you spend most of your time digging your truck out of the mud.

To avoid such a farming fiasco, design your fields and gardens to work in concert with the slope of the land: shed water from boggy or overly saturated areas and catch, store, and convey water to drier areas.

Proper water usage increases a farm's productivity and cost efficiency. Use flood irrigation for crops that don't demand as much water and carefully metered drip irrigation for thirsty plants. For example, when I properly irrigate kale plants, I can harvest twice as much from them. However, improper overwatering leaches hundreds of dollars of nutrients out of the soil.

Field Design

Growing annual veggies such as lettuce, broccoli, and tomatoes requires a different field design from that for long-lived perennial crops such as asparagus or fruiting trees and shrubs. Water harvesting and control are critical to both and ultimately guide the layout of fields and plantings. With both annual and perennial crops, the field or garden should catch rainfall but shed excess rain. At the Student Organic Farm (SOF) we use properly sloped raised beds or mounds of soil to catch rainfall and drain excess water while providing an ideal climate for root growth. Annual plants grown between young fruiting trees or shrubs make great companions. However, landscapes usually have ideal microclimates suitable for either annual vegetables or fruit trees, but not both. South- and southwest-facing slopes work best for annual vegetable production. The south-facing slope collects more solar energy during the winter, extending the vegetable growing season.

Another element to keep in mind is timing of spring bloom. If you plant fruit trees on a south slope, the trees may bloom early, which makes the flowers and trees susceptible to late freezes. South slopes also expose tree trunks to winter sunlight, creating extreme temperature changes that may lead to trunk cracking. Therefore, fruiting trees and shrubs grow best on north slopes. East slopes are also good for fruit trees and shrubs, as east-facing slopes dry quickly in the morning, which means less moisture and fewer disease problems.

Deep alluvial bottomland soils are also well suited for annual vegetable production. The thick topsoil produces strong, disease-resistant plants with little effort. However, pay careful attention to improving drainage and avoiding flooding. In contrast, ridges and upland areas with shallow soils are best suited for trees and perennial pasture and carefully managed annual crops.

Field Layout and Drainage

Farmers and gardeners know better than to work the soil when it's too wet. If your fields tend to stay wet, you can install drain tile, but that's an expensive proposition and usually requires hiring a specialized contractor. A cheaper, easier alternative is grading a field and building raised beds. Soil in raised beds dries out more quickly in the spring and after rains.

Farms build raised beds by mounding the soil using bed-shaping equipment. In home gardens use a shovel or walk-behind tractor with a furrower, bed shaper, or rotary plow attachment to build raised beds. For the basic backyard garden, form raised beds with rock or wood, and work the soil with hand tools. When properly sloped, the furrows between the raised beds shed excess water, enabling the field and gardens to dry quicker after a rain. When rainfall is scarce, the same furrows bring irrigation water to the crops.

Ideally, start by consulting a detailed elevation survey that shows the contours of land in 1- or 2-foot intervals. Though not necessary, an elevation survey helps identify ridges, direction of slope, and depressions in the landscape.

Access Roads and Paths

A ridge is a long narrow area in the landscape where water flows off in both directions. Roads and paths placed on ridges are easier to maintain because rainwater sheds off the road in two directions rather than accumulating and washing road material away (for more on this, see chapter 11).

Elements of the System

Save yourself money and time by considering factors such as sunlight and water when designing fields. Successfully employ drip and flood irrigation and organic no-till techniques while attracting beneficial insects and growing your own fertilizer, fuel, and feed. This chapter shows you how to:

► Design your fields for plant production, access, and drainage
► Select different types of irrigation for specific purposes
► Build swales, diversion channels, and terracing to capture runoff
► Grade fields and gardens with common types of equipment to harvest rain and shed excess water, resulting in productive fields resilient to extreme weather
► Use newly graded fields and gardens with flood irrigation systems to cheaply water plants and distribute nutrients
► Plant buckwheat to smother weeds and reduce mowing

The roads-on-ridges pattern establishes the main access route through the farm property and, thus, the position of the fields. The ideal is to grade the fields to create planting beds that extend out from the road at 0.25 to 1 percent slope. Rainwater will soak into the field because of the gentle slope. As long as the bed slope is continuous, excess water will make its way to the end of the bed. At the bottom a channel or collection ditch gently moves the water at a 0.25 percent slope into a retention pond, basin, or grassy waterway. I use the following section to establish the pattern and grade the field.

Collection (Diversion) Ditches

Once you identify the ridge and main access routes at the top of the fields, you can determine the location of the collection ditch at the lower end of each field. The upper road on the ridge forms one side of the field, and the lower collection ditch forms the opposite side of the field. The collection ditch should also have a shallow slope of 0.25 percent to slow the velocity and erosive potential of the water that collects in it. For small drainage areas less than 5 acres, dig a V-shaped ditch, but a flat-bottom ditch is best for larger areas. I built the ditches on our farm using a box scraper; I tilt the box scraper to make V-shaped ditches or use it flat to make larger ditches. To ensure that the ditch will drain the excess water, I excavate the collection ditch out of the land, digging the ditch lower than the field. I then deposit the spoils from the excavation into the low areas in the fields to improve drainage.

If you're working on a large scale, you may want to hire specialized contractors. But on a small scale, with careful persistence, a box scraper or rear blade will suffice. Every time I till the field for planting, I note low areas and resume with field grading. This usually equates to no more than four hours. Within a few years I can grade the field using simple farm implements, scraping higher spots to fill lower areas in the field. When I excavate collection ditches, I stockpile soil to fill depressions in the field. You will save time by carefully planning all soil movement to limit the distance soil moves.

The collection ditches carry water into an existing or built pond, basin, or grassy waterway for collection or disposal of the water. Dig your collection ditch large enough to convey the peak

Figure 10.1. The beds slope slightly downhill (from right to left in the image) to a collection ditch that slopes toward a basin at the far corner of the field near the trees. The stockpiled soil in the foreground will be used for more leveling work in the future.

flow of all the runoff from the field based on local soil and climate conditions. Reference the *National Engineering Handbook* or your local Natural Resources Conservation Service (NRCS) office for more information on this topic.

When installing collection ditches, consider what types of field equipment you'll want to drive across the ditch. It's easier to drive a tractor and implements across a wide ditch than a narrow one. As a rule of thumb, I make sure the bottom of the ditch is the same width as the tractor. If the ditch runs perpendicular to the beds, then the tractor and implements won't pitch and roll as they cross. I recommend a six-to-one ratio for all soil slopes on collection ditches to prevent erosion and enable the tractor to move easily across the ditch.

Raised Beds

The raised beds connect the road on the ridge to the collection ditch at the bottom of the field. An elevation survey helps in the orientation of the raised beds in relation to slope.

While I plan my general directions on paper, I use a transit or water level in the field to establish the final position of the raised beds. Starting at the road, I choose a slope across the field that continuously flows without dips or rises. The slope may be steeper in some areas and shallow in others, but it must continuously flow toward the bottom of the field, preferably at a 0.25 percent grade. If a hump or depression in the landscape prevents continuous flow, I level or grade the land.

If a depression is too large to fill with soil, I build a cross drain to remove water from the low spot. Cross drains run perpendicular to the raised beds, connecting with a furrow farther downslope. In really flat areas a cross drain may not work, and the only solution is to fill the low area with soil.

While I always favor bed orientation to minimize erosion and drain fields, the orientation in relation to the sun has important characteristics affecting crop growth. When beds are oriented on a north-south axis, uniform sunlight and heat affect the beds. When beds are on an east-west axis, the south slope of the bed warms and dries quickly and may extend the growing season, favoring low-growing early spring crops and fall crops. But east-west sloped beds cast shadows on each other as plants grow because sunlight is unequally distributed. If tall crops are planted on east-west sloped beds, it's not a problem in the middle of summer, but early or late in the season they shade crops to the north.

Field Irrigation

We irrigate crops in the field using two basic techniques: drip irrigation and furrow or flood irrigation. If water pressure is below 10 pounds per square inch but water storages are large or water is cheap, furrow irrigation is a good choice. I find furrow irrigation most useful for directly seeded

plants that don't need a lot of supplemental irrigation, such as potatoes, corn, and beans. All these crops can succeed without irrigation in our climate. However, they usually perform better with one or two irrigations when the tubers or fruit are developing if the weather is dry. Doing one or two flood irrigations for these crops is easier and cheaper than installing and removing drip irrigation. Since our furrows are widely spaced, they only work for irrigating mature crops. Water added to widely spaced furrows wouldn't expand out and up into the soil enough to reach the limited root systems of young crops.

Irrigating plants is based on soil moisture depletion or the difference between field capacity and the current soil moisture. With drip irrigation the idea is to maintain soil moisture in the top 10 inches of soil. With flood irrigation the soil moisture is allowed to deplete to the lower root depths before irrigation commences. During irrigation, if more water is applied than the amount depleted, water is lost to percolation. Irrigate crops when they reach their allowable depletion, which depends on the type of crop as well as the type of irrigation. In drip irrigation the allowable depletion is around 25 percent, so we water crops frequently to maintain field capacity. In furrow irrigation we allow soil moisture to deplete further and apply less frequent but larger irrigations.

Drip Irrigation

If water pressure is 8 pounds per square inch or greater and water sources are scarce or expensive, drip irrigation is a good choice. Drip tubing is made from polyethylene and has emitters placed along the tubing that deliver water directly to plants without watering the furrows and weeds between plants. Emitters on drip tape effortlessly water in transplants. Drip tape emitters also serve as a guide to transplant spacing when planting by hand since the emitters are spaced evenly down the bed. In addition, using harvested rainwater for drip irrigation systems allows me to conserve water because drip irrigation is highly efficient, with 90 percent of the water emitted being directly used by the plants. I also prevent plant disease because the leaves stay dry while I water.

COMPONENTS

Drip tubing comes in various sizes from as small as ¼ inch to as large as 1 inch to deliver different volumes of water depending on the size of the irrigated area. Within each tubing size, manufacturers have created slightly different diameter tubing. Some ½-inch tubing may have an outside diameter of 0.64 inches, and some may have 0.67 inches. Each configuration requires fittings corresponding with the correct diameter. I find it's easier to purchase drip irrigation parts from the same manufacturer so the tubing and fittings match up correctly throughout the system. Each type of tubing serves different functions.

Distribution tubing comes in sizes from ¼ inch to 1 inch. I use it to connect to the main water supply or the mainline coming from the rainwater storage and also as the sole distribution line from the rainwater storage if the size is appropriate for the application and the water is for irrigation. The smaller ¼-inch distribution tubing connects individual emitters or misters to the distribution line. Polyethylene distribution tubing lasts twenty years.

Oval tubing comes in larger diameters, and I use it in farming operations as headers to distribute larger flows to multiple rows of drip tape. The tubing lies flat, so tractors and vehicles can drive over it.

Emitter tubing comes in ½ inch and ¼ inch, with emitters placed at an even distance throughout the tubing. Emitter tubing comes with pressure-compensating emitters that deliver the same amount of water from each emitter regardless of hilly terrain. Different types of tubing deliver specific flow rates of water from each emitter. Some tubing has emitters that deliver ½ gallon per hour, while others deliver 1 gallon per hour. However, emitter tubing requires a minimum of 15 pounds per square inch to operate. Polyethylene emitter tubing lasts twenty years.

Soaker hoses have tiny pores allowing water to seep out slowly. Soaker hoses do not evenly distribute the water over the entire length of the hose; the first portion of the hose disburses more than the end. Most soaker hoses require a minimum of 15 pounds per square inch to operate. However, newer "rain barrel soaker hoses" operate under very low pressure. When connected to a rain barrel at ground level they drain all but the last 4 inches inside the barrel, meaning they require 0.14 pounds per square inch to operate. Soaker hoses only last a few years, clog easily, and are relatively expensive based on life span.

Drip tape is made from polyethylene and comes flat, making it less suitable for curves but better for straight rows. The tape comes in thin-wall versions that last a single season and thicker-walled versions for permanent or semipermanent crops that last up to ten years. Drip tape is available with different flow rates, and some run off pressure as low as 4 pounds per square inch. The tape uses a turbulent flow system to distribute water evenly from each emitter when the terrain is fairly level. Drip tape is the most cost-effective emitter system.

Drip emitters are attached directly to distribution tubing using a special barbed point on the emitter that pushes through the tubing or by using a punch to make a hole in the tubing. Different flow rates are available, and the emitters come in turbulent flow for low pressure and pressure compensating for hilly terrain and high-pressure systems.

Pressure reducers must be installed when high-pressure systems are attached to drip systems. The amount of pressure reduction is based on the type of drip system installed. Typically drip tape uses a lower 10 to 12 pounds per square inch pressure reducer, and drip emitters use 35 pounds per square inch reducers. Low-pressure gravity systems do not need a pressure reducer, and adding one may cause the pressure to be reduced too much.

Backflow prevention is required by code when attaching drip systems to city and well water systems used for potable water. The backflow prevention ensures that dirt, debris, and chemicals aren't sucked back into the potable water system, thereby contaminating it. When nonpotable rainwater-harvesting systems are used in drip irrigation systems, backflow prevention isn't necessary.

Control valves turn the water on and off for irrigation systems. Manual and automatic valves are available. Typically, I use solenoid valves, which run off low-voltage wire connected to a central timer. However, solenoid valves require a minimum of 25 pounds per square inch to operate properly. In low-pressure gravity systems I recommend using other automated options.

Hose bib timers connect directly to spigots, are battery operated, and have a timer and valve combined into a single assembly. Some of the hose bib timers are mechanically operated and don't require pressure to run (Toro and older Gilmour models). However, I've found they don't last very long. Another option for low-pressure gravity systems uses mechanical valves designed for zone control in hydronic heating systems. For example, the Taco Zone Sentry Zone Valve connects to irrigation timers and runs off the typical 24 volts. Because it's mechanically operated it won't require pressure to open and close.

Filters are required to prevent the small holes in drip emitters from clogging with debris. Usually 150-micron mesh is recommended for filtration. However, in low-pressure gravity-fed systems using inexpensive drip tape or cleanable turbulent flow emitters, a 100-micron filter may be warranted to keep pressure higher. I usually place the filter before the pressure reducer in high-pressure systems.

Fittings such as elbows, couplers, and tees ensure that the drip tubing attaches to all the various components and gets the water to where it's needed. Some fittings operate off compression, and you simply push the tubing into the fitting. Avoid fittings that you shove inside the tubing; they reduce the flow slightly in low-pressure systems. To install the fittings easily, let the tubing sit in the sun or submerge the tubing in soapy water to lubricate it a little. Endcaps or figure eights are placed at the ends of lines to stop the flow. Leave these visible to flush debris from the line as needed.

MANIFOLDS AND DRIP TAPE

Starting in the field, a thin drip tape conveys water to each plant, metered precisely through individual pores in the tape. The tape attaches to a larger manifold (header) line that carries water to the field. The manifold connects to valves, filters, and regulators that clean and control the flow of water.

I've installed drip tape in the field by hand and with a tractor. If I'm installing by hand I suspend the rolls between two sawhorses with rebar placed through the center of the roll. Starting at the road at the top of the field, I pull the tape to the bottom and secure it using sod staples. Maximum run lengths for drip tape depend upon the emitter spacing and flow rate and size of the tubing, with potential run lengths from 300 to nearly 3,000 feet.

Tractor-drawn drip tape applicators distribute drip tape on the surface of the soil or bury it up to 6 inches deep. Buried lines are not subjected to damage from tractors and cultivators. After harvest a drip tape remover carefully pulls up the tape with

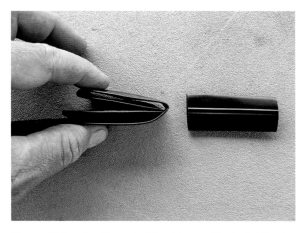

Figure 10.2. I plug the end of the drip tape line by folding the tape back on itself twice, then creasing and stuffing the folded line inside a short piece of drip tape.

automated or manual winders to coil the tape back into a roll.

Once I install the drip tape on the bed, I attach it to a header or manifold line that distributes water to the field. I use a 2-inch polyethylene tubing called oval tubing for larger fields up to ½ acre and smaller ½-inch tubing for gardens. The manifold line in the larger fields has plagued us with problems. Leaky connections favor weed growth and weeding around the manifold line. The zone 2 feet on either side of the manifold becomes a weedy dead zone representing the space of an entire bed in the field.

I've found a few solutions to fix manifold problems. I grow a crop in the manifold zone that quickly germinates to prevent weed growth and the hassle of removing it. I recommend a nonvining, loose crop to allow easy mowing and maintenance around the manifold. It's nice if the crop attracts beneficial insects or scavenges for nutrients. The perfect crop for this task is buckwheat. Buckwheat germinates quickly, smothering weeds. Flowers form within a month and attract a plethora of tiny micro wasps that fly into the field to parasitize plant pests.

Buckwheat also scavenges for phosphorous in the soil, making it available to other plants. We usually let the buckwheat go to seed so it replants itself for the next cycle; then we mow it down to extract the oval tubing. I would also like to try sunhemp in the manifold zone. This fast-growing nitrogen fixer would work well on the north side of

Figure 10.3. A sawhorse set up for installing drip tape in a field. Buckwheat was planted under the manifold line in the foreground and is now flowering. The buckwheat is just starting to germinate under the manifold line next to the sawhorse.

Figure 10.4. The same manifold line with buckwheat in flower. The buckwheat flowers attract beneficial insects that control plant pests, making it easy to extract the manifold line when crops are finished.

Figure 10.5. A manual valve at the manifold controls flow to each bed. We use valves with a fitting that tightens onto the manifold to prevent leaks.

the field, but tractor access would become difficult when the crop gets large.

To prevent leaks in the manifold line, I use connectors that tighten against the line using a threaded locking device. I also use connectors with manual valves. Since each of our fields contains a diversity of crops planted at various times, the valve enables us to turn rows on or off as needed.

VALVES, PRESSURE REDUCERS, AND FILTERS

The manifold line connects to a valve, a pressure reducer, and a filter. Drip tape requires a water pressure between 4 and 15 pounds per square inch. Higher pressures blow the tape and connectors apart, so a pressure reducer protects the equipment. We use a high-volume reducer to allow more flow through the line for feeding drip tape.

The pressure reducer attaches to a valve, controlling flow from the main line. At the minimal expense of twenty-five dollars, an automated valve is a worthwhile investment. We control automated

valves with a control timer. A 1-inch valve waters around 5,000 row feet, depending on the drip tape and pressure reducer. This equates to ¼ to ½ acre of land using drip tape, depending on row spacing. Valves over 1 inch in size are far more expensive. I install valves underground below the frost line inside protective valve boxes. The depth is dependent on climate; valves should be set deeper in colder climates. Draining lines during winter or blowing water out using compressed air is also an option. Stronger valve boxes are available, allowing tractors to drive on top of the boxes. If a field needs multiple valves, place the valve box in the middle edge of the field and run manifolds in opposite directions.

A flexible pipe extends from the valve box aboveground for attachment to the manifold line. The flexible nature of the pipe protects the valve assembly from breaking when aboveground pipes are moved, run over by the tractor, or hit by the mower.

To prevent emitters from clogging, drip irrigation requires filtration. Different types of filters are available for different types of water problems.

Figure 10.6. *Left,* a mesh filter with a pressure reducer attached; *middle,* discs for a disc filter; and *right,* the housing for the discs.

Figure 10.7. Automatic valves are protected from cold damage by installing them below freeze level in valve boxes. The top of the box is at ground level, and pipes in this location need to be 12 inches below ground level to protect them from freezing.

Wire mesh filters, disc filters, and media filters are common. Wire mesh filters work well with light loads or low-use systems. Disc filters work well with algae, and media filters handle higher volumes with organic debris. I prefer a single larger filter for the entire farm's irrigation system to streamline cleaning and maintenance. To simplify cleaning the mesh and disc filters, install an adjacent hose with a sprayer and stiff brush adjacent to the filter.

IRRIGATION SCHEDULING WITH DRIP SYSTEMS

Drip systems add small amounts of water daily to replace moisture lost from soil and plants (evapotranspiration, or ET). ET rates vary throughout the year and with different crops at different growth stages. Find average ET rates at your state climate office or local Extension office.

Instead of replacing depleted soil moisture, drip irrigation maintains soil moisture.[1] Soil can hold only a certain amount of water; this is called field capacity. Water that exceeds field capacity simply leaches below the roots of the plants. Shortly after moisture is lost, the drip irrigation system turns on to bring the moisture back. Research shows that soil moisture maintained slightly below field capacity (75 to 90 percent available soil moisture) maximizes plant growth.[2]

Table 10.1. Figuring Duration of Irrigation

Drip Tube Flow Rate		Row Spacing (feet)				
gph/100 feet	gpm/100 feet	4	5	6	8	10
11.4	0.19	21.9	27.3	32.8	43.7	54.7
13.2	0.22	18.9	23.6	28.3	37.8	47.2
20.4	0.34	12.2	15.3	18.3	24.4	30.6
27.0	0.45	9.2	11.5	13.9	18.5	23.1
40.2	0.67	6.2	7.8	9.3	12.4	15.5
80.4	1.34	3.1	3.9	4.7	6.2	7.8

Note: Hours required to apply 1 inch of irrigation water based on drip tube flow rate and row spacing.

With drip irrigation, frequent small irrigations are best for vegetables.

The total amount of water applied should equal the amount lost to ET. This may be as much as ¼ inch of water per day when it's hot and dry. The flow rate of the drip tape determines how long it takes to deliver the appropriate amount of water to replace ET. Table 10.1 shows the run time for various drip tape flow rates and spacing to achieve 1 inch of irrigated water.

The texture of the soil determines how much to water the soil. More clay or organic matter increases the soil's retention. Table 10.2 shows the fraction of water available for different soil types. For example, every inch of loamy sand soil holds 0.06 to 0.12 inch of water available for plant use.

Soil texture also determines how long to run the drip irrigation system. Soils that hold more water can be irrigated longer before water and nutrients leach below the root zone. Most plant roots, especially those of plants watered by drip irrigation, are in the top 10 inches of soil.

Let's look at an example of how to make use of the data in these irrigation tables. During June our vegetables at the SOF lose about 1 inch of water to ET over one week. Our drip tape flow is 0.45 gallons per minute for every 100 feet of drip tape, and

Table 10.2. Available Water for Different Soil Types

Soil Texture	Available Water (water inches/inches of soil)
Coarse Sand	0.02–0.06
Fine Sand	0.04–0.09
Loamy Sand	0.06–0.12
Sandy Loam	0.11–0.15
Fine Sandy Loam	0.14–0.18
Loam and Silt Loam	0.17–0.23
Clay Loam and Silty Clay Loam	0.14–0.21
Silty Clay and Clay	0.13–0.18

we space our rows 5 feet apart. The soil on our farm is a rich, loamy sand. How should we schedule the drip irrigation to replace water lost to ET without using excessive water or leaching water past the root zone?

Solution: From table 10.1 we see that for drip tape with a flow of 0.45 gallons per minute per 100 feet and a row spacing of 5 feet, we need to irrigate for 11.5 hours (690 minutes) over the week to replace the inch of water lost to ET. From table 10.2 we see that our loamy sand soil has about 0.10 inches (approximate median of values in table 10.2) of water capacity available for plants for

Table 10.3. Maximum Drip Time

Available Water (water inch/inch soil depth)	Maximum Number of Minutes per Application				
	Drip Tubing Flow Rate (gpm/100 feet)				
	0.2	0.3	0.4	0.5	0.6
0.02	20	14	10	8	7
0.04	41	27	20	16	14
0.06	61	41	31	24	20
0.08	82	54	41	33	27
0.10	102	68	51	41	34
0.12	122	82	61	49	41
0.14	143	95	71	57	48
0.16	163	109	82	65	54
0.18	183	122	92	73	61

Source: J. M. Kemble et al. *Southeastern U.S. 2014 Vegetable Crop Handbook.* Lincolnshire, Ill.: Vance Publishing Corporation, 2014.

Note: Maximum time for drip irrigation application on vegetable production based on flow rate. Available water is based on the soil texture. Based on 10-inch-deep root zone and soil water depletion of 25 percent.

every inch of soil. Using this info for table 10.3 we can see that for our flow rate of 0.4 (rounded down from 0.45 to account for time to pressurize the irrigation system) the maximum amount of time to water would be 51 minutes for our soil type before nutrients and water leach past the roots.

We then divide the amount of time to apply the weekly 1 inch of water (690 minutes) by our maximum irrigation time (51 minutes) to get 13.5 weekly irrigations (rounded up to 14) needed to apply one inch of water or approximately two irrigations a day for 51 minutes each. To get an exact time, divide 690 minutes by our rounded up number of 14 irrigations to get 49 minutes of irrigation needed twice a day. If we watered for 49 minutes at a time, we would have to irrigate 14 times over the week to replace the 1 inch of water lost. Therefore, we irrigate twice per day for 49 minutes each irrigation to water the vegetables 1 inch per week.

If I had course sandy soil, table 10.2 indicates a median of 0.04 inches of available water, and I would schedule my irrigation differently. Table 10.3 shows that using drip tape with a flow rate of 0.4 gpm/100 feet results in a maximum irrigation time of 20 minutes before water and nutrients are pushed beyond the root zone. I now divide the total irrigation time (690 minutes) by the maximum irrigation time (20 minutes) to get approximately 35 irrigations. Dividing my 35 weekly irrigations by 7 days shows I need 5 irrigations a day at 20 minutes each.

Furrow Irrigation

Flooding furrows between raised beds is an inexpensive but management-intensive solution for crop irrigation. The slope of the furrows and the slope of the existing landscape are important. The furrows should slope no more than 0.5 percent to prevent soil erosion. A minimum grade of 0.05 percent ensures that water drains out of the furrow after irrigation or rain. When the slope of the existing landscape is more than 0.5 percent, the furrows should cut across the slope to reduce the grade of the furrows to 0.5 percent. If the land slopes more than 3 percent, the risk of erosion from a breach in the furrows increases. Therefore, it's better to terrace land slopes over 3 percent to create a flat shelf before implementing furrow irrigation.

Building raised beds for good field drainage also sets up the field for furrow irrigation. However, more careful attention is needed to slope the beds properly, and curving the beds to follow contours may be necessary. To set up the beds for furrow irrigation, start at the upper part of the field. This is where the road or access path should preferably follow the ridge of the landscape. From here mark a line for the direction of the beds to travel. Using a water or transit level, lay flags on the ground at the required 0.25 to 0.5 percent slope to allow water flow.

In landscapes with continuous slopes you can line up all the beds in a straight line. In undulating landscapes the beds have to follow the contours of the land to conform to the shallow slope needed by the furrow. Use flags to mark a guideline for the furrows every 15 to 20 feet in undulating landscapes. If you'd like straight beds reset the beds or furrows at every guideline so the furrows maintain the proper slope. At the end of the field the furrows empty into a collection or diversion ditch, which conveys the water toward a storage or retention area, previously discussed in the section about field drainage.

The flow of water needed depends on soil type, the slope of the furrow, and the depth of the irrigation. More water flows over steeper sloped furrows and clay soils before being absorbed. In contrast, sandy soils need more water to compensate for percolation. In my experience a minimum flow rate is 5 to 7 gallons per minute for every furrow, but for a small garden even 1 gallon per minute may work. Shorter field lengths are more efficient, since less water soaks in on the way to the end of the field;

however, more runoff may occur because the end is reached sooner. Higher flow rates will also move water across the field faster, reducing percolation losses, but flow rates that are too high will cause erosion (table 10.4).

Efficient furrow irrigation involves replacing the amount of water lost without pushing water beyond the lower root zone of the plant. Different plants have different depths for root zones, and deeper roots have access to more soil moisture (table 10.5).

With flood irrigation, soil moisture is typically allowed to deplete up to 50 percent in the root zone of plants before irrigation commences.[3] However, I usually begin irrigation at 40 percent depletion to improve plant performance for the vegetables I grow. Based on the root zone of the plant and the soil moisture depletion, I use the following formula to determine how much water to add:

Available water based on soil type (table 10.2) × root zone depth (table 10.5) × 40% allowable depletion = amount of water needed

For example, we want to determine how much water we need to irrigate our potatoes. We have a loam soil with available water of 0.15 for every inch of soil.

0.15 (inches available water) × 24 (inches root depth) × 0.40 (allowable depletion) = 1.44 (inches of water needed)

We can then use evapotranspiration rates and crop coefficients to determine how frequently we need to irrigate (see chapter 6). For example, if 0.25 inch of water is lost per day in June to evapotranspiration, we divide the inches of water needed between irrigations by the daily amount of water lost to find the number of days between irrigations.

Table 10.4. Maximum Furrow Length

Furrow Slope (%)	Maximum Discharge into Furrow (gpm)	Clay		Loam		Sand	
		Net Irrigation Depth					
		2"	3"	2"	3"	2"	3"
0.0	47	325'	490'	195'	295'	100'	150'
0.1	47	390'	550'	295'	410'	150'	200'
0.2	39	425'	590'	360'	490'	200'	310'
0.3	31	490'	650'	425'	550'	250'	360'
0.5	19	490'	650'	425'	550'	250'	360'

Source: C. J. Brouwer. *Irrigation Water Management: Training Manual.* Rome: Food and Agriculture Organization of the United Nations, 1985.

Note: Approximate maximum furrow length based on slope, soil type, irrigation depth, and flow into furrow.

If the plants use 1.44 inches of water between irrigations, $1.44 \div 0.25 = 5.75$ days between irrigations.

I then use the following formula to estimate the depth of water applied to the field including losses to percolation.

$$(1{,}155 \text{ [coefficient]} \times \text{flow into furrow [gpm]} \times \text{time water applied [hrs]}) \div (\text{furrow length [ft]} \times \text{wetted furrow spacing [in.]}) = \text{gross depth irrigation (in.)}$$

For example:

$$(1{,}155 \times 5 \text{ gpm} \times 3 \text{ hours}) \div (200 \text{ [ft]} \times 60 \text{ [in]}) = 1.44$$

Therefore, I would need to apply 5 gallons per minute to each furrow for three hours to achieve 1.44 inches of irrigation.

The quarter-time rule guides the flow of water into the furrows and ensures an even water distribution across the field, increasing the efficiency of the irrigation. The flow should advance to the end of the furrow in one-quarter of the time required to apply the total irrigation. For example, if 180 minutes is needed to apply the total irrigation,

Table 10.5. Typical Rooting Depth of Selected Crops

Crop	Rooting Depth (inches)
Beans	20–27
Potatoes	16–24
Corn	40–67
Vegetables	12–24

Source: C. J. Brouwer. *Irrigation Water Management: Training Manual.* Rome: Food and Agriculture Organization of the United Nations, 1985.

the water should advance toward the end of the furrow in one-fourth the amount of time for the total irrigation or $180 \div 4 = 45$ minutes. The most efficient use of water requires a fast flow rate to get water through the field to the other end. Once water reaches the end of the field, slowing the flow rate using valves or adjustable gates at the header prevents excess water from flowing out the drain on the lower end, thus avoiding wasted water.

I use the same 2-inch header pipe for flood irrigation that I use for drip irrigation. However, flood irrigation uses at least four times more water than a typical drip line. As a result, a much smaller area is irrigated to concentrate the flow into a few furrows for flood irrigation. If I had a larger water

Figure 10.8. Furrow irrigation on potatoes during tuber set at the SOF.

supply enabling me to completely switch to furrow irrigation, I would use an 8-inch gated pipe with adjustable gates specifically designed for delivering large quantities of water for flood irrigation.

More on Water Management

Drip and furrow irrigation involve channeling stored water to plants, whether it's harvested rainwater or municipal water. We can also supply water to fruit trees and pasture through methods that capture rainfall as it happens. These methods include creating contour bunds and diversion ditches, structures that also help protect soil and plants from excessive rainfall that might otherwise cause flooding and erosion.

Figure 10.9. My backyard garden is set up for flood irrigation from the pond. The beds are sloped 0.25 percent away from the house. Water flows by gravity to fill the furrows between crops.

Contour Bunds (Swales)

Another technique for irrigating trees and pasture involves small earthen embankments — bunds — to stop overland flow, which occurs when the soil can't absorb the falling rain. Whatever water the soil can't absorb flows downhill as runoff. The bunds stop the runoff, forcing it to infiltrate and storing it in the ground. Then trees and pasture with deep roots planted near the bunds have extra water to use between rain events.

Contour bunds are most appropriate for slopes of 5 percent or less. The size of the watershed, the climate, the soil type, and the land use determine how much water flows off the land and into contour bunds. Small watersheds, sandy soil, and forested areas won't produce much runoff. Conversely, large watersheds, soils with clay and loam, and urbanized areas shed more water. The climate of the location also plays a part because some areas are more likely to experience intense storms with more runoff.

Make contour bunds by digging a ditch on contour and using the soil to form a bund or berm on the downhill side of the ditch. Build the bunds 8 to 12 inches tall, at least 30 inches wide at the base, and 5 to 7 feet long. A bund this size catches water from a watershed area of 100 to 500 square feet. If the soil is sandy and the rainfall in the area minimum, the larger area of 500 square feet is reasonable. If the soil has a lot of clay or the rainfall is high at the location, the smaller 100-square-foot catchment area makes sense. Runoff flowing from a watershed area larger than 100 to 500 square feet may wash out the bund, so I recommend building a larger bund or making the watershed smaller to adapt.

Bunds spaced 15 to 30 feet apart as you move up or down the landscape divide the landscape into a series of blocks, reducing the watershed size and permitting short bunds. Because every bund catches water as runoff flows downhill, all the bunds work together to infiltrate water evenly over the landscape. I reduce the watershed further by placing small earthen bunds, about 3 feet in length, on the uphill side of the bund perpendicular to the main bund. This prevents water from accumulating and moving along the length of bunds, in essence dividing the landscape into a series of small watershed blocks called microcatchments.

Sometimes, water from off-site flows into the bund area, potentially overtopping the bunds — a likely possibility if the bunds don't continue all the way to the high ridge in the landscape. To protect the bund area, build a diversion ditch to move water around the bunds (I describe diversion ditches later in this chapter).

Building contour bunds is simple. First, place grading stakes in the landscape at the location of the center of the bund. Lay the stakes out on a perfect contour using a leveling device such as a water level or a transit level. I place stakes every 10 to 20 feet depending on the curves in the landscape, then mark each stake with the height of the bund, adding 10 percent to account for settling of the soil.

For example, if I want to build a bund 12 inches high, I mark the grading stake 13.2 inches above ground level. Sometimes it helps to mark out the width of the base of the bund. For example, if the bund is 30 inches wide at the base I paint a line on the ground 15 inches on either side of the grading stake. Once I've outlined the bund, I dig soil from directly uphill of the bund location. This creates a shallow ditch, and I place the removed soil in the bund area. I tamp the soil every 6 inches to compact it. The side slopes of the bund and ditch should be as shallow as possible to stabilize the soil with a common ratio of one to one.

You can use various techniques to move the soil to the bund area. On the smallest scale, use hoes to pull soil down and form the bund. However, bund construction is well suited for machines. I find the rotary plow builds a quick and easy bund: Since the rotary plow throws soil in one direction, positioning the plow to throw soil in the downhill direction moves soil toward the bund. I start about 6 feet higher than the center of the bund and repeatedly pass in the same direction, working my way down toward the bund. Depending on soil conditions, three to nine passes with a rotary plow builds the bund. Next, I smooth the soil and cut angles with a hand rake. As soon as I'm finished, I seed and mulch the area to protect it from erosion.

Another technique involves using a disc harrow. First, I remove all the gangs of discs on the harrow, leaving a single set of discs angled to throw soil in one direction. Then I drive the tractor above the grading stakes, and discs throw soil downhill into the bund area. Sometimes it takes several passes to build the bund high enough.

A box scraper or rear blade on a tractor also forms bunds. First, loosen the soil with a chisel plow or tillage device above the grading stakes. Next, use the rear blade or box scraper to move soil into the bund area. Angling the rear blade allows soil to side cast into the bund area. Since box scrapers don't have the option of angling the

Figure 10.10. Place grading stakes every 25 feet on contour (or slope for diversions) at the point where the center of the bund will lie. Next, mark the height of the bund on the grading stakes, and bring the soil downhill to the marked height. The land slopes downhill from right to left; soil was brought from the uphill side to form the bund.

blade, either push the soil into the bund area while driving backward or pull soil into a pile and pull the pile into the bund area. Side casting with an angled rear blade is much easier.

A bulldozer with an angle blade also forms bunds. I angle the blade similarly to angling a rear blade on a tractor to side cast soil. Simply drive the bulldozer uphill of the bund, side casting soil into the bund area above the grading stakes. I've also pushed the soil downhill using the entire length of the blade if the blade is not set up to angle. As a last resort, use the bucket on an excavator or backhoe to pull soil downhill into the bund, but this process is tedious and leaves a rough surface requiring additional handwork. If all you have is an excavator to work with, it's usually easier to use the blade on the front of the excavator to push soil downhill than using the small bucket on the arm.

Diversion Ditches

Diversion ditches are giant earthen gutters placed across the landscape to harvest and move water in a manner similar to rain gutters on a house. Like contour bunds, diversion ditches harvest runoff from large rain events. Unlike contour bunds, diversion ditches are designed to move water safely through the landscape. Building diversion ditches is comparable to building contour bunds. Excavated soil from the diversion ditch is placed on the downhill side, forming a bund to support the depth of the channel. However, the bund is slightly off contour at an average slope of 0.25 percent over the entire length. Diversion ditches have the following uses:

- To harvest rainwater for pond and soil storage
- To protect gardens, cultivated fields, and water-harvesting structures from inundation and erosion

- To protect buildings and roads from overland flow that may cause flood damage
- To divert water away from areas undergoing erosion

The size of the watershed, the climate, the soil type, and the land use determine how much water flows off the land and into diversion ditches.

The NRCS recommends agricultural diversions have a minimum size to handle the peak discharge from a ten-year-frequency, twenty-four-hour-duration storm. Diversions for buildings, roads, and more important structures should be larger and capable of handling the capacity from a twenty-five-year, twenty-four-hour-duration storm. In addition, the velocity of water inside the diversions should be slow enough to prevent erosion. This requires sloping the diversion across the slope at a shallow 0.25 percent. For more information on sizing diversions, refer to the *National Engineering Handbook*.

The shape of the diversion may vary. I typically build V-shaped diversions on smaller watersheds and flat-bottom trapezoidal shapes on larger watersheds. Cutting shallow slopes prevents erosion and allows tractors or mowers to ride over the diversion ditch. A four-to-one ratio for all slopes works well.

To build a diversion, start at the outlet — a pond, retention basin, swale, or some other area with the capacity to hold and safely release the harvested water. Peg a grade line uphill from the outlet at the required slope, usually a shallow 0.25 percent unless you're using rock or erosion-control fabric for extra protection. Mark the required height for the berm on the grading stakes, as well as the required width to achieve the four-to-one slopes. Next, using techniques similar to those described for building contour bunds, move soil downhill to form the berm and compact the soil every 6

Figure 10.11. A small diversion ditch is protected from erosion through mulching, log check dams, and keyline pattern cultivation uphill.

inches. Protect the diversion ditch from erosion by applying seeds and mulch immediately after construction. I also place small earthen, rock, or log check dams every 20 feet to prevent water from flowing in the diversion ditch until vegetation takes hold. Once the vegetation is a few inches tall, I remove the check dams.

Using Bunds for Flood Irrigation

Contour bunds catch and store runoff water in the soil. When slightly sloped, bunds can also move water through the landscape to flood-irrigate trees and pasture on landscapes with a slope of 5 percent or less. When bunds are used to irrigate, they zig-zag through the landscape, with each level of bund sending water in the opposite direction. Careful design protects the bunds from overtopping and eroding by inhibiting bunds from accumulating excess water. I suggest making this type of bund less than 300 feet long to prevent erosion. Also, place each bund within 30 feet of the uphill bund to limit the size of the watershed flowing into the lower bund. During flood irrigation, water will contact more of the landscape because of close spacing.

Next, carefully consider the slopes of the bunds for improved water flow. I like to place the first third of the bund at a 0.5 percent slope, the second third of the bund at a 0.25 percent slope, and the last third of the bund on contour. The steeper start gets the water flowing before a large volume

Figure 10.12. The bund at the top of the incline slopes gently from left to right across the contour. The bund below it slopes gently from right to left. These newly formed bunds have been seeded and mulched so they will not erode.

Figure 10.13. The same bunds (photographed from a different perspective) are flooded with water from a pond at the top of the property to irrigate pasture and young fruit trees hidden in the tall grass.

accumulates. The middle and end of the sloped bund slows the water down. The average slope over the entire bund should not exceed 0.4 percent to prevent erosion.

To move water to the next bund downhill, I build a spreader bank on the final third of the bund placed on contour. The spreader bank is a perfectly level area about 5 feet in length that is slightly depressed from the level of the rest of the bund. It allows water to overflow safely in a thin layer over the top of the bund. Once the water flows over the bund, the next bund downhill catches the water and moves it in the opposite direction across the slope. I plant permanent vegetation to protect the soil between the spreader bank and the bund below.

Flood-irrigate the landscape by applying a large volume of water to the upper bund, preferably with a 4-inch pipe connected to a pond or large storage tank. The water slowly flows through the landscape in a zigzag fashion as it works its

way downhill. Because of the shallow grade of the bunds, water has time to soak into the soil, irrigating the adjacent trees. I recommend keyline plowing (described later in this chapter) if you're growing pasture between your bunds.

Unfortunately, bunds don't run parallel at an equal distance to each other. Since each bund slopes in the opposite direction, the bunds move toward each other and eventually connect at a single point if the bunds are long enough or close enough. The nonparallel nature of the bunds complicates spacing of trees and plants.

Terracing

Terracing slows the downhill flow of water and soil while creating level areas suitable for cultivation. When most people think of terracing they picture the stepped-bench terracing with a near vertical backside held in place by stone or permanent vegetation. Such terraces are usually built with bulldozers. But other types of terracing require less effort and allow cultivation of all parts of the terrace if the land slope is reasonable.

A graded terrace is simply a contour or sloped bund with a wider channel uphill of the bund. The slope of the land determines the width of the channel. The NRCS recommends a minimum channel width of 8 feet on slopes less than 5 percent, 7 feet on slopes 5 to 8 percent, and 6 feet on slopes greater than 8 percent.[4] Channels should have a depth of 12 to 18 inches to contain runoff. Design the channel on contour to retain water or slope the channel to drain water at one end. In either case, water from the channel needs a safe outlet to prevent erosion. Shallow grassy slopes or stone waterways make great outlets for channels.

The climate, slope, and soil determine the distance or interval between terraces. Areas with less intense rainfall, erosion-resistant vegetation, and permeable soils permit more widely spaced terraces. The NRCS uses the following equation to calculate the vertical interval between terraces:[5]

$$\text{Vertical interval} = x \times S + y$$

where

x = rainfall factor; S = slope (percent); and
y = soil and cropping factor (1 for erodible soil and 4 for soil protected with vegetation)

The rainfall factor (x) varies between 0.4 and 0.8 inches depending on the intensity of rainfall. In the Deep South $x = 0.4$, and the number increases to 0.8 as you move north.

For example, at my property in South Carolina I have slopes of 5 percent and permanent pasture and fruit trees in the landscape. The rainfall factor for South Carolina is 0.5.

$$\text{Vertical interval} = 0.5 \times (5) + 4$$
$$= 6.5 \text{ feet}$$

Figure 10.14. Invented in 1885 by Priestly H. Mangum, the Mangum terrace is a type of graded terrace with a wide base slightly sloped to slow the flow of water. The terraces were the first farmable terraces designed to allow equipment to traverse them. Over a hundred years later, they're still part of the landscape in the South—harvesting rainwater.[6]

Therefore, the maximum vertical distance between my terraces is 6.5 feet to prevent too much water from accumulating on the terrace. To determine the horizontal distance between each terrace I use the following formula:[7]

$$\text{Horizontal interval} = VI \div S \times 100$$
$$= 6.5 \div 5 \times 100$$
$$= 130 \text{ feet}$$

Consequently, I space my terraces no more than 130 feet apart. The actual horizontal distance varies slightly from the distance measured across the surface of the land, which is usually slightly longer than a distance measured horizontally across space, but the difference is negligible unless the slope is steep.

Irrigation Ditches

I first learned about the irrigation ditch from reading P. A. Yeomans's *Water for Every Farm*. Yeomans used a series of irrigation ditches connected to ponds to water his pastures. The ponds flooded into the ditches, and the ditches carried pond water to the pastures on the farm. Using a sheet of metal or cloth "flagging," he also dammed the irrigation ditch to overflow into the landscape. With a metal chain sewn into the bottom of the flagging and a rope sewn into the top, he staked the chain to the bottom of the ditch and the rope to the top to create a temporary movable dam for the ditch. The irrigation ditch serves the purpose of a pipe, delivering water to flood-irrigate hilly areas with slopes up to 5 percent.

To build an irrigation ditch, start at the drain of your pond or other type of water storage. I peg a line with markers from the drain at a 0.5 percent slope downhill; this represents the lower lip of the ditch. Place markers every 10 feet to achieve a perfect 0.5 percent slope. Sometimes I paint a line on the ground with marking paint to simplify the process. Make sure the lower lip below the line remains undisturbed during excavation. If any part of the lower lip varies from the 0.5 percent slope, water won't spill evenly over the lip when the ditch is dammed. During excavation place soil on the uphill side of the ditch and leave the lower side undisturbed.

Yeomans dug his irrigation ditches a maximum width of 4 feet at the top and 2 feet at the base.[8] He used a delver, a ridger, a lister, or a large single-furrow plow to break the soil, then a hand shovel to move the soil uphill.[9] Yeomans concluded, "There is no suitable implement that will construct an irrigation drain" because soil always gravitates downhill.

However, I don't think Yeomans had a rotary plow. This implement is ideal for constructing small irrigation ditches. Small versions of the irrigation ditch can flood-irrigate small pasture areas of approximately 1,200 square feet, or about the size of a pasture for a small flock of chickens.

Even a few irrigations per year made a noticeable difference in my pasture.

The rotary plow looks like a giant corkscrew pulled through the ground by the tractor. The corkscrew spins and throws soil toward the side as it excavates the ditch. The rotary plow is usually used to till soil and build raised beds, but its throwing action makes digging irrigation ditches a breeze. One or two passes of the rotary plow forms a ditch with the water-moving capacity of a 4-inch pipe. If the ditch fills in with soil, cleaning it out is as simple as making another run through the ditch with the rotary plow. The entire process of installing a ditch with the rotary plow is nearly as quick and easy as installing layflat tubing for drip or flood irrigation.

Always place irrigation ditches below roads and access paths so water flowing from the ditches

Figure 10.15. A small irrigation ditch is flooded to irrigate pasture. When the ditch is dammed, water spills over a 25-foot section.

during flood irrigation won't pass over the road. Long ditches should also have flagging left in place to prevent water from flowing down the ditches during rains and causing erosion. The irrigation ditch works well in conjunction with keyline pattern cultivation to distribute the water evenly over the land.

Keyline Pattern Cultivation

Taking advantage of water's natural tendency to flow downhill, keyline pattern cultivation harvests rainwater and builds rich, fertile topsoil. Normally, water flows from ridges into valleys. The ridges stay dry, and the valleys accumulate moisture. In keyline pattern cultivation you place shallow rip lines in the soil in a direction that brings the water from the valley toward the ridges to distribute water evenly over the land and increase infiltration. The net effect is that the rip lines hold water for infiltration instead of the water running down the slope. If water moves at all it moves toward the dry ridges. With more water in the soil, plant growth and soil microbes increase. The increased plant and root growth builds soil as plant matter decays to feed soil life.

P. A. Yeomans, who invented the technique, and Bill Mollison, who promoted the technique, claimed that keyline pattern cultivation was the quickest way to build soil. Keyline pattern cultivation intrigued me for years, but I first had to implement the technique before becoming a convert. My holdup was the specialized equipment I thought was necessary to take on the task. Normally, keyline or chisel plows carry out the pattern cultivation. Similarly to a chisel plow, the keyline plow has narrow, stiff tines that scratch lines in the landscape. The specialized narrow tines are designed to keep soil layers intact without turning the soil or bringing subsoil up to the surface. Some keyline plows have cutting disks in front of the tines so the soil surface is left smooth and level after the cultivation. Tine spacing on keyline plows ranges from 18 to 24 inches apart.

A closer examination of Yeomans's work revealed that before the invention of the keyline plow and before the chisel plow was widely available, Yeomans used rippers commonly found on earthmoving equipment. Bulldozers and box scrapers for tractors commonly employ ripping tines to loosen the soil, enabling the blades to move the loosened soil. After reading about

Figure 10.16. I use the scarifiers on a box scraper to cultivate in a keyline pattern fashion.

Yeomans's earlier techniques I realized I could use ripping devices that attached to my box scraper. By shortening the top link on the box scraper as far as possible, I engaged the ripping devices without having the blade of the box scraper touching the ground. I then used the simple, and widely available, box scraper to implement keyline pattern cultivation.

With rippers Yeomans recommends a spacing of 12 inches between ripping tines. The tines leave a rougher appearance in the landscape in comparison to the specialized keyline plows, but I've gotten excellent results.

To implement keyline pattern cultivation, mark the grade lines on the landscape toward the ridges at a slope of 0.25 percent. Approximately 50 feet downhill or uphill of the first grade line, install an additional grade line at the 0.25 percent slope toward the ridge to guide the tractor. Using the rippers on a box scraper, keyline plow, or chisel plow, put shallow rip lines of a consistent depth into the landscape, following the slope of the marked grade lines. The maximum depth is 2 to 3 inches before engaging the soil-moving blade with rippers on the box scraper. Keyline plows and chisel plows can penetrate much deeper. However, Mollison recommends a depth of no more than 4 inches for the first treatment.[10]

Before implementing the keyline pattern cultivation, I amended the soil with the necessary lime and phosphorous to bring the pH and nutrient levels within recommended limits. Keyline cultivation helped incorporate the material into the soil and prevent its runoff. The disturbance from the rip lines also creates a nexus for seeds to germinate for pasture renovation and overseeding.

The results of keyline pattern cultivation have amazed me. Before I implemented the cultivation, a 1-inch rain event filled up all 600-plus feet of contour bunds on our property. After implementing a single keyline pattern cultivation, we've had two five-hundred-year-intensity rain events, and the contour bunds failed to fill. The pattern cultivation has permanently changed the water-holding ability of the soil. Where soil and plants were thin and sparse, thick vegetation now grows. With the additional moisture in the soil, I can now grow water-hungry forage crops such as white clover favored by my chickens.

Growing Fuel, Fertilizer, and Feed

At the SOF, most of our fertilizer comes from growing cover crops and green manures. We grow these crops to supply subsequent crops with nutrients, prevent weed growth, control erosion, and build soil organic matter. Nitrogen-fixing legumes such as crimson clover, vetch, cowpeas, and sunhemp can add over 100 pounds of nitrogen per acre and several tons of organic matter. Two thousand pounds of organic fertilizer equals the amount of fertilizer grown by the cover crops per acre.

We use two techniques to harvest the nutrients and soil-building properties of cover crops. For early spring and midsummer vegetables we mow the cover crop to the ground using a flail mower, then till the material into the soil (see chapter 11). However, tilling destroys the life in the soil and turns precious soil organic matter into carbon dioxide. Cover crop roller crimpers kill cover crops and leave the residue on the soil surface. The residue protects soil from wind and water erosion and prevents soil moisture from evaporating. The thick blanket of residue also acts as a weed-suppressing mulch. Roller crimpers save time, since all you have to do is put the plants in the ground, then come back and harvest. It almost feels as if I'm cheating when the system works well.

However, to kill it with the roller crimper, the cover crop must be mature. If you roll a young cover crop it will stand right back up and start growing again. Crimp after flowers are well formed on the cover crop but before viable seeds form to become weeds. For crimson clover wait until the flower petals are withering on up to 80 percent of the head on the individual flower heads. With ryegrass pull an individual seed out of the seed head and give it a squeeze. The seed should be milky or doughlike for the ryegrass to die with crimping. At the SOF we have a two- to four-week window in late spring when the cover crop is mature enough to use the roller crimper, allowing us to plant almost half our farm using the no-till practice.

Early- and late-maturing varieties extend the crimping window.

We plant the early spring crops and all directly seeded crops into tilled cover crop fields. Once the overwintered cover crop reaches maturity, we roller crimp the cover crop and plant tomatoes, peppers, eggplant, winter squash, summer squash, sweet potatoes, and melons through the thick mulch made by the roller crimper. The field must meet the following requirements for successful organic no-till production:

- Thick, dense mature cover crop.
- No perennial or biennial weeds in the field. Till fields to get rid of perennial weeds,

Figure 10.17. Preparing a no-till field with raised beds for planting with our roller crimper.

then convert to no-till once the weeds are eliminated.
- Good field drainage with no low areas to pool water.

Planting through the thick mulch is far more difficult than planting into tilled soil. We use a posthole digger to plant larger 4-inch transplants and a bulb planter to plant smaller transplants. To plant sweet potato slips, we use a dibble or trenching shovel. I've heard of farmers modifying mechanical waterwheel transplanters to plant through no-till residue. We add a drip line on top of the mulch to help mark the rows and water the plants.

For cover crops we use grain rye mixed with crimson clover as our first maturing winter cover crop for no-till. Vetch matures later than the crimson clover but can also be used with grain rye. The vetch allows a second crimping two weeks after the crimson clover to help spread out the planting time. The best weed suppression happens in the first month after crimping. Therefore, it's imperative to plant within a week after crimping to take advantage of the weed suppression. We eliminate any weeds that penetrate through weak areas by using the flame weeder while it's raining to prevent the mulch from catching on fire. For long-season crops such as peppers and eggplant, I add a layer of leaf mulch recycled from the city's fall cleanup to the top of the crimped mulch using the manure spreader. The extra layer of mulch gives longer weed control than the crimped cover crop alone.

I've wanted to plant fall vegetable crops into summer cover crops killed by the roller crimper but haven't been able to get the cover crops weed-free or dense enough to take the risk. The crimper kills sunhemp and millet but not cowpeas. The sunhemp we planted was a large dense crop, but once the residue was laid down, gaps were present that would have allowed weeds to grow. I've never been able to get a good weed-free stand of millet to grow. A mix of both millet and sunhemp works well if you can keep it weed-free.

Crimping Raised Beds

One issue we faced was using the roller crimper with raised beds. Traditionally, roller crimpers are used on flat ground to access the full crimping action of the roller. Raised-bed roller crimpers are available to crimp the furrows between raised beds. Recently I've been able to crimp raised beds with a traditional crimper.

First, I form the raised beds before planting the cover crops. Leaving just enough space between the raised beds to fit the tractor tires, I then broadcast the cover crop seed onto the raised beds and run the bed shaper back over the beds to help bury the seeds. I've noticed the cover crops perform better on raised beds and become mature earlier. When it comes time to crimp, the tires of the tractor provide enough pressure to kill the cover crop in the furrows and the roller crimper kills the cover crop on top of the beds.

While most of our nutrients come from cover cropping either tilled into the soil or crimped using no-till techniques, some of our heavy-feeding crops require additional nitrogen and other nutrients. Currently, we use seed meals to compensate for the difference. Inspired by the work at California-based Farm Fuel Inc., our latest endeavor aims to grow our own seed crops and press them to make meal for fertilizer and oil for fuel.

After visiting Farm Fuel's operation in Santa Cruz, I brought home 15 pounds of mustard seed to plant on our farm. I broadcast the 15 pounds of mustard seed over 1.5 acres of land in late September. I was concerned the mustard wouldn't make it through our winters (which are colder than California winters) because cold tolerance varies

Gold in Those Hills

Driving along Highway 1 through California's central coast, everyone notices the yellow fields blanketed in mustard blooms. For most, this springtime display gets no more than a passing glance. However, a small group of California farmers and scientists saw gold in those hills.

Farm Fuel Inc. is a "by the farmers, for the farmers" manufacturing and research company located in Santa Cruz. Focused on eco-conscious sustainable solutions, it turns domesticated mustards into soil amendments while providing oil to produce biodiesel, an alternative to petroleum-based fuels.

After harvesting and crushing the plants' seeds, the company processes the oils into biofuel to use on their farms and sells the remaining seed meal as mustard meal.

The Organic Materials Review Institute has certified Farm Fuel's Pescadero Gold Mustard Meal fertilizer for use on farms and in gardens with a nitrogen-phosphorus-potassium ratio of 4.5-1.5-1.15, comparable to nonorganic commercial fertilizers such as cottonseed meal. While it's not labeled for weed control or pest control, research shows that mustard meal acts as a biofumigant.[11] Mustard meal contains glucosinolates, the compound responsible for the pungent flavor of the condiment. The breakdown of glucosinolates in the soil prevents seeds from germinating but doesn't affect transplants added to the soil. The meal suppresses soilborne diseases from bacteria, fungi, or nematodes.

"If we're going to grow healthy foods from healthy plants, then we need healthy soil," said Stefanie Bourcier, CEO and research coordinator for Farm Fuel. "In order to have healthy soil we have to use products that don't wipe out soil biology, but rather preserve and work with the soil's biology."

With this in mind, Farm Fuel has recently developed another product called Pajaro Valley Gold as part of an anaerobic soil disinfestation (ASD) treatment. ASD is a preplant soil treatment developed in Japan and the Netherlands as an alternative to methyl bromide, a soil fumigant and pesticide that is being phased out because of its effects on the ozone layer.

Pajaro Valley Gold is tilled into the soil, watered thoroughly, then covered with an oxygen-impermeable tarp. These anaerobic conditions control pathogens and restore a healthy balance to the soil. Tested on strawberries in coastal California and on peppers and eggplants in southeast Florida, Pajaro Valley Gold has produced positive results.

Farm Fuel works with teams of researchers from the University of Idaho and the University of California–Santa Cruz to test the effects of the mustard meal and ASD treatments. They've had exceptional results, particularly in strawberry cultivation, with a 50 percent increase in strawberry yields.

Figure 10.18. A field of mustard flowers next to the farm.

among varieties. Like most other fall-planted cover crops, the plants were off to a slow start. We had a particularly harsh winter for our mild southern climate, with temperatures dipping to 12°F. The plants took a little beating but came back strong. By March our field was flourishing with gorgeous yellow flowers that lasted five weeks.

A surprising benefit of the mustard was the timing of the flowers. Early spring nectar sources are hard to find in our area. As a result, our bee colonies suffer without nectar and pollen. The March mustard flowers helped build strong hives, preparing the bees for major nectar flows following the mustard in mid-April. The mustard honey has a tendency to candy, so I advise caution if it's robbed from hives instead of being used for feeding the bees for early hive growth. Another bonus: The mustard flowers helped beneficial insects get off to an early start, aiding with pest problems on our vegetables.

The mustard seed is mature when all the pods start to turn brown and the cotyledons inside the seeds turn from green to yellow. In our climate plants are ready in May. Converting mustard seeds into fertilizer and oil involves a little machinery magic orchestrated by David Thornton, organic and biofuels project coordinator for Clemson University. Thornton designed and built a portable biodiesel lab on a trailer, which looks more like it belongs in a NASA facility than in a composting lot. The lab teaches students how to convert grease from the school cafeteria into biodiesel used by the university's maintenance fleet. Thorton has keenly researched growing fuel for the university, looking at such crops as sunflower, mustard, and canola to provide the raw material. Canola is similar to mustard but contains lower quantities of erucic acid and glucosinolates, making the meal a good animal feed and oil crop. Sunflowers produce high yields of native oil and feed crop. Thornton harvests the seeds and crushes them in an oil press. Oil flows out the bottom of the press while pellets of fertilizer or feed fall from the end. Thornton also developed a technique using his oil press to extract oil from black soldier fly larvae. After dehydrating the larvae, he combines them with dried sunflower seed in a 30 percent black soldier fly to 70 percent sunflower seed mix (see chapter 9).

Thornton wanted a small plot combine to harvest seed crops for his operation. When he started shopping around, most combines he found were around thirty thousand dollars and needed to attach to a tractor. Additional Internet searches turned up a small walk-behind, diesel-driven rice and wheat combine available through the online shop Alibaba (Zhengzhou Amisy Trading Co., Ltd).

You've probably stumbled upon the Alibaba site when shopping online but shrugged off making a purchase because the products are made in distant places, such as China. As it turns out, small walk-behind combines are not made in the United States but proliferate on small farms overseas. As a result, small farms in the United States don't grow grains beyond a small novelty amount harvested by hand. The mini rice and wheat combine harvests at a rate of 0.16 to 0.25 acres per hour using an efficient 9-horsepower diesel engine.

Thornton bravely took the plunge and ordered the walk-behind combine from China for the 3 acres of grain we wanted to plant. The combine cost thirty-two hundred dollars plus another nine hundred in shipping and tariffs. It arrived in a box and required serious assembly. So far our experience with the combine has been limited, but it's done well with wheat. We also tried it on sunflowers with some success, but the plants needed longer to dry in the field. Combining mustard and crimson clover produced fair results. I tried the combine on ryegrass but experienced poor results because the tall rye tended to clog the machinery. The height of the cutting bar on the combine can be raised slightly, but the machine is designed for

Figure 10.19. David Thornton and his walk-behind combine, seen from the front (*left*) and the rear (*right*).

low-growing crops. The best crop so far has been buckwheat. Planting crops densely, using defoliants, or mowing grasses before they set seed may reduce crop height and improve the performance of the combine. In addition, it helps to cut the crop with a cutter bar, and leave it in heaps in the field to dry, then hand feed it into the combine.

Yields for mustard in our area average around 1,200 pounds of meal an acre and about 100 gallons of oil. We're hoping also to use it to grow and harvest our own cover crop seeds. Andy Pressman, sustainable agriculture specialist with Appropriate Technology Transfer for Rural America (ATTRA), has also used the small plot combines from China and found an importer in the United States at www.eqmachinery.com. His experimentation with wheat and barley produced fairly good results. While I'm still a novice at growing grains, I'm looking forward to having a small tool for our farm that makes it possible.

SUMMING UP THE FUNCTIONS

1. Sloping the field properly helps keep beds and pathways dry.
2. Properly sloped furrows flood-irrigate beds.
3. Cover crops grow nutrients for vegetables.
4. Graded fields reduce erosion.
5. Shallowly sloped furrows and diversion ditches harvest rainwater.
6. Buckwheat planted under manifold lines attracts beneficial insects.
7. Raised beds reduce disease.
8. Buckwheat planted under manifold lines reduces mowing under irrigation tubing.

PART TWO

CASE STUDIES IN LIVING BIOSYSTEMS

CHAPTER 11

The Right Connections
Bio-Integrating the Farm Core and CSA

Now more than ever, consumers want to know where their food comes from. They want to meet the farmers, view the gardens, perhaps even harvest some of their own fruit and vegetables. Farms located close to customers have a huge advantage: Customers prefer to come to the farm. This saves the farm time and money that's usually spent packaging and delivering produce. In return, customers experience the farm that grows their food and ultimately feel better about their purchases.

Every farmer's dream is selling his produce before it's planted. The community supported agriculture (CSA) program makes that dream a reality. Before planting the first seed in the ground, we've already sold all our vegetables at retail prices — guaranteeing our labor will not go to waste. However, there is a downside. A difficult season — too much rain or not enough — can quickly transform that dream into a beast of burden. The pressure's on to produce, and the clock is ticking. So our most important tools are efficiency, organization, and planning.

Often, we humans organize and plan without considering all the factors influencing our results. A farm or garden's success depends on how well we work in concert with the natural elements of sun, water, wind, and soil. One of the greatest

farm designers of all time, P. A. Yeomans, devised a process to guide the design, called the Keyline Scale of Permanence. Yeomans examined the permanent parts of the landscape — climate and landforms — before determining the placement of water storage, roads, trees, buildings, and fields. I applied Yeomans's principles while designing our CSA at the Student Organic Farm (SOF).

Ridges and Roads

First in the design process we need to understand landforms to consider the effects of other elements. Landforms are divided into valleys and ridges. The ridges represent the highest point in the landscape, and the valleys are the areas between the ridges. Water runs off the ridges and down through the valleys, eventually flowing into rivers, then oceans. Landscapes have main ridges separating watersheds and secondary ridges or shoulders descending from the main ridge. Even land that appears completely flat has subtle elevation changes that create ridges and valleys. When rain falls, water that doesn't soak into the soil moves in a direction perpendicular to the ridgeline into the valleys on its way downhill.

Why do some roads wash out, flood, and require continual maintenance while others stand strong?

Figure 11.1. Although the terrain at the Clemson Student Organic Farm is only gently sloped, this road follows the ridge, and thus it stays dry during rain.

Elements of the System

Each bio-integration pattern is like a piece to a puzzle. The ultimate goal is fitting all of the pieces together to form a working, functioning farm. In my experience farmers and gardeners aren't philosophers, they're doers. They want to know why — but most importantly they want to know how. In this chapter you'll learn the best layout for your farm and the best tools for each task. You'll learn:

► To design roads and paths to reduce maintenance and facilitate field access.
► To use the proximity of farm components to locate farm buildings.
► To configure parking, access, and market buildings corresponding to public impact.
► How to facilitate a farmers' market or CSA program.
► The ins and outs of farm tools and equipment.

Placing a road in the proper location saves countless hours of maintenance and headaches later. Moving an improperly placed road usually pays for itself over time. Two rules guide the placement of roads and paths.

First, place main roads and paths on ridges. Ridges naturally stay dry during rain events, making them easily accessible during all times of the year. From a ridge, gravity takes water in both directions away from the road, preventing water from accumulating on the road and washing away road material.

The second rule concerns roads or paths running across the landscape perpendicular to ridges: Place these roads or paths above or below diversion ditches. Diversion ditches (described in chapter 10) are wide ditches sloped at a shallow grade of 0.25 percent or a 1-foot drop for every 400 feet of length. The diversion channels are like giant gutters that catch water flowing over the land and direct the water into ponds or ground storage areas. The roads and diversion channels run parallel to each

other, with water storage intricately connected to the road. Either the water storage dictates where the road is placed or the road dictates where the water storage is placed. Either way, the shallow slope of the road is protected from the erosive forces of water.

One farm I redesigned had a road that ran straight down a valley. Every time it rained, a river of water ran down the road and gravel eroded away as water flowed off the ridges and into the valley. At the base of the valley, sediment, muck,

Figure 11.2. This road blocks the natural flow of water across the landscape and channels it into the corner of the field. A culvert needs to be built here to let water pass through without washing the road away. If the road were placed 100 feet to the right, on the ridge, a culvert wouldn't be needed and establishing field drainage would be easier.

and water stood for days after the rain, making the road impassable. The farmer spent countless hours every year pulling his truck out of the mud with a tractor. We ended up rerouting the road along a higher ridge. The new road stayed dry year-round and was passable during and after rainstorms.

Following these two rules for road placement also guides field layout and size. Whenever fields are forced into a particular shape or standardized field size, trouble follows. A common mistake is making fields perfectly square in undulating landscapes for easier planning. However, the result is often poor field drainage and a constant need for road maintenance.

Another road, on the same farm, dissected the landscape to form a square field, with a disregard for landform and water flow. The road created a dam that directed water into the square corner formed by the field (see figure 11.2). Water, dammed by the road, would pool, then wash out the road with every rain. The farmer placed a

culvert at the corner to allow water to pass under the road to prevent the road from washing out. However, the culvert quickly filled with debris and silt from the shallow slope, and the problem reoccurred. We eventually rerouted the road along a ridge that forced the field into an odd shape. But without a road damming it up, the field drained earlier and became more profitable. Additionally, the road on the ridge rarely needed maintenance, saving the farm time and money.

Relative Location of Buildings

Older farms hark back to an era when the land intertwined with a family's income and walking was the prominent mode of transportation. Buildings were constructed nearby out of necessity — to save time and energy. On newer farms workers use golf carts, gators, and pickup trucks to travel between buildings spread great distances apart. All of those automated vehicles add to the expense of running a farm, from the purchasing to the maintenance. As the farm hires more employees the expenses compound.

Rather than creating rural sprawl on your farm, follow the permaculture principle of "relative location" by placing components in close proximity. Not only will the components assist each other, they will also assist the farm workers by making the farm more efficient. Throughout the following sections I've grouped buildings according to what the designer Christopher Alexander calls circulation realms. This concept prioritizes what buildings and components on a farm need to be close together so they can assist each other.

When I first started working at the Clemson Student Organic Farm (SOF), we spent hundreds of extra hours a year walking between farm buildings. The toolshed was located a three-minute walk away from the produce-washing shed and the

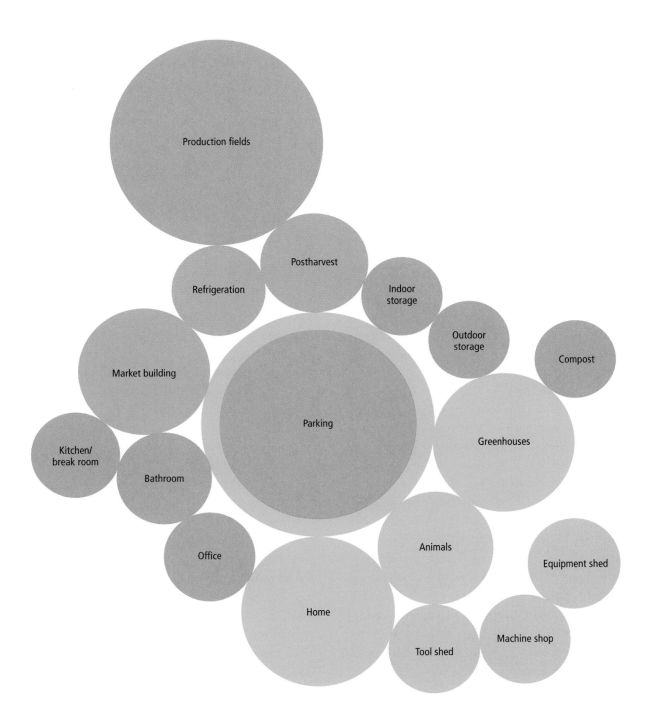

Figure 11.3. Connected farm components for a mixed production farm. The proximity of circles to each other represents the importance of placing the components close together. Colors represent circulation realms.

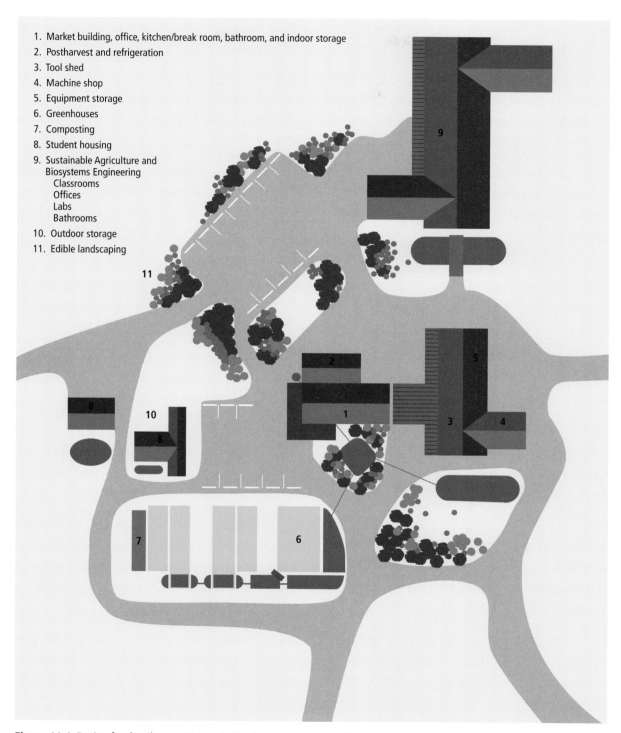

1. Market building, office, kitchen/break room, bathroom, and indoor storage
2. Postharvest and refrigeration
3. Tool shed
4. Machine shop
5. Equipment storage
6. Greenhouses
7. Composting
8. Student housing
9. Sustainable Agriculture and Biosystems Engineering
   Classrooms
   Offices
   Labs
   Bathrooms
10. Outdoor storage
11. Edible landscaping

Figure 11.4. Design for the Clemson University Student Organic Farm, with buildings arranged according to circulation realms and importance of proximity.

tractor shed. This apparently innocuous distance had huge ramifications. Since each of our four employees took a minimum of three trips per day between the buildings, collectively they spent seventy-two minutes per day walking between the distant shed and other farm buildings. Over the entire 248-day growing season that amounted to 297.6 hours of wasted labor, or $4,464 in labor expenses calculated at $15 per hour. Needless to say, moving the shed at a cost of $300 paid for itself rather quickly. On a farm every step costs money and time — two precious resources of a farmer.

Place buildings as close together as possible to form building complexes, and locate the buildings near main roads and below water storage. Since every 50 feet of gravel road costs about $2,500 to build properly and requires approximately $250 of yearly maintenance, locating buildings next to main roads saves time and money.

A common mistake is placing buildings a great distance from municipal roads to achieve privacy. This makes sense when municipal roads are busy, but for buildings next to rarely traveled roads, privacy can be achieved by planting a few evergreen trees by the roadside. So let municipalities do the work of maintaining the public roadway, and keep your farm roads short.

Yeomans recommends the age-old courtyard as the basis for the perfect building complex — all the buildings positioned closely together around a central courtyard. At the SOF our courtyard functions as a parking lot so all the buildings are easily accessible. To apply Yeomans's system, analyze every farm component and how it connects to other farm components. Then either combine connecting components or place them as close to each other as possible. If fire safety is a concern, space the buildings 50 feet apart to prevent fires from spreading.[1] See figure 11.3, which is a diagram of common farm components and their associated applications.

Parking and Access

When parking areas are not easily recognizable, people park all over the landscape. This is a problem because parked vehicles compact the soil and limit the usefulness of the land if not contained. A weed-free gravel area with concrete parking bumpers works well to define direction and location of cars. Keeping the parking area weed-free organically can become a challenge. Burn-down organic herbicides are available, but it costs about forty dollars to treat our parking area, and the herbicides don't kill the weeds — they only turn the tops brown. If weeds are large I burn them down with the organic herbicide, then flame weed them a week later, burning all the dead material and providing a permanent kill. However, it's much easier to hand pull or flame weed small weeds before they become large and out of control. I also recommend grading parking areas to minimize low areas and draining water to retention basins.

It's easy to determine how many parking spaces you'll need for a home or business by looking at your local ordinances. The county where I reside requires two parking spaces for every dwelling, one space for each 200 square feet of retail space, and one space for every 100 square feet of home business. For nursery and greenhouse operations the county requires one parking space for every 300 square feet of floor area used for sales, plus one space for each employee and a space for every business vehicle.

However, it's difficult to determine the number of parking spaces necessary for farm markets or CSA programs. Unlike most retail businesses, the customer base of farm markets and CSAs is confined to a short sales period. Most retail businesses are open every day all day, which distributes customers over a long period of time. With our on-farm CSA and market programs,

I prefer to have customers come during a short time period to limit the amount of time the farm crew spends selling or distributing the produce. This means that our large customer base visits during a short span of time, which requires more parking spaces.

To lessen the impact, we offer two pickup days a week and split our customer base into two groups accordingly. Customers pick up their share any time over a three-hour period on their assigned day. Offering longer pickup times or more pickup days would reduce our parking needs, but it would also decrease productivity on the farm.

CSAs differ in how they distribute their goods. Some operate like cafeterias, where CSA members package their own boxes, while others hand off the produce in prepackaged boxes. At the Clemson University farm we use a cafeteria style setup for our 125 members, and they pack their own boxes. Cafeteria style obviously takes the customer longer than picking up a prepackaged box. Our eight parking spaces have proven inadequate with our current pickup system for our 125 CSA members. I recommend one parking space for every five members when pickup is distributed cafeteria style over a three-hour period. I created the following formula to determine the amount of parking spaces needed for CSAs based on pickup time, the number of CSA members, and the number of employees present:

$$\text{Number of parking spaces needed} = M \div (C + H) + E$$

where

M = number of CSA members on pickup day; C = a variable factor: 5 for cafeteria style CSA and 10 for pickup with prepackaged boxes; H = number of hours beyond three that pickup lasts; and E = number of employees present during pickup

Parking and Access Connections

Parking lots are connectors between roads and buildings. Ideally, they're connected to all the buildings in a courtyard or linked fashion. However, priorities sometimes dictate that only one building has a parking lot. In situations where not all buildings can connect to a parking lot, use the following list to guide choices of which buildings should connect. For example, a market building has a higher priority for parking than kitchens and bathrooms.

1. Market building
2. Home
3. Greenhouses
4. Office
5. Kitchen/break room/bathroom/ handwashing

Here's an order of priority for roads and access:

1. Parking and all associated parking connections
2. Fields
3. Produce wash station
4. Refrigeration
5. Animal housing
6. Toolshed
7. Tractor shed
8. Machine shop
9. Composting
10. Outdoor storage
11. Indoor storage

For example, if 50 members are picking up CSA shares on Tuesday and 50 members are picking up shares on Friday over a four-hour period each day (cafeteria style) and two employees will be present, how many parking spaces are needed?

Number of parking spaces needed = 50 ÷ (5 + 1) + 2
Number of parking spaces needed = 10.3

Parking lots ruin the landscape, creating an unsightly jungle of multicolored metal. The designer Christopher Alexander recommends small parking lots, "serving no more than five to seven cars, each lot surrounded by garden walls, hedges, fences, slopes and trees, so that from outside the cars are almost invisible."[2] So if you need more than seven parking spaces, design a series of smaller lots instead of one large lot to prevent the landscape from becoming a "sea of cars."

Flow in a Market Building

On-farm sales operations usually distribute produce within a market building. Ideally, market buildings are enclosed and climate controlled. An open-air market is a romantic idea, but it wreaks havoc on the quality of the produce. Open-sided buildings are exposed to an onslaught of birds, rodents, dirt, wind, rain, sunlight, and theft, leading to the deterioration of everything inside the building. An enclosed, climate-controlled building is more suitable for food storage since it protects produce from heat, cold, and animals.

The market building is the face of the farm. Create a flow path guiding people from the parking area through the building. Enhance the customers' experience by beautifying the transition between the parking area and the building with a pleasant view of flowers or ponds. Even if you have an open-sided market building, install

Prioritizing Market Building Connections

Locating the market building as close as possible to other farm components that use it saves time, and ideally all these components would be inside one building. However, this is not always the case. Use the following list to prioritize which components should be placed closest to the farm building.

1. Refrigeration
2. Parking
3. Produce wash station
4. Office
5. Bathroom
6. Kitchen/break room
7. Indoor storage
8. Home

a sheltering gable roof that provides a covered entrance and protects produce and customers from the sun and rain.

CSA markets have a check-in at the entrance, and retail markets have a checkout near the exit. A linear flow through the tables of vegetables from entrance to exit moves customers through all the produce efficiently. Alternatively, produce is prepackaged and delivered. Efficient delivery requires stackable boxes and bins. Since boxes cost $1.50 each, it would cost our farm $4,200 to buy boxes for every one hundred CSA members over a twenty-eight-week season. The costs of purchasing and storing boxes would have to be passed on to the

Figure 11.5. Customers picking up vegetables at our cafeteria-style CSA pickup. A white board lists the amount of produce available for each customer to pick up.

customers. By picking up and packaging their own produce, our CSA members not only save the farm money, they also save themselves money.

We request that our customers bring their own plastic bags for packaging produce, but we also provide bags if needed. We've spent as much as five hundred dollars per year for plastic bags to supplement what the CSA members bring. To encourage members to bring their own bags, we occasionally offer a limited-supply produce item as a special reward only to members who bring their own bags. This provides incentive for customers to reuse and saves the farm hundreds of dollars per year.

Maintaining good records requires keeping tabs on the amount of produce coming in and the amount going out. For organic production, record keeping is required. Commonly, items are recorded by weight, bunch, item number, or number of leaves, with each technique requiring a different amount of time during the harvest and record-keeping process. For example, weighing a bin of tomatoes is a lot quicker than counting all the tomatoes placed in the bin. Therefore, when harvesting tomatoes it's much quicker to weigh the tomatoes to keep track of records. However, for distributing the tomatoes, counting small individual quantities (say, six tomatoes) is faster than weighing a bunch of specific small amounts (two pounds of tomatoes), which would take a while.

Many hands make light work, so splitting the labor among many CSA members saves the farmer a considerable amount of time and money. Provide plenty of scales for the customers, and let them do as much weighing as possible. The farmer weighs produce as it arrives at the cooler or market building and records a total; then the produce is passed along to the customers, who weigh the individual lots.

Another way our farm cuts costs is by breaking the bunching habit. Previously we bunched all our greens for distribution. After I realized how many boxes of rubber bands we were using, I started thinking of new techniques to reduce our material consumption. Instead of bunching we now count leaves. Rather than recording the number of bunches harvested, we record the number of leaves harvested. We then offer CSA members a specific number of leaves to take during distribution. The CSA members count their leaves and package them into their bags. There's no need for rubber bands — it's just another way to pass the savings on to our customers.

Some items are so tedious to harvest they inevitably end up netting the farmer very little income. CSA members can harvest some of these items themselves when they visit the farm to pick up produce. While I would never make my customers stoop down and pick beans, I've found they do enjoy harvesting flowers and cherry tomatoes. I plant these items close to the market building so the customers don't have to walk far. The cherry tomatoes are placed in cages to enable an easier harvest, and we maintain access around the crops by mowing. With just these two crops alone we save six hours of labor for each CSA pickup and enhance the farm experience for our customers — everyone wins.

Efficiency of Harvest and Postharvest

Harvesting, washing, and moving produce into refrigeration takes more labor time than growing the food, so making the process as efficient as possible saves money. Every farm needs a building or an area dedicated to cleaning and storing produce, as well as the harvesting equipment. To efficiently harvest produce out of the field, we use the following flow: hand washing, harvest bin cleaning, harvesting, produce cooling and washing, labeling, and refrigeration. We connect all these components into one circulation realm.

Hand Washing

First and foremost, it's necessary to have a hand-washing station. Ideally, multiple faucets allow several workers to use the sink at the same time. A pump-style soap dispenser is more sanitary than bar soap. I like to place sunscreen in this area as well, because I wash the sunscreen off my hands after application.

Figure 11.6. A two-person handwashing sink located next to the postharvest area encourages frequent hand washing.

Harvest Bins

For food safety purposes and to comply with Good Agricultural Practice (GAP) guidelines, we clean and sterilize our bins before using them to harvest in the field. If you use new wax-covered boxes to harvest, skip this step.

Specialized harvest bins are an integral part of the farm and are worth the extra expense. Purchase tough plastic bins that you can clean, sterilize, and stack. Our harvest bins have slits to enable ventilation and drainage. Either they come in a collapsible form or they nest inside each other to save space when storing. The nesting kinds last a little longer since they incorporate fewer moving parts. We can stack the bins on top of each other in the fridge, which eliminates the need for shelving and functionally doubles the capacity of the refrigerator. Each type of bin is rated for a certain weight and type of produce; larger bins for leafy vegetables and smaller bins for heavy root crops and fruit. Exceeding the weight limit can damage the bins and make a large stack unstable and unsafe. We purchase all our bins from the same manufacturer to guarantee they will stack properly if we mix them together in the cooler.

Storing unused bins for a large CSA takes up a lot of space. I prefer to store the bins under the tables used for produce washing. To comply with GAP guidelines, we must raise the bins off the ground. I accomplish this by placing them on top of square or round metal tubing.

We dunk bins in a washtub for cleaning and sterilizing, adding food-grade bleach (which meets the standards of NSF International) to bring the chlorine levels to 150 parts per million. A stiff brush is kept next to the tub to help remove any caked-on dirt. We've converted 100-gallon plastic stock tanks into washtubs for bins and produce. Doing the conversion well is important: The first stock tank I converted was horribly inefficient to

Figure 11.7. Harvest bins are dunked in sanitizing water, then stacked on pallets for rinsing. The pallets slope slightly forward, facilitating removal of debris.

clean. It took seven minutes to properly remove dirt and debris accumulating on the bottom. Since we clean the tub five days a week over our twenty-eight-week harvest season, we spent 16.3 hours a year cleaning the tub. Now that I've created a better way to convert the tank, we spend only one minute cleaning the tub, totaling 2.3 hours per year. The amount of time saved easily pays for the cost of properly converting the tubs and highlights the importance of making systems as efficient as possible.

Stock tanks are not designed for debris removal the way sinks are. To make the stock tank easy to clean, slope the bottom toward a recessed drain. First, use a hole saw to cut an opening into the tub as close as possible to the edge on the end of the tub. I make the hole large enough to insert a sink-style drain. The bottom of the stock tank is laced with reinforcing ridges that you'll need to avoid while drilling the hole to allow the drain to seat properly. Next, use a chisel, angle grinder, or Dremel tool to shave away the plastic on the tank around the edge of the hole, allowing the rim of the drain to sit level with the bottom of the tank. Then install the tank on top of cinder blocks supported by a 4-inch gravel base. Adjusting the depth of the cinder blocks within the gravel base, carefully slope the tank slightly toward the drain on the side of the tank. To clean the tank, place a coiled-style hose with a spray nozzle nearby to use to wash the debris toward the drain.

The drainpipe then slopes continuously at 2 percent so debris in the pipes properly discharges into a gray water or municipal system for treatment. You can also discharge water into the landscape; however, chlorine levels must be reduced to 0.1 milligrams per liter to comply with EPA regulations for discharge of chlorine. Letting the water sit overnight helps evaporate the chlorine.

Once we've dunked our bins into the washtub for cleaning, we stand them on their sides on top of pallets. The pallets slope slightly forward so debris easily flows out the front when we spray them with water. We use a high-pressure hose nozzle to force the debris out. A wall behind the bins supports them, preventing them from being tipped over by the force of the water.

Organizing for Harvesting

The first step toward efficiency is organization. We store all our harvest equipment together so we're not running around at the last minute gathering tools. Here's a basic list of equipment necessary for hand-harvesting tasks: harvest bins, knives, rubber bands, gloves, trowels, shovels, scissors, and baby-greens harvesting equipment.

The second step toward efficiency is preparation. Before harvesting we write a harvest list, then clean the bins and harvest equipment. Instead of counting rubber bands out in the field, we count and group them ahead of time for bunching produce. Rather than making multiple trips, we load all our harvest equipment into a clean truck or cart and drive it out to the field. Here's a little trick of the trade: Start harvesting produce a distance from

your truck or cart, then work your way back toward the truck. You'll only have to carry the heavy full bin a short distance to your destination.

To reduce stooping and limit lifting, I prop a bin against my leg so I can harvest while standing. When I'm ready to move down the row, I drag the bin along with me. Some vegetables require cutting rather than picking. We use specialty knives and tools to make the job easier. Lettuce knives cut by pushing the knife forward; broccoli knives work in a similar way but have a rounded head to prevent damage to emerging side shoots. We use a serrated-blade green harvester with an attached bag to harvest baby lettuce quickly. Some harvesters have motorized blades powered by cordless drills to harvest baby greens effortlessly.

One of the most important skills to develop on a farm is using both hands. Two hands hold more fruit than one — that's what I tell our rookies at the farm. Using two hands cuts the harvest time in half. When we're picking small fruits or vegetables like cherry tomatoes and okra, we strap 1-gallon buckets over our shoulders. This allows us to fill the buckets without having to stoop to move them. Once we fill the buckets, we dump the produce into a larger bin.

Moving large fruit such as watermelons and cantaloupe takes teamwork. We use our "bucket brigade" technique by forming a chain and passing the fruit from person to person until the fruit finally reaches the truck. Before rolling watermelons off the truck and into the market building, we hose them down in the truck bed. Larger operations use watermelon boxes and pallet jacks to move the melons. We place cantaloupes in bins, where they are washed and hydrocooled before refrigeration. The watermelons can remain in the market building at room temperature.

To give our customers a longer-lasting product, we harvest our produce during the cool morning

Figure 11.8. Chelsi Crawford harvests lettuce into bins using a lettuce knife.

hours. I usually start with the most perishable products, such as leafy greens and lettuce, then move on to less perishable crops, such as beans, tomatoes, and okra. Harvesting the dirtiest crops last keeps the wash water clean as long as possible.

Plant diseases are spread easily during harvesting. Since older plants are more prone to disease, we always harvest from the younger plants first and move on to the older plants later. I've also split staff into separate groups to harvest in different areas if I'm concerned about transferring disease from one field to another.

Produce Washing

Within hours of harvest, vegetables begin losing their quality. So food preservation should start immediately. Different vegetables require different handling; some need hydrocooling, while others need curing. It's important for every farmer and gardener to know how to extend the life of the harvest.

Basically, all crops fall into one of two broad categories: Either they like cold storage temperatures or they don't. Vegetables sensitive to the cold may develop skin lesions, decay susceptibility, or shriveling caused by a disorder known as chilling injury. Tomatoes, peppers, cucumbers, and Irish potatoes are but a few of these sensitive vegetables. We wash our peppers and cucumbers and store them in our cooler above 45°F. We also wash the tomatoes but store them at room temperature. The Irish potatoes like it dirty; they tend to rot once washed. So we simply stick them unwashed in the cooler inside stacked bins.

Some veggies thrive from a quick dunk in ice water. This cold submersion, known as hydro-cooling, drastically improves the quality and shelf life of produce such as greens, broccoli, and corn. When I sold at farmers' markets I was shocked by all the wilted ugly veggies displayed for sale. By dunking the produce in cold water, you can quickly remove the field heat — the first line of defense against spoilage.

Ideally, your produce-washing area is enclosed and connected to the market building. Enclosed buildings prevent birds, rodents, and vermin from damaging your harvest and also control the climate for better food preservation. But even a hoop house will suffice. Just cover it with an 80 percent shade cloth to convert it into a wash facility. For vegetable washing and hydrocooling, set up one table for stockpiling incoming produce from the field alongside a washtub to clean and cool the produce, and another table for draining vegetables and labeling bins before refrigeration. Not all vegetables can tolerate icy cold water, so we just clean those in cool water.

The tables and washtub are commonly set up in a linear fashion. This way, as we bring produce out of the field, we can move it in a straight line on its way into the cooler. To maximize what little space we have, our system is set up using a "keyhole design." With a keyhole design, access is through the keyhole area in the center with tables and tubs placed around the access point on three sides. One person can stand in a single location and move produce through the design taking very few steps (see figure 11.9).

We've used tables as basic as untreated 2 × 4s or metal greenhouse benches. Roller conveyors can substitute as tables to move the bins easily while preventing wear and tear from sliding. The washtub can be as simple as a stock tank, as described earlier in this chapter, or as sophisticated as a custom-built stainless steel sink. Adding ice to the water aids in cooling the vegetables, or you can connect the washtub to a milk chiller for cooling. Automated commercial washing stations are even better but get pricey. Commercial systems use brushes, sprayers, or bubbles to clean a wide variety of products efficiently. Prices range from four thousand to a hundred thousand dollars. I've found the best deals on www.alibaba.com.

Some veggies are fussy — they don't like a bath or the cold. Crops such as sweet potatoes and garlic require a different process, called curing. We cure our sweet potatoes inside a walk-in cooler without the refrigeration running, stacking them in bins to promote airflow. A small electric radiator heater warms the temperature to 85°F inside the cooler. After three to five days we move the sweet potatoes to a well-ventilated cool area in the north side of the market building.

To cure garlic we lay it on top of benches in our large propagation greenhouse for about six weeks. Covered with an 80 percent shade cloth and equipped with a large fan, the greenhouse provides perfect conditions for curing the garlic.

Labeling and Weighing

On a table next to the washtub, we label the bins as they drain. The label includes the harvest date of the produce as well as the field where it was harvested. This information helps us monitor the age of the produce and where it grew in case any problems occur. We use painter's tape to make our labels; then we simply tape the labels onto the bins. For larger scale operations a tagging gun similar to the ones used in grocery stores to mark prices on produce would work well. The tagging guns come with rolls of removable labels.

Before storing the produce, we weigh it or count it. Weighing is much quicker. Maximum bin weight is around 50 pounds, so we use a 100-pound

Figure 11.9. A "keyhole design" produce-washing station to efficiently clean and cool produce before moving it into refrigeration. A table collects the produce before hydrocooling, and a table drains the produce while it is labeled for storage. Bins slide onto a scale for weighing. Lifting is reduced through the whole process.

scale to accommodate all bin sizes. We try to position our scale so we can simply slide the bins onto it without any heavy lifting. Our backs thank us at the end of the day.

Refrigeration

Different vegetables require different storage temperatures, from room temperature to cool to cold. Ideally, a farm should have two coolers with adjustable temperatures. One of the coolers stores produce suitable for cold temperatures, and the other stores the sensitive summer vegetables at a warmer temperature. If you're using only a single cooler, start the season with cold storage at 32°F before you harvest summer vegetables. Raise the temperatures to 45°F once summer vegetables are ready for harvest. Even though some vegetables keep longer at 32°F, the warmer 45°F temperatures will suffice.

In our coolers we organize produce based on its age. Typically, we stack bins of similarly aged produce together. But if one stack includes a range of differently aged produce, we always place the oldest on top so it will be used first.

Figure 11.10. Stackable bins inside the cooler fit a lot more produce than bins placed on shelves. The cooler slopes slightly toward the door to facilitate cleaning and sterilizing.

Circulation Realm Connections

In order of priority, harvest and postharvest circulation realm connections are as follows:

1. Truck access
2. Market building
3. Office
4. Bathroom
5. Home or kitchen/break room
6. Indoor storage
7. Compost
8. Production greenhouses
9. Fields

For convenience and efficiency position your cooler inside or next to the produce washing room so you're not moving the washed vegetables a great distance.

For large operations a walk-in cooler is a worthwhile investment. Purchasing a new cooler is slightly more expensive than refurbishing an older cooler with a new refrigeration unit. You can custom design a new cooler, making good use of available space. If space is limited, you can place walk-in coolers outdoors and protect them with a thermoplastic polyolefin (TPO) roofing membrane. Make sure the membrane is white to reflect sunlight and keep electricity costs down.

Walk-in coolers are relatively simple to build. When purchased, they come broken apart into panels and pieces that are hooked together like a giant jigsaw puzzle. We put our cooler together

ourselves, saving thousands of dollars. Once the cooler was in place, we hired an HVAC technician to make the final connections for the refrigeration unit and charge the system with coolant.

Coolbot systems convert window air-conditioning units into refrigeration units. However, coolbots don't work well with temperatures under 36°F or if the cooler door opens frequently. With the abundance of produce going in and out of our cooler daily, our conventional refrigeration unit barely keeps up with our cooling needs, so a coolbot would not work for us.

Walk-in coolers come with or without a floor attached. The floorless coolers are situated on top of an insulated concrete slab. We built a slightly sloped slab for our cooler. It has a 1 percent slope in two directions toward the door, keeping the base of the cooler in one plane similar to sloping greenhouse platforms. The sloped floor enables easy cleaning. We simply use a hose to wash dirt and debris out through the cooler door. As long as both the cooler floor and the produce-washing floor are both concrete and at the same level, they'll be appropriate for using a pallet jack or hand truck to move produce between them.

Sheds, Garages, and Shops

The next circulation realm on our farm includes the toolshed, tractor shed, and machine shop. It's best to combine these components into a single building, either attached to the market building or placed between the market building and the fields or greenhouses, since they are so interconnected.

A small farm may combine these components into a small shed or garage, and a large farm may need larger buildings to house its equipment and tools. Regardless of the size of the farm, tractors always require tools for maintenance and repair, and fields always require equipment to grow crops. So placing components as close as possible saves time.

Toolshed

Our toolshed houses all the hand tools we use to cultivate crops in the field, as well as our tools to maintain the tractor and equipment. If the toolshed entrance and windows face south, the shed stays warmer in winter and cooler during summer. A south-facing entrance exposed to the winter sun also remains free of snow and ice and

Figure 11.11. Hangers for tools self-sort high-use tools to the outside and increase storage space.

takes advantage of natural daylight. The daylight exposes any mess accumulating in the shed to motivate cleaning.

When our wall space becomes limited, we build hangers to hold multiple tools. The tools we use most naturally sort themselves to the outer portion of the hanger. To build the hanger, use wooden dowels as the supports. Next, drill holes the size of the dowels into a clear piece of 2 × 8-foot lumber. Use a drill press to ensure all the holes are oriented perpendicular to the board. Position the wooden dowels inside the holes, and mount the hanger on the wall. Finally, shim the bottom of the board so all the wooden dowels rest at a slight upward angle.

I'm often asked about tools and equipment, so I've compiled a list of those I've found most useful throughout my years of farming. The list is constantly evolving as I learn about new tools and techniques. The following hand tools are useful for small and medium-size farms:

Collinear hoe. If I could only own one farm and garden tool, this would be the one. The collinear hoe is a lightweight tool with a sharp blade used for cutting small weeds under the soil. The long handle allows you to stand upright and hold the hoe with both thumbs pointing up. With practice, the hoe becomes an extension of your body, and small weeds are quickly eliminated. The collinear hoe works best on weeds 2 inches or smaller.

Wheel hoe. Slightly quicker and more efficient than the collinear hoe, the wheel hoe attaches a sharp stirrup-style blade to human-powered handles and a wheel to hold the weight. A rhythmic forward-and-back motion moves the blade under the soil, slicing the roots of small weeds and quickly eliminating them. Wide blades will take out larger areas of weeds, or use shorter blades to straddle narrow rows of vegetables, cutting weeds on both sides at the same time. The wheel hoe works best

on weeds 2 inches or smaller, but you can use it on weeds up to 4 inches with more effort.

Flame weeders. Using the power of propane, flame weeders kill small weeds with heat. The flame weeder is less successful on grasses unless the grass is less than ½ inch high. You can eliminate most broadleaf plants at the three-leaf stage. In addition to weed size, effectiveness also depends on a smooth soil surface, as irregularities can deflect the flame and prevent heat transfer. With our raised beds, I find the furrows between the beds create angles that make flame weeding not as effective. But the flame weeder is successful on the tops of the beds when the soil is smooth.

The flame weeder does not scorch the plants; rather, the cell walls of the plants rupture from expansion. The plants won't look different until the day after being flamed. To make sure your flame weeder is working, press the leaf of the plant between your fingers after you've flamed it — if the leaf turns dark, the flame weeding was effective. The speed of the flame weeding depends on the type of weed and the power of the burner on the weeding device. The hotter the burner, the faster you can move through the field flaming weeds. Flame weeding has the advantages that it doesn't disturb the soil or bring new weed seeds to the surface like traditional cultivation does. And the price is right, since a 20-pound propane tank weeds about 1 acre of land.

I find flame weeding useful in these situations:

- When the field is too wet to cultivate with tractors or hand tools.
- To kill weeds that emerge before seeds I've directly planted in the ground. For example, carrot seeds take a long time to germinate, so the weeds sprout first. I pass the flame weeder over the bed to kill the weeds, while the carrot seeds lie protected under the soil.

- When I want to plant seeds that germinate quickly, I give the weeds a head start by delaying my planting a few days after it rains. The weeds sprout before my seedlings, and I use the flame weeder to eliminate the weeds. My seedlings then emerge into nearly weed-free soil.
- To use a stale seedbed technique. For example, I prep a field to plant within two weeks. However, rain falls before I can finish planting. So I let the weeds germinate with the rain, then use the flame weeder before putting out transplants or seeds.
- To implement organic no-till weed control. Weeds inevitably find their way to the surface in organic no-till or mulched areas and compete with crops. Usually you'd hand-pull these weeds, disturbing the mulch and creating future weed problems. Although flame weeding over mulch would normally ignite the mulch and cause disastrous results, flame weeders work well even when it's raining. So if weeds emerge in flammable mulch, we kill them with the flame weeder during or directly after a rain.

Flame weeders come in units small enough to attach to a backpack or large enough to mount to a tractor. At our farm we use a single-burner handheld unit as well as a six-burner unit attached to wheels that we push over the beds. Reasonably priced at a little over five hundred dollars, the six-burner unit quickly paid for itself in the amount of time it saved us. A comparably sized tractor-driven model would cost thousands more dollars. Our six-burner flame weeder fits perfectly between our emerged plant rows when we use a single row per planting bed. If the field is too wet to cultivate with the tractor, we run the flame weeder between the rows and quickly eradicate the weeds. The flame weeder has rescued our CSA program from complete failure during wet years when the field never dried out. Using flame weeders in fields too wet for tractors is a major benefit for our farm, so a small handheld unit makes sense.

Large hoe. Though large hoes aren't my first choice for weed management, I find them useful in moving loose soil. We plant our 4-inch pots and potatoes inside furrows dug down the middle of our beds. Using the large hoe, we pull the soil over the potatoes or transplants. Large hoes work well to move soil short distances such as when digging ditches or fine-tuning bunds and pond edges.

Manure fork. A hybrid between a hoe and a pitchfork, the manure fork moves loose material a short distance when a large hoe won't easily penetrate the material. For example, a hoe would pull manure or mulch off a pickup truck, but it doesn't sink into the manure. A manure fork sinks in, then pulls the manure off the truck. The pulling motion uses your body weight, requiring less energy than lifting a shovel or pitchfork.

Shovels. When I have to move heavy materials, I grab a shovel, and the type of shovel I choose depends on the task. For digging trenches I use narrow trenching shovels, but if I'm spreading compost over a bed I select a flat-top transfer shovel.

Pitchforks. If you're moving something loose or lightweight such as manure or wood chips, a pitchfork works best. The tines easily penetrate the material so you don't have to use as much effort.

Rakes. We use garden rakes to smooth the soil and mulch, especially inside greenhouses or small gardens, where we can't use large tractor-drawn implements.

Backpack sprayer or small sprayers. When smaller areas require spraying, we reach for a backpack sprayer. Battery-powered and compressed carbon dioxide sprayers handle medium-size spraying jobs. We also use small hand-pump sprayers for mixing up chlorine solutions for sanitizing coolers and other equipment.

Seeders. Small farms need walk-behind seeders. They accomplish the following tasks with ease:

- Meter seeds out evenly, while conserving seed
- Place the seed at the appropriate depth using an adjustable plate
- Cover the seeds with soil
- Firm the soil over the planted seeds to guarantee good seed-to-soil contact

Earthway and Jang are the two most common brands of walk-behind seeders. I've only used an Earthway seeder, but I've heard the Jang seeder singulates seed better, conserving seeds. I recommend running the seeder over a tarp before planting to gauge seed placement and make adjustments by using different seed plates or by taping holes in the seed plate. It's important to push down on the back wheel when walking behind the seeder to ensure the soil is pressed firmly over the seeds. It takes about four hours to plant a diversity of crops over an acre with a walk-behind Earthway seeder.

Some small rotary seeders come with bags or containers that hang over your shoulder. I use these when I'm planting cover crops. I fill the bag with seed and a hand crank whirls the seeds out in an even pattern. A hand-crank bag seeder can apply over 100 pounds of seed an hour, depending on seeding rate.

Wheelbarrow. Every farm and garden needs a wheelbarrow. When you need to move soil amendments, potting soil, tools, or heavy material, a wheelbarrow makes the job easy. Not all wheelbarrows are designed equally, however. Industrial models with metal buckets and puncture-proof tires last longer and require less maintenance.

Toolshed Connections

Tools are an essential part of farm work. Site your toolshed for close connections to the places where tools are most likely to be needed. Here's a list of those places in order of priority.

1. Fields
2. Machine shop
3. Equipment shed
4. Greenhouses
5. Home
6. Animals
7. Parking

T-post driver and puller. Keep your tomatoes off the ground by using a T-post driver to build a Florida trellis system. Drive T-post-style fence posts into the ground every 10 feet, and attach twine between the posts to support the tomato plants. At the end of the season, use a T-post puller to remove the fence posts.

Hatfield transplanter. Sets transplants without stooping by making a hole and dropping transplants through a tube.

Irrigation equipment. We organize and store all our parts and tools for irrigation systems in the toolshed (see chapter 10 for more details on irrigation parts).

Miscellaneous tools. Owning tools to fix, build, and maintain equipment is always helpful. The more you can do yourself, the less you have to pay or barter for with other people. Circular and reciprocating saws, hammers, mallets, drills, socket sets, and pipe wrenches are necessities. Compressors, welding equipment, table saws, and presses are good tools to scale up to if you have available space.

Equipment Shed

Larger tools such as tractors and tractor implements also need a designated location on the farm. If this area is located under a pole shed protected from the rain, tools will last longer and will not rust. Ideally, equipment and tractors are stored under a long, narrow shed. Open on all sides, the shed gives access for attaching implements stored on both long edges.

Usually, you add new equipment to your farm over time. Locate your equipment shed in an area where you can add length to the end of your shed to accommodate growth. If space is limited, store expensive power take-off (PTO)–driven tools under the shed and simpler, cheaper tools in the open. Here's a rundown of common equipment stored in the equipment shed for a 2- to 10-acre organic row crop farm.

TRACTORS

Before I became a farmer, only one image came to mind when I thought of a tractor. But tractors come in many different sizes, from the ride-on-top to the walk-behind, and have hundreds of implements to make the job easier. Here is a list of some tractors and tools I've used.

Walk-behind tractor. For farms less than 2 acres, a walk-behind tractor is all that's needed. High-end gear-driven models such as BCS and Grillo with PTO attachments are best. Almost any implement used by a larger ride-on-top tractor can be found for walk-behind tractors, even hay bailers. However, finding used equipment is difficult, and new implements are nearly as expensive as large tractor implements. I had to search on Craigslist for six months before I found a decent used BCS walk-behind tractor. I don't

recommend belt-driven walk-behind tractors with limited attachments. You can purchase any of the equipment listed in this section for a good walk-behind tractor.

35- to 50-HP tractor, 2WD or 4WD with loader. For farms over 2 acres I recommend purchasing a ride-on-top tractor. A front-end loader attached to the tractor is useful when composting is part of the farm operation and also helps with moving fertilizer, soil amendments, manure, potting soil, and anything heavy. You can also use the loader to move soil during field grading operations. A tractor canopy protects you from the sun on hot days.

Flail mower. Since cover cropping is the foundation of sustainable agriculture, tools to manage cover crops are a must. The flail mower chops up the cover crop into small pieces and evenly distributes the material over the soil. The

Figure 11.12. A flail mower cuts cover crops close to the ground, evenly distributing the material over the soil. The close cut and small pieces are easily tilled in, quickly preparing the field for planting.

cutting action of the flail mower reduces the cover crop to ground level. The low cut and small chopped-up pieces kill the crop and give a quick turnover of the field from cover crop to prepared planting bed.

Before I used a flail mower, I used a rotary mower. The mower cut cover crops high and left debris in large pieces clumped around the field. Incorporating the material into the soil using tillage equipment took several weeks and twice as many passes in comparison to the flail mower. The time we've saved not having to till the field as much and the increased production we acquired from quick field turnovers has quickly paid for the flail mower.

Rotary mowers. Using a single blade similar to that of a lawn mower, rotary mowers come equipped for high-weed mowing or finishing mowing.

Sickle or cutter bar. A series of blades move along a shaft to cut grass or saplings up to 1 inch in diameter. The cut material is left whole on the surface. I use the cutter bar to cut grass for making hay and feeding grains to chickens.

TILLAGE EQUIPMENT

Once the cover crop or previous crop is mown using the flail mower, tillage equipment incorporates the plant material into the ground. The material is decomposed, and nutrients are released for the next crop. Some tillage equipment leaves material in clumps, and some pulverizes into fine pieces. Tillage equipment that leaves clumps protects soil structure and prevents organic matter from breaking down too fast and being converted into carbon dioxide.

Reciprocating spaders incorporate mown cover crops in a single slow pass. Shovel-like tines lift soil in clumps, preserving soil. Additional equipment creates a fine seedbed.

Rotary tillers pulverize soil and organic matter into small pieces using sharp blades. Several passes are needed to incorporate cover crops. They are best used at shallow depths and medium speeds to incorporate soil amendments or fertilizers in the top layer or to prepare the top layer for planting seeds.

Rotary plows are used with walk-behind tractors to till the soil deeply in a single pass. The plow consists of a giant rotating corkscrew that's pulled through the soil. Because rotary plows throw soil to the side of the tractor, they also make raised beds. The side-casting nature of rotary plows allows them to build irrigation ditches and diversion channels as well. The rotary plow will also dig a shallow trench (see chapter 10 for rotary plow techniques).

Disc harrows are useful with 35-horsepower tractors or bigger. Adjacent discs cut through and chop soil into clumps, preserving soil structure better than a rotary tiller. Additional equipment is needed to create a seedbed for planting small vegetable seeds. Depending on the size of the disc harrow and the amount of residue in the field, four to eight passes and adequate time are generally needed to prepare the soil from a cover crop after mowing. Disc harrows work at high speeds. Tilting some forward slightly gives deeper tillage. Adjusting all the discs to run straight helps smooth out the soil surface for cover crop planting. The straight-running discs also lightly bury the seeds to improve germination.

Roller crimper. A large metal drum with dull blades rolls over mature cover crops, killing them and leaving a thick layer of mulch to suppress weeds and conserve moisture (see chapter 10 for more information).

Box scraper and rear blade. Grade fields and build ditches, diversion channels, and ponds by attaching scrapers and blades to the rear of your tractor.

Surprisingly, a box scraper can dig a pond. When you attach rippers in front of the blade on the box scraper, they loosen the soil enough to move large amounts of soil in a single pass. The feat is accomplished by shortening the top link as far as possible on the three-point hitch to engage the scarifier tool to full effect. Scarifiers also perform keyline pattern cultivation if needed. By adjusting the drawbars on the tractor to angle the blade, you can dig V-shaped ditches with the scraper.

Box scrapers collect soil within the sides of the box and can then spread or deposit it in a pile for moving with the front-end loader. A box scraper lacking the sides to hold soil back is called a rear blade. The open-ended blade casts soil to one side as the tractor moves forward. Rear blades are ideal for making and maintaining swales, ditches, and diversion channels.

Bed shapers. Tractor-drawn bed shapers form raised beds. Discs pull soil into a center mound; then shaper pans form the flat-top raised bed. The furrows in between the raised beds help drain the field. Properly sloped and graded fields used in conjunction with raised beds prevent water from pooling in the field and permit quick access to crops after a rain event. The furrows between raised beds also guide tractor-drawn cultivation equipment for precision weeding with tractors. You can add drip tape layers to the bed shapers to bury drip tape while you form the raised beds.

Specially designed primary bed shapers can form raised beds in high-residue conditions, allowing earlier access to fields for final bed forming. Finishing bed shapers work best with well-tilled soil.

We added row markers to our bed shaper to make small furrows for guiding planting activity. Each raised bed is marked with three rows spaced 12 inches apart. Coiled springs bolted to equipment make good row markers.

TRACTOR CULTIVATION EQUIPMENT

In our tilled areas we accomplish most of our weeding with tractor-drawn cultivation equipment, which uses various knives and cutting devices to kill weeds between plants. Older tractors specially designed for cultivation use belly-mounted equipment attached to the middle of the tractor. Belly-mounted equipment is not subject to the large movements that rear-mounted equipment is prone to, making precision weeding possible.

Rear-mounted equipment uses gauge wheels or shaper pans to guide the cultivation equipment along raised bed furrows. As long as the plants are in the correct place on the raised bed, cultivation equipment locks into the furrows between beds and passes within 2 to 4 inches of the crops without killing them. Various types of weed-killing devices are available.

S-tines. Cutting under the surface of the soil, S-tines kill small weeds and loosen crusting topsoil. The flexible spring steel forming the shaft of the S-tines creates a shattering action and bends around rocks.

Shovels or sweeps. Using a stiffer shaft and wider blade, shovel attachments cut under compacted soil and control larger weeds.

Beet knife. Cutting close to small seedlings without moving the soil, the beet knife is good for the first cultivation.

Crescent hoe and side knife. The angled side of raised beds needs an angled cultivation knife to cut under emerging weeds. Crescent hoes and side knives fill this important niche.

Row spiders. Using an angled blade on spinning wheels, row spiders take advantage of the forward motion of the tractor to rip small weeds out of the ground. The wheels are angled to throw soil away from smaller plants or toward larger plants to bury weeds underneath.

Figure 11.13. Tractor-drawn cultivation equipment controls weeds without hand labor. Sweeps kill weeds in furrows, and side wings on sweeps maintain furrows for field drainage (*left*). Spider wheels throw soil at the base of the potato plants, burying small weeds (*right*).

Figure 11.14. Here, the spider wheels have been removed, and an S-tine is placed in the middle for cultivating between two rows on a raised bed.

Furrower. Cultivation equipment destroys the shape of raised beds, leading to poor drainage after cultivation. The furrower cultivates and rebuilds the furrow between raised beds, which permits good drainage after cultivation. The furrower can also help guide the cultivation equipment by locking into the furrows.

Shaper pans and row crop shields. Shields protect small plants from getting buried by soil thrown during cultivating. Shaper pans rebuild the edges of raised beds and help guide equipment.

MISCELLANEOUS EQUIPMENT

There are tools that don't fit into a particular above category but are very useful on the farm. Below is a list of equipment every grower should have for scaling up their operation.

Fertilizer spreaders. Moving fertilizer and soil amendments into the field takes considerable effort. Cone spreaders attached to tractors make the job easier with a large hopper positioned over a spinning disc. The PTO of the tractor or wheels of the spreader cause the disc to spin the material out in an even pattern. Bander kits attached to the spreader confine the material to an even band over the row. Since the pathways between beds are not fertilized with banded applications, one-third less fertilizer is used. Another technique uses fertilizer spreaders attached to cultivation toolbars. The hoppers use gravity and vibration to move material through chutes and next to plants for side dressing or applying fertilizers to the tops of beds.

Manure spreader. Manure spreaders are attached trailers with chain drives on the floor that push the manure toward the back of the trailer, where beater bars mix and throw the material out into the field. Some chains and beater bars are driven by the motion of the wheels (ground drive) on the manure spreader, and some are driven by

the PTO on the tractor. PTO-driven models can be parked and used to mix compost ingredients, as shown in figure 8.6. Additionally, the manure spreader can deposit mulch evenly onto planting beds. Simply fill the manure spreader with the mulch, then slowly drive over the bed, and the chain drive pushes off the mulch.

We stockpile leaves from the city recycling facility, then use the manure spreader to distribute the leaves evenly over the beds. The beds we know we want to cover with leaves are spaced slightly farther apart to accommodate the width of the manure spreader. This technique works great with garlic. After planting the garlic shallow, we place a thick layer of leaves on top using the manure spreader. The garlic pushes up through the mulch, but the weeds can't make it.

Potato digger. Various devices are designed to remove potatoes and sweet potatoes from the ground. We use a PTO-driven chain digger attached to the back of the tractor. Soil falls through the chains, and root crops are deposited on the surface. Simply mow the crop as close as possible with a flail mower, then run the chain digger through to remove the root crop. The size of the crop and the type of soil determines the pitch of the chain. Coulters or cutting discs added to the front cut through sweet potato vines missed by the mower on both sides of the row, enabling easy harvest of vine crops.

Larger sprayer. Sometimes our best efforts to prevent pests fail, and we resort to an approved organic pesticide to rescue our crop. A large PTO-driven tractor sprayer turns a long, difficult task into an easy one. In our area caterpillar pests are a major threat to most brassica crops. The frequent spraying of bacteria-based pesticides such as Bt to control caterpillars is quickly completed by the ease of a tractor sprayer. The tractor sprayer also facilitates quick applications of bacterial inoculants to fields after planting. Our sprayer is separated into three booms. We can turn each boom on or off in any combination for precise applications.

The tractor sprayer also comes equipped with a hand-operated nozzle for spot treatments and for fruit trees, shrubs, or plants too tall for the booms.

SUMMING UP THE FUNCTIONS

1. Roads placed on ridges stay dry and reduce maintenance.
2. Parking lots and roads control and direct cars to designated locations.
3. Placement of building complexes improves efficiency and movement of people, tools, equipment, and supplies.
4. Customers weigh, count, and box produce for distribution.
5. Customers harvest produce during pickup.
6. Location of components maintains freshness of food.
7. Equipment facilitates production of produce.

CHAPTER 12

The Mini-Innie

A Small-Scale Bio-Integrated Greenhouse

In the hilly, steep terrain of the Appalachian Mountains, the roots of agriculture lie in subsistence farming. Early natives and settlers foraged chestnuts, acorns, and all sorts of berries on the steep slopes. The few flat areas available were readily cleared for annual crops. In these small clearings the natives probably first experimented with plant combinations in an effort to maximize the available space. Without enough space for separate plots of corn, beans, and squash the natives likely planted them together, much closer, to see how they worked in concert.

The steep hills of Appalachia also gave birth to the next bio-integrated pattern. As a farmer living in the mountainous but urban terrain of Asheville, North Carolina, I had to develop maximum integration methods for small confined areas, so I designed the bio-integrated greenhouse with an interior chicken coop and pond.

A pond encompassed the entire greenhouse floor, storing the sun's daytime heat, then releasing that heat at nighttime when it was needed most. A bridge crossed the greenhouse, providing access to wicking hydroponic plants that lay along the edges of the pond absorbing nutrients. A suspended chicken roost fertilized the pond and kept the greenhouse warm during the night. And the

minnows grew in the water below — future feasts for the roosting chickens.

Studies and models have shown a drastic improvement in greenhouse heating performance in these encapsulated pond greenhouse systems. Normally, a greenhouse built over soil and covered with a single layer of plastic provides only 0° to 1°F of protection from nighttime low temperatures. During the day the soil heats up and transfers the

Figure 12.1. The outside of this greenhouse looks ordinary, but it harbors a hidden pond.

319

Elements of the System

Efficiently and effectively pool your resources in a bio-integrated greenhouse with an interior chicken coop and pond. In the age of rooftop and vertical gardens, urbanites are always exploring options in their search for sustainability. Not everyone has an acre or 2 or 10 — sometimes you have to make the best of what you've got. This chapter shows you how to:

▸ Create a greenhouse environment where the entire floor is composed of a pond
▸ Add bridges above the pond for better access to the interior of the greenhouse
▸ Nest a heated pond inside the larger pond for raising transplants and over-wintering fish
▸ Attach a chicken coop to a greenhouse

heat into the greenhouse, creating a warm micro-climate. But soil's heat capacity or ability to retain heat is limited. The heat quickly dissipates, and temperatures plummet.

Water, however, possesses a high heat capacity, over five times greater than soil, storing more heat for use at night. With the pond encompassing the entire floor of the greenhouse, I observed nighttime minimums inside the greenhouse 10°F warmer than outside minimum temperatures. I conducted my experiments with only a single layer of plastic over the greenhouse and no insulation beneath the pond. The addition of both insulation

and a second layer of plastic would improve the performance greatly.

A greenhouse enclosing a pond works in any climate with an available water supply to keep the pond full. The size of the greenhouse and pond is dependent on the slope of the site and the ability to create a flat area. Since a pond must be perfectly level, the greenhouse platform needs to be level as well. Sloped sites require terracing to create level platforms, and the terrace size limits the size of the greenhouse. Refer to chapter 2 for basic pond construction techniques. Chapter 5 shows many features of this type of greenhouse/pond system, with some useful construction techniques illustrated. Also see chapter 10 for the technical aspects of terrace construction to create flat areas needed for ponds.

Building the Pond and Greenhouse

I spent several weeks designing and building the greenhouse in Asheville. Because I was working with a tiny lot on steep terrain, every square inch mattered. Luckily the site was already terraced with a level location for the greenhouse. We formed the frame from galvanized steel tube bows, then covered the greenhouse with a single layer of greenhouse plastic. The greenhouse, measuring 10 feet wide and 24 feet long, was part of a larger water-harvesting system involving multiple houses and lots. The system harvested overland flow and directed water into a storage pond in the upper lots. The elevation of the storage pond allowed gravity to feed water into the greenhouse pond and into the landscape of the lower lots. We connected the greenhouse pond to a recirculating system, including a waterfall and pond outside the greenhouse. The entire system drained into retention basins that also harvested water off the roof of the houses (see the appendix for an illustration of this system).

This chapter describes the construction details of and considerations in covering an entire pond with a greenhouse, combined with a chicken coop.

Site Layout and Excavation

This is a two-in-one project. Building both a pond and a greenhouse requires forethought and strategy. Since the greenhouse floor is the pond, the greenhouse should sit perfectly level to ensure even depth throughout the pond. The water in a pond always assumes a perfectly level position. So any discrepancy from level of the greenhouse will be obvious once you fill the pond, as described in chapter 5.

Ideally, construct the pond before building the greenhouse. First, mark the greenhouse's corners, then dig a perfectly rectangular pond inside those corners. Reuse the excavated soil to create a perfectly level rim around the pond for the base of the greenhouse. You can also use the excavated soil to form an embankment, creating a ditch or vegetated waterway on the downhill side of the greenhouse. When I built the Asheville greenhouse, I used the vegetated waterway on the downhill side to convey rainwater runoff from the greenhouse and landscape into the pond inside the greenhouse. Water first flowed into the vegetated waterway; then a riser allowed water into or out of the greenhouse pond.

If you're retrofitting an already existing greenhouse, first make sure the greenhouse sits reasonably level. Remove the plastic so you can reach between

Figure 12.2. A vegetated waterway outside the greenhouse feeds water into the greenhouse if needed and connects to a waterfall and outdoor pond for recirculation. Nutrient-rich water in the greenhouse drains into the basin below to flood-irrigate pots for nursery plant production.

Figure 12.3. The same vegetated waterway outside the greenhouse with established plants.

the bows with the digging arm of an excavator to carefully remove soil to excavate the pond.

Pond Slope and Depth

As explained in chapter 2, when building any pond, you must consider the effects of erosion. If the sides of the pond are too steep, you may end up with a big hole in the ground full of mud — which is certainly not the goal. I recommend a slope with a two-to-one ratio for the sides of the pond if you don't cover the pond liner with soil. You should use a ratio of three to one if you cover the pond liner with soil, which will result in a shallower pond.

The slope of the pond's sides and the size of the greenhouse determine the depth of the pond.

Typically, the larger the greenhouse, the deeper the pond. A flat, level area at least 20 inches wide on the pond floor is essential to provide space for piers to support structures (such as a bridge). Sloping this flat area to a low point makes draining the pond easier.

I've also used a cinder block wall to make a vertical edge around the perimeter of the pond, maximizing the pond's volume. For instructions on building a cinder block wall for a pond, see chapter 5.

Once you've excavated the pond, install the ground stakes for the greenhouse structure and attach the greenhouse bows to the ground stakes. Similar to the ponds built inside greenhouses in chapter 5, the pond will need a baseboard made from treated or engineered wood to attach the pond liner material around the edge of the pond. Since the pond will encompass the entire greenhouse, install a baseboard around the entire perimeter. I've learned the best location for the baseboard is inside the greenhouse bows.

Many factors affect where the water level of the pond surface lies. First, consider the height of the greenhouse and the height of the greenhouse door. If you install a standard-size door on your greenhouse, the size of the door may dictate the level of the pond surface. In a small greenhouse you can only raise the door so high before it butts against the roof of the greenhouse. Usually, I cut out the wooden greenhouse baseboard to allow enough depth for a standard-size door to fit on a small greenhouse. Thus, in small greenhouses, the door determines the maximum water height since the door should hang above water but below the roof. I recommend designing the pond's water height to fit a standard-size door height of 81 inches. Typically, metal storm doors are used for greenhouse entrances. This height allows installation of long-lasting metal doors without the hassle of cutting doors to accommodate a higher water level.

The wooden baseboard of the greenhouse also affects the maximum height of the water level. The top of the baseboard is the highest level at which to attach the pond liner. Above this level, water simply spills over the liner and leaks out. Thus, the baseboard dictates the maximum water level if doors don't.

Finally, consider the shallow area around the edge of the pond. Shallow and emergent areas around the edge of ponds support wicking hydroponic systems for self-watering plants. Alternatively, building the pond to the highest height possible with a deep edge maximizes volume and depth. However, the deeper water on the edge won't support wicking hydroponic systems. Hydroponics is still possible by pumping water from the pond into commercially available or homemade hanging hydroponic systems. In addition, the floating hydroponic systems presented in chapter 5 are also a good option and help reduce evaporation and condensation issues.

You can control water height by connecting an adjustable standpipe to a drain or by creating a limited low area along one edge of the pond. If you install a standpipe and drain, place the drain line before installing the liner, and leave 12 inches protruding into the pond. In Asheville we designed a low area on the edge of the pond to control water height. We attached the pond liner to the baseboard but dipped it down below the baseboard at one location to allow water to exit under the baseboard and into the vegetated waterway outside.

Installing the Pond Liner

Installing the pond liner is similar to the techniques used to install the liner in the interior pond described in chapter 5. The only difference is that in this case the baseboard encircles the entire greenhouse. Follow the steps in chapter 5 to insulate and seal the pond.

Creating Access

Access is in the mind of the designer. We installed a bridge across the length of the pond connecting the doors at both ends of the greenhouse. The bridge became a stable foundation for attaching the chicken coop and provided necessary access to the plants along the edges of the pond. However, a bridge is not completely necessary.

Roll-up sides could give access to the pond edges along the length of the greenhouse, while short bridges across the width could create support for the addition of chicken coops or other structures. A simple stepping-stone placed inside the door

Figure 12.4. A bridge crosses a pond inside a greenhouse, with wicking hydroponic systems on the edge.

could provide keyhole style entry. You could easily move floating hydroponic rafts using a pole to push or pull them into reach.

If you plan to build a bridge across the pond, you may need piers to support the bridge, depending on your bridge material. We used black locust, harvested on-site, cut 3 inches thick and 12 inches wide. The locust is rot resistant and chemical-free.

To build our bridge, we first poured small concrete pads measuring a few feet square under the door and protruding into the greenhouse about 4 inches on both ends. We placed the bridge on the 4-inch section of the concrete pad where it extended into the greenhouse. The height of the concrete pad was placed 1 inch above the final height of the water, bringing the base of the bridge

in close contact with the water. The concrete pad then butted up against the pond liner.

Since the strength and length of the wood was inadequate to span the entire length of the greenhouse, we added a pier to support the bridge. Before we laid the pond liner, we prepped the foundations below the planned pier by excavating the soil and adding 6 inches of compacted gravel. The gravel provided support to prevent the pier from shifting. Then we covered the gravel with a layer of sand to protect the liner from potential punctures. To enable the pier to contact the solid bottom surface, we removed insulation from the pier area.

Next, we supported the bridge with cinder block piers laid on top of the pond liner. We located the cinder blocks on top of the areas we had previously prepared with gravel and sand beneath the liner. Next, we placed several pieces of scrap pond liner on the liner to protect it from damage, and set the cinder blocks in place. Stacking the cinder blocks up to the height of the concrete pads, we laid them directly on top of the pieces of protective pond liner. We positioned the bridge in place and used shims to fill the final gaps between the bridge and the pier. Since a single wooden plank wasn't long enough to span the entire length of the pond, we used two pieces that met at the concrete pier in the middle. Metal plates hidden underneath fastened the pieces together to secure the connection. At least one additional bridge or support structure should extend out from the side of the bridge and connect with the wooden greenhouse base to give the bridge lateral stability (see figure 12.5).

In hindsight, I would probably spend a little more money and use metal I-beams to suspend the entire length without a pier for support. Then, I could use thin pieces of wood to provide a deck on top of the I-beams. I'm sure a metal fabricator skilled in building bridges for docks could also construct a nice custom pond bridge.

Connecting Chickens to the Greenhouse

Before implementing this project, I'd read a lot about connecting chicken coops to greenhouses — and in theory it works. But as with all theories, the application itself presented challenges. In a nutshell, chicken coops connected to the greenhouse are both beneficial and problematic.

Chickens provide heat and carbon dioxide to greenhouses, both of which are limiting factors for plant growth. They also fertilize the pond. However, chicken manure produces ammonia gas, which is toxic to plant growth. And elevated greenhouse temperatures on hot summer nights may also deter chickens from coming home to roost. Ultimately, isolating the chicken coop from the rest of the greenhouse and filtering the chicken coop air before it enters the greenhouse may present the best option.

Coop Design

The coop I built inside the greenhouse measured 4 feet wide and 4 feet deep. It was made with a simple wooden frame and wrapped in galvanized poultry netting to keep the chickens out of the greenhouse and the rodents and predators out of the chicken coop. The bottom of the frame extended over the pond inside the greenhouse and connected to the bridge in the center. Inside the coop I installed four roosting bars made of 2 × 3-inch wooden boards, turned vertically and extending the entire width.

The chickens could enter and exit the coop through a framed opening equipped with two cabinet-style doors at the bottom edge of the curved greenhouse wall. To retain heat inside the greenhouse while the doors were open during the day, I added overlapping plastic strips to seal the door. With a little training the chickens learned they could push through the plastic strips to enter or exit the coop.

We affixed the laying boxes to the top of the coop frame, and the chickens accessed the laying boxes from the outdoors. To keep the entrance and laying boxes dry when it rained, I attached greenhouse plastic to the backside of the laying boxes. In case we had early layers, I added a laying box to the interior of the coop. The hens laid eggs in the interior box until we opened the doors of the coop, but they preferred to lay in the outer boxes because of the heat inside the greenhouse. I could have installed an automatic door set to open early in the morning and close at the end of the day after the chickens returned to the roost. Opening the door

Figure 12.5. A simple wood-framed chicken coop covered with poultry wire provides heat for the greenhouse at night when the chickens return to roost.

Figure 12.6. The open laying boxes above the doors to the coop are quite large and would benefit from partitions at 1-foot intervals.

early would have prevented chickens from laying inside the coop.

The bottom of the coop contained a sliding plywood drawer. Manure from the roost fell into the drawer, and we pulled the drawer out to remove the manure easily. If the drawer remained pulled out, manure fell through the poultry netting on the bottom of the coop and into the pond to fertilize the pond and plants.

Just inside the opening, we suspended an automatic waterer over the plywood drawer. In hindsight, I realize we should have sawed a hole in the drawer to drain excess water, thereby extending the life span of the wood. We placed a feeder

Figure 12.7. *Left*, the automatic waterer is just inside the coop. *Right*, I insulated the water lines that feed the waterer to protect them from freezing during the winter. Prevent chickens from eating the insulation by fastening aluminum flashing with cable ties around the insulation.

opposite the water, inside the opening, for easy access and filling.

Heat Exchange and Temperature Management

How much heat does a chicken produce? Research indicates a ten-week-old broiler produces around 375 kilocalories per day.[1] This equates to about 18 watts or 1,500 Btus per day per chicken. Consequently, sixty-one chickens provide the same amount of heat in a day as burning 1 gallon of propane. That's why farmers equip large chicken production houses with huge fans and evaporative coolers to lower temperatures.

Chickens inside greenhouses can be a heat source, but the heat is only necessary at night, and the high temperature inside a greenhouse during the day may compromise chicken health. The ideal temperature range for poultry is 65° to 75°F. Heat stress begins when temperatures climb above 80°F, and emergency measures and survival concerns

occur at temperatures over 100°F.[2] Consequently, in warm climates, greenhouses are not a good place for chickens during the day. But a greenhouse can offer good nighttime roosts for chickens that are allowed to forage outdoors during the day. Chickens return to the roost inside the greenhouse at nighttime to provide heat to the greenhouse when it's needed most.

During the summer, temperatures in the greenhouse may remain so high in the evening that they deter chickens from coming back to the roost. You'll need to shade and ventilate the greenhouse to encourage chickens to return to the roost at night. Use roll-up sides to provide easy natural ventilation during the summer. Otherwise, you can set up an alternate summer roost outdoors.

In our mild southern climate, greenhouse temperatures may become too hot for chickens without automated ventilation even during winter days, so we have to open the chicken coop in the morning to let the chickens out. Oversleeping on a warm winter day could kill the chickens if they remain trapped in the coop. During the summer the door can simply stay open, if predators aren't a problem. I've experimented with automatic doors and ventilation systems; though helpful, they are also harmful if the electricity goes out, creating a dangerous situation for the flock. Managing ideal temperatures in a greenhouse for chickens is more difficult than managing temperatures for plants. In my opinion, colder climates are more suited for this pattern, and I recommend paying close attention to greenhouse temperatures and ventilation. But if your conditions are suitable, then connecting chickens to your greenhouse is an efficient way to use a small piece of land.

Carbon Dioxide Concerns

How much carbon dioxide does a chicken produce? Chickens provide carbon dioxide through respiration. Carbon dioxide is a limiting factor for plant growth in greenhouses (see chapter 8). Studies indicate a chicken produces around 75 liters of carbon dioxide per day or 2.65 cubic feet per day.[3] When farmers apply carbon dioxide to stimulate plant growth, on average they apply 0.002 to 0.004 cubic feet of carbon dioxide per hour per square foot of greenhouse floor area.[4] Therefore, every chicken supplies enough carbon dioxide for 27 to 55 square feet of greenhouse floor area on an hourly basis.

However, plants only need carbon dioxide during daylight when photosynthesis is strongest. But chickens want to go outside during the daytime, not stay confined in a hot greenhouse. To maintain adequate temperatures for chicken health, I recommend venting the greenhouse before optimum temperatures for greenhouse heating and plant growth occur. That's why it's more beneficial to permanently keep chickens in greenhouses in northerly climates, where daytime temperatures in greenhouses don't reach high levels. Chickens confined to greenhouses in cooler climates only experience minimal heat stress during winter, and the greenhouses can also remain closed most of the day, retaining carbon dioxide for plant growth. Warm winters in southern climates require ventilation in greenhouses to maintain chicken health, resulting in a loss of carbon dioxide. Greenhouses in warmer climates make better chicken roosts, capturing heat during the night but allowing chickens to forage outdoors during the day.

Another more practical option is to join the coop with the greenhouse using a shared wall. Placing the coop on the north side of the greenhouse insulates the greenhouse from the cold without shading the greenhouse from winter sun.

Managing Manure

Wherever chickens roam and reside, manure accumulates. Manure can be used to fertilize ponds and plants when applied as a fertilizer. However, carefully manage manure to prevent toxic ammonia gas from entering the air and injuring plants. It's easy to overfertilize plants and ponds, so use special care when applying manure.

Ammonia is a gas resulting from the decomposition of bird droppings. When chickens are housed in a greenhouse, ammonia may reach toxic levels for plants, birds, and humans. The gas affects the respiratory system and damages the eyes in animals. The Occupational Safety and Health Administration (OSHA) sets a maximum safe limit of 25 parts per million for ammonia gas for humans. Most people can smell ammonia at concentrations between 20 and 30 parts per million, so using your nose can be a good gauge.

Plants are far more sensitive to ammonia gas. Studies have found that broccoli leaves are injured at 0.86 parts per million. Symptoms in plants include black spots, necrotic areas on leaf margins, increased susceptibility to cold injury, and reduced plant growth.[5] Cultivated plants most sensitive of plastic. You can wrap the chicken roost in greenhouse plastic during winter to prevent gases from entering the greenhouse. Since chickens are more tolerant of ammonia than plants are, concentrations inside the chicken coop can remain higher than concentrations inside the greenhouse, thereby limiting maintenance. During the summer you can unwrap the roost and roll up the greenhouse walls to allow airflow and gas dispersion.

Dust

Chickens produce a lot of dust. The amount of dust increases with the number of chickens and their access to dusty material to scratch inside the coop. Carefully consider the types of vegetables you grow in the greenhouse, since your vegetables are subjected to this dust. For example, a lettuce plant whose leaves are eaten raw is probably less safe than a cooked vegetable if both are covered in chicken dust. In either case, it's a good idea to prevent chicken dust from settling on your plants. Eliminating scratching areas inside the roost helps. Also, isolating the coop with vines growing around the coop or plastic helps block or screen the greenhouse from dust.

Lessons Learned

Although I designed the coop to fit twelve roosting hens, we tested it out with three hens during a cold winter and a hot summer. The hens enjoyed the accommodations, foraging outdoors even on the coldest days and returning to the roost nightly. Again, in hindsight the nesting boxes were a bit large and should have been reduced to 1 square foot each to prevent fighting over the same box.

Once, we left the drawer out for a few days so the chicken manure would drop directly into the pond to fertilize it. The water quickly turned green, and a dense population of mosquito fish flourished. However, water and manure from the surrounding

Figure 12.8. A drawer below the roosting bars slides out for easy manure removal or to allow manure to fall into the pond as fertilizer.

to ammonia toxicity include tomatoes, potatoes, peas, beans, beets, and strawberries. Cultivated plants most tolerant are in the *Allium* genus; for example, onions and leeks.[6]

To prevent ammonia gas from accumulating, keep droppings dry and remove them frequently. Moisture speeds the decomposition process and the release of toxic ammonia gas. By keeping droppings as dry as possible, you will stunt ammonia production. To prevent moist droppings, design areas around watering devices so that they drain away from manure-collecting areas. The addition of wood shavings under roosts also helps absorb moisture. Keeping fewer chickens or simply removing the droppings more frequently also prevents ammonia buildup.

An alternative solution is to isolate the chicken roost from the rest of the greenhouse with a barrier landscape also flowed into the pond, as did the water running off the greenhouse roof, so additional fertilization didn't seem necessary. I didn't observe problems with ammonia toxicity, but we were farming a small number of chickens. We also kept the manure dry and removed it frequently.

In our Asheville greenhouse the chickens minimally benefitted the greenhouse pond system. We could have easily hand delivered the small amount of manure needed to fertilize the pond with a bucket or shovel. Since the chickens didn't spend much time inside the greenhouse during the day, the amount of carbon dioxide generated for plant use was minimal. On the other hand, manure accumulated under the roost and required frequent removal to prevent ammonia toxicity for plants. The amount of heat generated by the chickens was equivalent to the amount of solar heat that could have been stored if we had replaced the coop with thermal mass.

Confining the chickens during the day in a separate climate-controlled coop and blowing air from the coop into the greenhouse to deliver carbon dioxide might have been more successful. We could have filtered the air to remove ammonia, eliminating the maintenance of frequent manure removal. However, we would have had to confine the chickens constantly, creating a less humane environment for the birds.

Heated Propagation Pond

Originally, I thought I'd heat the entire pond inside the greenhouse because I wanted to start seedlings in the warm water. However, after I realized the expense associated with heating the entire pond, I decided to try another idea: a pond inside a pond.

Nested inside the greenhouse pond, a heated propagation pond proved to be an excellent place to raise transplants in floating seed trays, similar to those described in chapter 2. The added warmth

quickened seed germination and seedling growth. Isolating and insulating a smaller pond within the bigger pond meant less energy required for heating.

I've used several techniques to isolate and insulate a pond inside another pond. With the first technique, I suspended a wooden frame above the water. To keep the wooden frame in place, it must be supported by the surrounding framework of the greenhouse, bridge, or walkways. I affixed the wooden frame to two layers of pond liner extending from the wooden frame down to the bottom of the pond. Between the pond liner layers, I placed extruded polystyrene insulating panels the exact depth of the pond. I then secured the pond liner sandwich to the bottom of the pond by placing cement over the excess pieces of liner, burying the edge on the bottom. The connection wasn't completely watertight but was sufficient to supply some insulation to the inner pond.

Building a cement-based wall is another way to isolate an inner pond. I combined a lightweight concrete mix composed of one part cement and six parts perlite with enough water to create a pasty consistency. The lightweight concrete acts more like wood when hardened; I could cut it with a saw and drill screws into it easily. The lightweight concrete reduced the weight of the wall, thus preventing excessive pressure on the liner. The mix also provided some insulation value, preventing heat loss from the pond. You can affix additional extruded polystyrene insulation to the sides of the concrete using stainless steel screws and washers.

Before making the concrete mix, first build plywood forms on the outer edges of where the wall will sit, leaving a 3- to 4-inch gap between plywood sheets. Secure the forms to prevent the weight of the concrete from pushing them out. Place concrete wire mesh between the forms to provide additional stability to the concrete mix. Be sure to suspend the wire mesh above the liner to prevent

Figure 12.9. A corner of the greenhouse pond, isolated by a bridge, forms an insulated pond for starting seedlings in floating "speedling trays."

Figure 12.10. Insulation and an aquarium heater provide more heat.

Figure 12.11. A roof made of window sash holds heat in at night. A dip net stands ready to harvest minnows for chickens.

Figure 12.12. Seedlings get a head start in the heated pond. A plant-eating fish added to the pond prunes the roots of the plants.

puncturing, then fill the gap between the plywood sheets with the concrete mix.

Besides insulation, cement-based walls also provide the benefit of durability and structural integrity. You can use a hole saw to cut through the lightweight concrete wall, which makes it easy to install pipes. However, the walls are permeable, and water will move through the wall and the seams, slowly allowing some heat to escape.

I've also tried to create a completely watertight seal by placing a smaller pond liner inside a larger pond. The water level of the inner pond must be kept higher than the water level of the outer pond to provide the necessary pressure to keep the liner on the bottom. Otherwise, the inner liner becomes a floating piece of plastic in the outer pond. Heavy liners composed of EPDM perform better in this situation than lightweight HDPE liners. You can also add an inner liner to the previously mentioned wall systems, such as the lightweight concrete walls, to create an impermeable layer with insulation. However, the water level must remain higher in the inner pond than in the outer pond to provide enough pressure to keep the liner submerged.

The principle behind the "pond inside the pond" concept was to cut costs and conserve energy. Since our small propagation pond held less than 50 gallons of water, we simply used an aquarium heater to keep the water warm. It's important to size the heater for your pond. Larger ponds require pond heaters, and expenses increase dramatically with size. I've also used heat exchangers coupled with hydronic heating systems to heat ponds, as described in chapter 5.

Floating Transplant Trays

We started seedlings in the inner heated pond. We'd fill Styrofoam trays with a modified potting mix, making the mix lightweight by increasing the perlite content of the soil to 75 percent. As the trays floated on the surface of the warm water, the plants wicked up the moisture. Meanwhile nutrients in the water from fish or chicken manure helped fertilize the plants.

You can order commercial floating trays called "speedling trays." However, at sixteen dollars per tray, this becomes pricey. Roots also grow out from under these trays and down into the water. Most of these dangling roots fall off during transplanting, leading to plant shock. I recommend moving transplants before the roots grow long.

To cut costs and prevent plant shock, I've used a different technique, as presented in chapter 2. I float traditional transplant trays on top of ¾-inch extruded polystyrene insulation sheets. Submerge the floating raft ¹⁄₁₆ to ⅛ inch deep, allowing the bottom of the transplant flat to wick up water. Some roots still grow out under the flats but fewer than from floating speedling trays.

I craft the floating rafts by first filling the transplant trays with a lightweight potting soil mix; then I leave the trays along the edge of the pond with just their bottoms submerged for about an hour or until saturated. Once the trays are fully saturated I assess the weight of the trays to provide proper sinkage of the insulation sheet. I prefer making floats for only one to two flats at a time to simplify the process of depth adjustment.

Start by cutting a piece of insulation slightly larger than the flats, place the flats on the float, and check the depth. Remove strips off the float or add rock weights until the flats sink to a depth of ⅙ to ⅛ inch. It's important to check the flats regularly the first week and adjust as needed. I prefer to use Winstrip flats with open slits along the edges to provide additional aeration and root pruning.

Covering the transplants with a row cover material or a glass screen helps retain the heat, ensuring good germination and growth. I used a window sash hinged to a frame on top of our heated pond. Old window sashes are usually cheap or free. You can easily find them at Habitat for Humanity ReStores or salvage yards. We propped our sash open during the day to promote good ventilation. At night we lowered the sash to retain the heat. You can also purchase a solar-powered louver opener similar to the openers used for the greenhouse vents in chapter 4 to automatically raise and lower your sash.

While most vegetables germinate and grow in floating systems, not all plants perform well in these conditions. I don't recommend this technique for growing watermelons, cucurbits, and probably many other plants that prefer moist but not completely saturated soils. Test small batches before committing a plant variety to the system.

Wicking Hydroponics

Soil wicks water like cloth wicks oil in a kerosene lamp. That's why the shallow edges of ponds create the perfect microclimate for potted plants. Normally, the edge of a pond is far too moist for vegetable production. The saturated soil limits growth for everything except adapted plants such as watercress (*Nasturtium officinale*). However, if you containerize the plants in pots, you raise the level of the soil. The bottom inch of the container

Figure 12.13. Wicking and floating hydroponic systems in the pond.

remains submerged in muck or water, but the rest of the container rises above the moisture. Meanwhile the water subtly sneaks its way up through the soil.

To grow a wider diversity of plants in this microclimate, I enhanced soil aeration by using a highly aerated potting soil mix with 75 percent perlite. For even more aeration, a mix of 25 percent vermiculite and 75 percent perlite does the trick. The vermiculite absorbs moisture and nutrients while also providing aeration. I've grown basil, chard, tomatoes, chives, groundnuts, and brassicas using this technique.

Wicking hydroponic containers can be placed on the steep upper edge of ponds by securing the containers to the greenhouse framework to prevent them from tipping over into the pond. You can use galvanized aircraft wire or rope to secure the containers to eyebolts mounted in the greenhouse frame. Save money by using the eyebolts as hardware to secure the greenhouse bows to the wooden baseboards. Space the eyebolts several feet apart. A single wire can then support several pots.

A steeply sloped container will lose some soil volume as it spills to the side. You can compensate by leveling the container. Simply fold small pieces of old carpet and use them like shims beneath the downhill side of the container.

Containers sunken into soil on top of liners will wick moisture even if the pots are several inches above the water level. If the top of the liner is bare, you can still place plants above the water level and use thick nylon ropes to wick moisture up into the containers. Before you fill the container with soil, loosely coil the nylon rope in the bottom of the container, then let it dangle out of a hole in the bottom of the pot.

With any wicking hydroponic system, you'll need to initially water the plants for a week to establish the wicking action. You may also need to add soluble nutrients every few weeks to maintain good growth. This can be as simple as using a watering can dipped into the fertile pond water to irrigate the potted plants.

SUMMING UP THE FUNCTIONS

1. Collected rainwater outside greenhouse reflects sunlight into greenhouse during winter.
2. Collected rainwater inside greenhouse stores solar heat for use at night.
3. Greenhouse captures heat from chickens to assist in plant growth.
4. Greenhouse provides protection and shelter for chickens.
5. Chickens provide carbon dioxide for plants.
6. Chickens provide nutrients for pond.
7. Pond provides water and nutrients for plants.
8. Greenhouse provides season extension for plants and fish.
9. Pond provides habitat for predators of plant pests.
10. Pond grows minnows for supplemental chicken feed.

CHAPTER 13

Concrete Forest

The Bio-Integrated Parking Area, Driveway, and Patio

Although I'm a farmer, I love the city life. I've spent some of my best years living in small towns and cities. With friends, shops, and work all within walking distance, I rarely drove my car anywhere. But city living does have its drawbacks — mainly limited space for growing food and relating with nature.

Near the coast of South Carolina, Stephanie and I lived in a charming small city about 50 miles inland from Charleston. Our home in Walterboro offered me an opportunity to create a memorable harvest-only design. This pinnacle of perfection in permaculture is a design that requires no maintenance — the only work is harvesting. When all parts of the landscape are truly connected, any action in the landscape provides sustenance for another part.

Every solution begins as a problem. Living in Walterboro, I faced similar problems that urbanites all over the world face — massive runoff from roofs, driveways, and parking lots and a constant downpour of leaves and organic debris (which usually ends up in landfills). Of course, water and organic debris are the fundamental building blocks of all life on Earth. To turn problems into solutions it's simply a matter of stacking those blocks into life-giving patterns.

Acorn Harvesting Curb

Large oak trees hung over the long concrete driveway and patio beside our house. When we first took ownership, the previous (and absentee) owners had long abandoned the property. Leaves on the driveway had decomposed into a rich, thick soil and a lawn of baby oak trees had taken root. On the downhill edge of the patio, rain-washed leaves coated the ground and a thick mass of earthworms writhed under the surface. Mushrooms abounded in the fertile layer of detritus coating the ground. Every autumn, leaves and acorns fell like clockwork, coating the driveway — nature working to rebuild a forest floor.

I recaptured the driveway and patio by harvesting the thick layer of detritus forming on top of the concrete. The rich soil provided a great amendment for the garden. Once I cleared the debris, I faced the ongoing task of keeping the surface clear. Using a blower, I gathered the mass of branches, leaves, twigs, and acorns into a pile, then lugged it to the garden or the compost pile. The huge pile of leaves would decompose into a small pile of soil. Frustrated by the amount of time I spent handling this monotonous chore, I soon devised an easier way to manage the leaves.

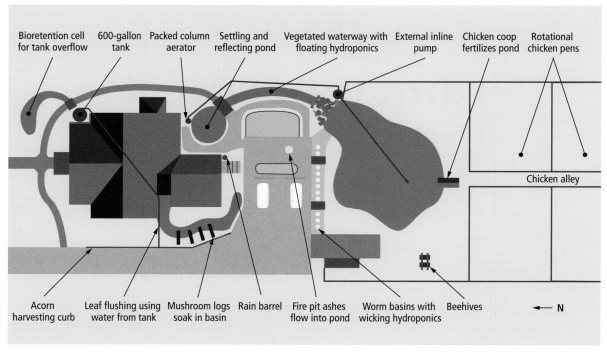

Figure 13.1. This landscape plan shows an overview of the Walterboro property once I'd added rainwater-harvesting ponds and leaf-collecting basins.

My first step was separating the leaves from the acorns. High-protein jewels that fell from the trees, acorns supplied the perfect feed for my six laying hens ensconced in a coop in my backyard. I noticed that the stream of air from the electric leaf blower moved leaves faster than acorns, so I devised a technique to separate leaves from acorns by building a rounded concrete curb. A straight-edged curb will block the motion of fallen leaves, but leaves can easily blow over a rounded curb. As the leaves blow over the curb, the acorns settle into a pile at the base, too heavy to make the journey.

To use the acorns as a chicken feed, I first drove my car over the acorns. The tires crushed the acorns, releasing the precious nutmeat. After I pulverized the acorns with a few passes of the car, I scooped up the acorns with a shovel and threw them to the chickens as feed.

Not all acorns are created equal. While roughly six hundred different oak tree species exist worldwide, the oaks are loosely divided into a red oak group and a white oak group. The red oak group produces acorns every other year, and the white oak group produces acorns every year. Red oak acorns are also notorious for a high tannin content, which gives them a bitter or astringent quality. The tannins inhibit protein assimilation in animals, reducing the nutritional benefits. This is a natural defense mechanism: The astringent tannins protect the acorns from consumption so they can sprout. The white oak group has fewer tannins and a sweet nutty flavor. Chickens prefer acorns from the white oak group. Research using acorns from the Persian oak (*Quercus brantii*) in the red oak group showed that incorporating acorns as 20 percent of a laying hen's diet has no soaking the cracked acorns in a pond for a few days should leach out the tannins.

Capturing Leaves in Worm Basins

One good idea leads to another. The leaves blown over the curb and off the patio built nests of decomposing organic matter. It didn't take long for the earthworms to get wind of the new feeding site, and suddenly I had an army of annelids to handle my composting needs. The earthworms shredded the leaves into rich, fertile soil. Instead of lugging the leaves to the curb or compost pile, I could instead harvest earthworms for chicken feed and potting soil for plant starts.

First, I dug a shallow basin next to the patio, which was plagued with leaf accumulation. The basin acted as a storage reserve for the leaves. The size of the basin needed will depend on your leaf load, but I found a width one-tenth the width of the patio was adequate under heavy leaf loads. For safety and stability I sloped the sides at a shallow three-to-one ratio (for more about digging basins with sloped sides, see chapter 3). With this ratio, a 6-foot-wide basin will have a maximum depth of 1 foot; a 3-foot-wide basin will be 6 inches deep because each side of the basin is sloped. Make sure the bottom of your basin is perfectly level so that water and leaves will distribute themselves evenly in it.

It's also important to design the basin so it will retain some water (earthworms can't live in a pile of dry leaves) but also allow excess water to drain away. Adding a liner provides for water retention, and creating a low spot on the downhill berm of the basin provides an overflow route for water. The crest of the lower edge sat between 1 and 2 inches above the bottom of the basin. Next, I lined my

---

Elements of the System

Leaf litter plagues parking areas and driveways. As soon as you clean it up, it starts accumulating again. Rather than slaving away at this dreaded chore, take advantage of this valuable resource by using these simple techniques. This chapter shows you how to:

- Separate acorns from leaves using specially designed curbs and use the acorns as chicken feed
- Design leaf-harvesting basins next to driveways and parking areas that attract earthworms to shred the leaves into fine castings
- Grow wicking hydroponic plants and mushrooms in the leaf-harvesting basins
- Use harvested rainwater to flush leaves and cool patios

---

serious adverse effects on performance.[1] As long as hens are not forced to eat the acorns, they will take what they need and ignore the rest, naturally balancing their diet.

My friend Vaughan Spearman taught me to look at the tips of the leaves to determine the difference between the two groups of oaks. Generally, the red oak group has a sharp spine on the end of the leaf, and the white oak group has a smooth leaf tip. Some of the red oaks have deciduous spines that fall off as the leaf matures. And live oaks are an exception to the rule — they are red oaks with low-tannin acorns. To make the red oak acorn group more palatable, basin with pond liner material. The liner held in moisture and facilitated easy harvesting of earthworm castings and earthworms. Also, without the liner, the roots of any tree within 50 feet could have found the thick layer of leaves accumulating in the basin. If roots had spread throughout the castings, they would have made extraction of earthworms and potting soil nearly impossible — comparable to breaking through a thicket of branches rather than digging in loose sand.

When it rained, water flowed off the patio and into the basin. However, since the basin only held 1 to 2 inches of water, earthworms could easily crawl up into the upper strata of leaves to escape the flooding water. The shallow pool of water under the collecting leaves attracts and retains earthworms since the area remains moister than the surrounding soil. However, always keep the water covered with leaves to prevent mosquitos from taking up residence. Initially, I used leaves from other areas to keep the water covered, but once the system began catching leaves off the driveway and converting them into the rich worm castings, the water remained covered.

To keep the system working, it's important to harvest worm castings every year or so. If the worm castings accumulate in the basin, leaves have nowhere to go and will once again require being hauled to the curb or the compost pile — which defeats the purpose of the worm basin. It's far easier to haul off a small amount of worm castings after the earthworms shred the leaves than to carry away a large cache of loose leaves.

You can easily harvest earthworms from the basin once you remove the worm castings. The liner prevents the worms from escaping because they can't dig any deeper. So it's easy to sift through the loose castings without cumbersome roots if you need a few worms for fish bait. I'd harvest a large amount of worm castings by shoveling a pile of soil out of the basin, taking care not to puncture the liner, and placing it atop a piece of plywood in the chicken yard. The chickens pulled all the earthworms out, leaving me with the worm castings for adding to potting soil or feeding to plants.

For the system to work properly, it's imperative to prevent roots from intruding. I secured the pond liner to the concrete patio using an engineered

Figure 13.2. The liner in this basin will keep moisture in and tree roots out.

Figure 13.3. In the completed basin, earthworms feed on the leaves, turning them into rich soil. Logs and leaves grow edible mushrooms, and flooding stimulates fruiting. The rounded curb facilitates easy separation of acorns from leaves.

wood strip to seal out the roots. The opposite edge of the basin, away from the patio, had a lip protruding up a few inches to prevent roots from crawling over the edge of the basin and into the castings. To hide exposed plastic of the protruding lip, I sandwiched the protruding lip with rocks or bricks on either side.

When the basin was full of leaves, its surface sat level or slightly lower than the patio. This posed a safety hazard, since the basin looked like solid ground covered by a layer of leaves, but actually contained a thick deep mass. Someone could have unknowingly stepped into the basin and sunk, possibly causing injury. Therefore, I recommend adding a small barrier fence to prevent intrusion or, better yet, adding wicking hydroponics or mushroom logs to define your basin.

Earthworms for Basins

Earthworms are categorized into three basic groups, depending on the regions they inhabit in the soil. Most people are familiar with the epigeic group, the leaf litter or manure worms such as red wigglers, commonly grown to compost food waste. The epigeic group lives on the surface directly under leaves or decomposing waste. Another group, the endogeic, live in the topsoil a few inches below the surface digging horizontal burrows. The anecic group makes deep vertical burrows, coming up to the surface only to gather plant material.

Research shows earthworms accelerate plant litter decomposition.[2] Most earthworms present in a landscape are not native; earthworms have been moved around the world by humans for centuries. Nonnative earthworms may actually pose a threat to some ecosystems. These exotic earthworms are capable of "significantly affecting soil profiles, nutrient and organic matter dynamics, other soil organisms, and plant communities."[3] Worm lovers will be glad to know that red wigglers are probably

Figure 13.4. Earthworms will not eat branches, so I bundle them up into wattles and use them as mulch and habitat for birds, ground-nesting spiders, and ground beetles—all predators of plant pests.

okay, even though they're native to warm areas of Europe. Since the red wigglers don't survive through cold winters unprotected, they never have the opportunity to invade the ecosystem outside warm climates. Problems occur when fishermen release European night crawler worms used for bait at pristine ponds and lakes, seeding earthworm invasions into new areas.

The best earthworms for an earthworm basin depend on what earthworms are already in your local area. I don't recommend moving earthworms except within the local area, and red wigglers don't readily consume leaves in basins. The basins I've built attract whatever native or nonnative earthworms are in the area, but favor the endogeic group. If your immediate area lacks worms, scour the neighborhood for deposits of deep organic matter, and dig local worms to seed your basins. In colder climates the earthworms overwinter by hibernating underground or as eggs near the surface, depending on the species. Once temperatures warm up, the earthworms return to the favorable habitat provided by the basin. Extremely dry climates may lack native or nonnative earthworms, limiting the application of this pattern. However, within a year's time in most locations, you'll have a teeming population of earthworms in your basin shredding away your leaves.

Wicking Hydroponics

Stack another function by growing plants in a wicking hydroponic system in your worm basin. Place large pots or nursery containers, at least 1 foot tall, inside your basin. The soil inside the pots and containers will wick up moisture from water pooling beneath. The bottom of the pot should sit on the bottom of the basin, submerged in 1 to 2 inches of water. Larger pots can handle deeper water, but submerge smaller pots in 1 inch of water at most. Use a highly aerated potting soil mix containing 75 percent perlite and 25 percent worm castings or peat moss in the pots. Once you fill the pots, insert the plants. Apply water from the tops of the pots for the first week to initiate the wicking action. After that, the soil inside the pots will wick water pooling inside the basin into the pots to irrigate plants.

If it's hot and it doesn't rain for a week, you may need to add more water to the basin to irrigate the wicking hydroponic plants. Every few weeks add an organic fertilizer to the pots to support plant growth. This can be as simple as adding urine to the bottom of the basin and letting rainwater distribute the fertilizer or sprinkling your favorite complete organic fertilizer on top of the pots and watering it in.

Growing Mushrooms in Basins

Not many plants tolerate shady locations, but mushrooms thrive in shade. The blewit mushroom

Figure 13.5. Basil grows in wicking hydroponic pots inside a leaf-harvesting basin. When fertilizer (such as urine) is applied to one end of the basin, it flows throughout to fertilize all the pots.

(*Clitocybe* sp.) makes a fine edible when cooked, and it grows readily on leaf litter. To grow blewit mushrooms in worm basins, first buy bags of spawn or mycelium. I recommend buying spawn grown on sawdust instead of grain to limit the number of pests potentially feeding on the spawn. Add the spawn after the earthworms have accumulated a layer of castings deep enough to cover the water pooling on the bottom. Evenly spread the spawn throughout the basin, under the leaves but on top of the castings, then water it in.

Alternatively, hunt your own wild blewits in a nearby forest during fall. After proper identification, harvest chunks of the mycelial mass below the mushrooms and spread them into your worm basin in a similar manner as spawn.

I inoculated logs with shiitake mushrooms, placing one end of the log inside the basin where it naturally wicks up moisture when it rains. (For more on mushroom cultivation, see chapter 3.)

Figure 13.6. The leaf-catching basins that border the patio overflow from one to another and finally into the pond.

Add mushroom logs to your basin only after they are fully colonized by the mycelium. I placed one end inside the basin and propped the upper end on top of a brick or concrete block to prevent ground contact. The lower part of the log soaked in the water for a day after a rain while the upper part, separated from soil, was protected from competitive soil fungi.

Since worm basins are shallow, they only hold a small amount of water. Flow from impervious areas entering my worm basins inevitably caused the basin to overflow. To account for the overflow, I created three stacked worm basins. Each one caught leaves from the driveway, and overflow

Figure 13.7. A chicken coop parked over the pond fertilizes the water for fish production.

from the upper basins flowed into the lower basins. The overflow from the last basin flowed into a pond. The basins filtered the leaves and debris from the runoff before it entered the downhill pond. A final basin without a pond liner could also serve to harvest the runoff and place water back into the ground where it's needed.

Leaf Flushing

I liked my system of blowing leaves to separate out acorns, but at other times of the year, I wanted to avoid using the blower. So I thought about it and realized that rainwater could move leaves for me.

Using harvested rainwater to move leaves offers a time-saving solution to an age-old problem. Water is powerful enough to move cars when concentrated into a fast-running current. I released my harvested rainwater through a 2- or 4-inch pipe, sending a pulse of concentrated flow to push the leaves into the worm basins. With wicking hydroponic systems and mushroom logs incorporated into my basins, the pulse of water also provided irrigation to the plants and logs.

Collecting water in tanks and cisterns uphill of parking and roads reserves a large weight of water available with the turn of a valve. For a successful system, consider the following variables. First, you'll need a large enough reserve of water and pipe delivery system to move the amount of leaves you have. The slope of your driveway or patio, the depth of the flow of water, and the duration of the flow determine the distance the leaves will travel and the size of the area the water will clear.

I'm certain some complicated engineering formulas could solve this problem. However, I can only give advice from my limited experience and engineering skills. I hope a real engineer will

Evaporative Cooling

While emptying my 55-gallon rain barrel across the patio one day, I noticed a cooling effect in the air. As water evaporates it cools surfaces and air; this is known as evaporative cooling. For each kilogram of water evaporated, approximately 890 Btus of heat are transferred. That means that the water in a 55-gallon drum can remove an amount of heat equivalent to burning 2 gallons of propane.

The water from the rain barrel flowed across the patio and into the basins to help irrigate the potted plants.

If you want a cooler patio on a hot summer day, simply unload a pulse of water from your leaf-flushing system onto your patio. As the water evaporates it takes the heat with it, leaving your patio much cooler.

conquer the task of formulating an appropriate equation to enable better designs. Until then, experiment with temporary systems until committing time and money to permanent pipes.

Here are a few suggestions to help guide the design:

- A 2-inch pipe with low pressure from a gravity-fed system typically delivers 55 gallons of water per minute. On a shallow slope 55 gallons may clear an area 4 feet wide, pushing leaves a distance of 20 feet.
- Water flowing over a shallow slope spreads out evenly over the surface and moves a wider swath of leaves.

- Water on steep slopes stays in a concentrated area, moving a smaller width of leaves. On steep slopes a raised ridge a few inches high perpendicular to the flow of water helps spread out the flow.
- Water moving over a slope of 1 to 2 percent tends to float material downhill. On steeper slopes water moves around leaves instead of pushing them in its current.
- Curbs on the downhill edge of drives and parking accumulate leaves. A pulse of concentrated flow directed into the curb moves the accumulated leaves through cuts in the curb or into nearby basins.

SUMMING UP THE FUNCTIONS

1. Rounded curb separates acorns from leaves, allowing easy harvest of the acorns for chicken feed.
2. Basin next to the parking area and driveway collects leaves.
3. Basin filters water flowing off parking area and driveway before it enters pond.
4. Parking area and driveway harvest rainwater for pond.
5. Earthworms in basin shred leaves into potting soil for plants.
6. Earthworms in basin feed chickens.
7. Earthworms in basin feed fish.
8. Mushrooms grow on logs and leaves in basin.
9. Basin sustains wicking hydroponic plants.
10. Harvested rainwater flushes leaves into basin.
11. Harvested rainwater cools patio.
12. Harvested water in ponds reflects sunlight in the wintertime into the house and patio.

ACKNOWLEDGMENTS

First and foremost, I would like to thank my wife, Stephanie Jadrnicek, for her enduring support of all my home projects and her help in putting all the designs and concepts into the written form of this book. My daughter, Sage Jadrnicek, also helped with editing, photo selection, and project creation with my home projects and did a substantial amount of cooking and cleaning to afford us the time to write the book.

The first book I ever read about gardening was *Rodale's All-New Encyclopedia of Organic Gardening* edited by Fern Marshall Bradley. So it seemed fitting that serendipity brought us together again. Fern's editing skills elevated this book to the next level because she could always see the big picture when Stephanie and I felt bogged down with the details. I would also like to thank Makenna Goodman and Margo Baldwin and the wonderful team at Chelsea Green for giving me the opportunity to present bio-integration to a larger audience in a beautiful package.

I would like to thank Charles (David) Thornton for his collaborative efforts in soldier fly rearing, compost heating, and small grain production at Clemson University. His wit and sense of humor inspired the innovative names for most of the chapter titles. I'd also like to acknowledge Scott Davis, my aquaculture mentor, who passed on his wisdom before retiring from managing the Clemson University aquaculture facility. I'm grateful to all the students and staff at Clemson University who helped build projects and offered advice when creating the Clemson University Student Organic Farm designs, with special thanks to Charles (Alex) Pellet, James Black, Carly Basinger, Carly Cox, Chelsi Crawford, Will Dukes, Chance Lawrence, Tradd Cotter, David Haines, Robert Bennett, and David Robb.

With the Ashevillage Institute project, I would like to thank Janell Kapoor, Lloyd Raleigh, Jason Boyer, Steveo Brodmerkel, and all the other volunteers and paid workers for help with the design and construction of the site.

At the Urban Permaculture Institute of the Southeast, I would like to thank Vaughan Spearman, Mikel Mckee, Thomas Angel, and the Clemson University Cooperative Extension Service for their support and help.

I would like to thank Dr. Philip Crossley for providing the photographs for the chinampas chapter. His research on the subject helped inspire my desire to design, and his review of the chapter increased its accuracy.

Finally, I would like to thank my advisors at Clemson University, Dr. Geoff Zehnder, Dr. Caye Drapcho, Dr. Dara Park, and Dr. William Bridges, for their support of my research and innovation efforts. I also acknowledge the support of Clemson University, USDA NIFA Hatch Project #SC-1700459, USDA SARE Professional Development Project #GS13-126, and Clemson Creative Inquiry for the financial and in-kind support of creation of the bio-integrated designs described in this publication.

APPENDIX

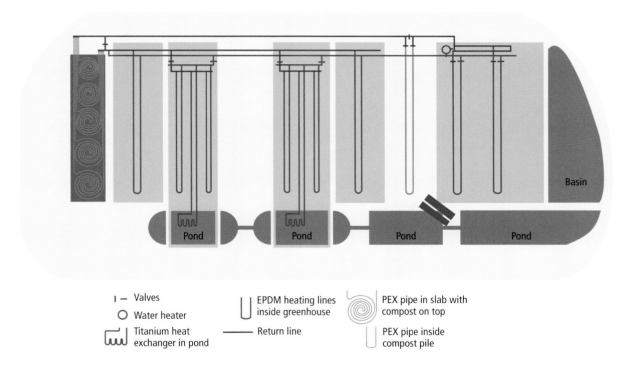

Figure A.1. Hydronic Heating System Layout

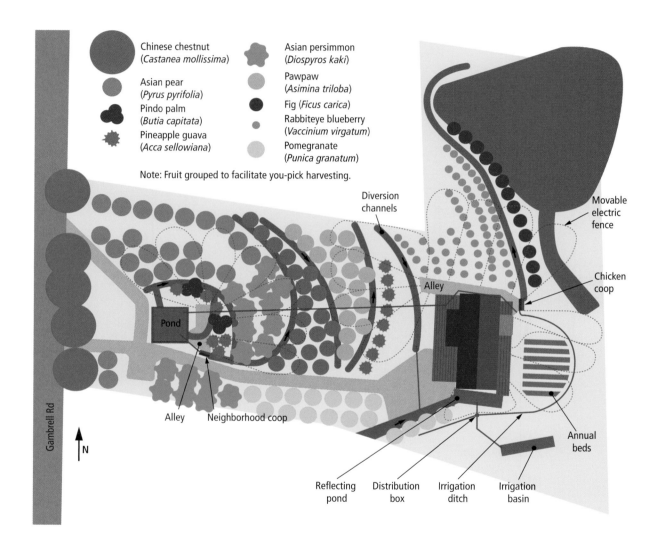

Figure A.2. Jadrnicek Homestead Edible Landscape and Pasture Pens Plan

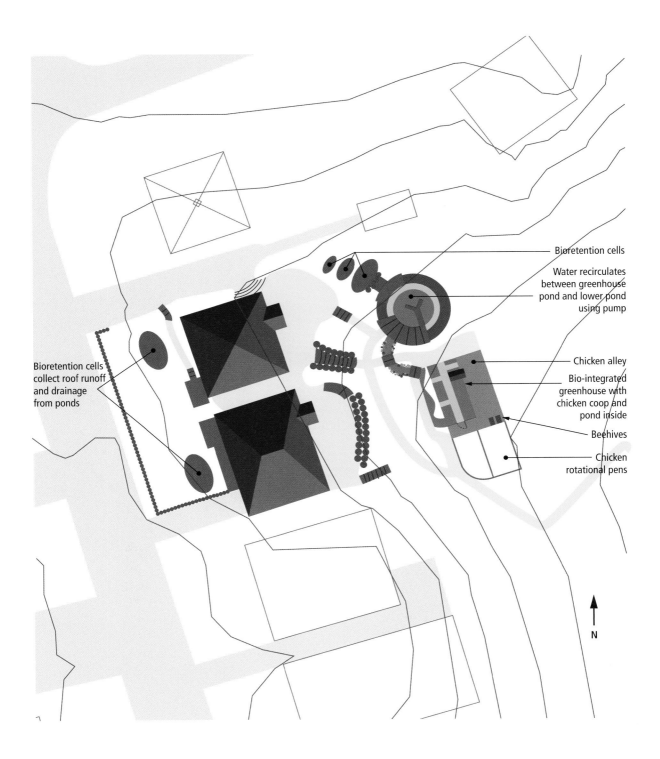

Bioretention cells

Water recirculates between greenhouse pond and lower pond using pump

Bioretention cells collect roof runoff and drainage from ponds

Chicken alley

Bio-integrated greenhouse with chicken coop and pond inside

Beehives

Chicken rotational pens

N

Figure A.3. Ashevillage Institute Landscape Plan

NOTES

Introduction

1. Bill Mollison, *Permaculture: A Designers' Manual* (Tasmania, Australia: Tagari Publications, 1988).
2. David Holmgren, *Permaculture: Principles and Pathways beyond Sustainability* (Hepburn, Australia: Holmgren Design Services, 2002).
3. Mollison, *Permaculture: A Designers' Manual*.
4. Ibid.

Chapter 1: The Chinampas

1. Dave Jacke and Eric Toensmeier, *Edible Forest Gardens* (White River Junction, Vt.: Chelsea Green, 2005).
2. Michael D. Coe, "The Chinampas of Mexico," *Scientific American*, July 1964: 90–98.
3. Phil Crossley, "Sub-irrigation and Temperature Amelioration in Chinampa Agriculture," (PhD diss., University of Texas at Austin, 1999).
4. Emory Dean Keoke and Kay Marie Porterfield, *Encyclopedia of American Indian Contributions to the World: 15,000 Years of Inventions and Innovations* (New York: Facts On File, 2005).
5. Michael D. Coe, "The Chinampas of Mexico."
6. Jeffrey R. Parsons, "The Role of Chinampas Agriculture in Food Supply of Aztec Tenochtitlan," in C. Cleland, *Cultural Change and Continuity: Essays in Honor of James Bennet Griffin* (New York: Academic Press, 1976).
7. Phil Crossley, "Sub-irrigation and Temperature Amelioration in Chinampa Agriculture."
8. Ibid.

Chapter 2: A Pool of Resources

1. University of Oregon Solar Radiation Monitoring Laboratory (http://solardat.uoregon .edu/SoftwareTools.html).
2. Build It Solar: The Renewable Energy Site for the Do-It-Yourselfers (http://www.buildit solar.com/SiteSurvey).
3. Weather Underground (http://www.wunder ground.com). Directions to find a personal weather station close to your site: Simply visit the site and search for the town you live in. After the website locates the closest town or city, click on "change station" to search for the nearest weather station to your location. Then click on the station name next to the change station icon to get the personal weather station data. Next, scroll down and change the daily mode to the yearly mode using the drop-down tab. Click on "View," and scroll down to display graphs depicting wind direction and speed, as well as rainfall patterns for every month of the year.
4. This technique might work with other types of trays, but I've only used Winstrip. Van Wingerden International (VWI) holds a patent on Winstrip flats. They order more every few years and sell the extra stock. The trays are presold months in advance, and you must get on a prospective buyers' list through Scott Arrington, who brokers the sales. His contact info is 404-509-8338, or torahisgoodnews@ gmail.com.

Chapter 4: Greenhouse with an "Outie"

1. Christian von Zabeltitz, *Integrated Greenhouse Systems for Mild Climates: Climate Conditions, Design, Construction, Maintenance, Climate Control* (Heidelberg, Germany: Springer, 2011).
2. Ibid.
3. Ibid.
4. Personal communication, Delta T Solutions: Engineering, Integrating, Innovating Systems, 27711 Diaz Rd., Suite B, Temecula, CA 92590; http://www.deltatsolutions.com.

Chapter 5: Greenhouse with an "Innie"

1. W. C. Dickinson et al., "The Shallow Solar Pond Energy Conversion System," *Solar Energy* 18.1 (1976): 3–10.
2. Baughan Wisely, John E. Holliday, and Ray E. Macdonald, "Heating an Aquaculture Pond with a Solar Pool Blanket," *Aquaculture* 26.3–4 (1982): 385–87.
3. K. Bradley Fox, Robert Howerton, and Clyde S. Tamaru, "Construction of Automatic Bell Siphons for Backyard Aquaponic Systems," University of Hawai'i College of Tropical Agriculture and Human Resources, *Biotechnology*, June 2010, BIO-10.
4. Odd-Ivar Lekang, *Aquaculture Engineering* (Oxford: Blackwell, 2007).
5. Ibid.
6. Ibid.
7. Ronald Malone, "Recirculating Aquaculture Tank Production Systems: A Review of Current Design Practice," *Southern Regional Aquaculture Center* Publication No. 453 (2013).
8. John A. Hargreaves, "Biofloc Production Systems for Aquaculture," *Southern Regional Aquaculture Center* Publication No. 4503 (2013).
9. Hargreaves, "Biofloc Production Systems for Aquaculture".
10. Lekang, *Aquaculture Engineering*.
11. James E. Rakocy and Andrew S. McGinty, "Pond Culture of Tilapia," *Southern Regional Aquaculture Center* Publication No. 280 (1989).
12. Craig Upstrom, "The Necessity of Regional Nurseries for the Success of Freshwater Prawn Farming" (presentation at USFPSGA meeting, Aquaculture of Texas, Inc., 2004).
13. Ibid.
14. C. E. Boyd, "Advances in Pond Aeration Technology and Practices," *INFOFISH International* 297. (1997): 24–28.
15. Michael P. Masser, James Rakocy, and Thomas M. Losordo, "Recirculating Aquaculture Tank Production Systems: Management of Recirculating Systems," *Southern Regional Aquaculture Center* Publication No. 452 (1999).
16. Thomas Popma and Michael Masser, "Tilapia Life History and Biology," *Southern Regional Aquaculture Center* Publication No. 283 (1999).

Chapter 6: The Big Flush

1. Christopher Kloss, *Managing Wet Weather with Green Infrastructure Municipal Handbook Rainwater Harvesting Policies* (EPA-833-F-08-010, December 2008).
2. For more information, visit the NSF International website for a list of approved roof coatings under Protocol P151 entitled, "Health Effects from Rainwater Catchment System Components."
3. J. Mechell et al., *Rainwater Harvesting: System Planning* (College Station: Texas AgriLife Extension Service, draft version September 2009).

4. Brett Martinson and T. Thomas, "Quantifying the First Flush Phenomenon," 12th International Rainwater Catchment Systems Conference, November 11, 2005, New Delhi.

5. D. A. Cunliffe, *Guidance of Use of Rainwater Tanks* (EN Health Council, Department of Health and Ageing, Australian Government, 2004).

6. Mechell, *Rainwater Harvesting*.

7. Art Ludwig, *Water Storage: Tanks, Cisterns, Aquifers, and Ponds* (Santa Barbara, Calif.: Oasis Design, 2007).

8. Ibid.

9. Ibid.

10. Ibid.

11. Mechell, *Rainwater Harvesting*.

12. Ibid.

13. *2012 Green Plumbing and Mechanical Code Supplement* (International Association of Plumbing and Mechanical Officials, 2012).

14. Mechell, *Rainwater Harvesting*.

15. E. Myre and R. Shaw, *The Turbidity Tube: Simple and Accurate Measurement of Turbidity in the Field* (Michigan Technological University, 2006).

16. Ibid.

17. NOAA, "Atlas 14 Precipitation Frequency Estimates," http://hdsc.nws.noaa.gov/hdsc/pfds/.

18. David M. Hershfield, *Rainfall Frequency Atlas of the United States* (Department of Commerce, United States of America, Technical Paper No. 40, May 1961).

19. Mechell, *Rainwater Harvesting*.

20. Kloss, *Managing Wet Weather*.

21. Mechell, *Rainwater Harvesting*.

Chapter 7: Chicken No Tractor

1. M. W. Priest, D. J. Williams, and H. A. Bridgman, "Emissions from In-Use Lawn-Mowers in Australia," *Atmospheric Environment 34.4 (2000)*.

2. Anders Christensen, Roger Westerholm, and Jacob Almén, "Measurement of Regulated and Unregulated Exhaust Emissions from a Lawn Mower with and without an Oxidizing Catalyst: A Comparison of Two Different Fuels," *Environmental Science & Technology* 35.11 (2001): 2166–170.

3. Harlan E. White and Dale D. Wolf, *Controlled Grazing of Virginia's Pastures* (Virginia Cooperative Extension, May 1, 2009).

4. C. Sheppard, "House Fly and Lesser House Fly Control Utilizing the Black Soldier Fly in Manure Management Systems for Caged Laying Hens," *Environmental Entomology* 12 (1983): 1439–442.

5. SDB-200 Protex Wall Mount Drop Box.

Chapter 8: Harnessing Heat

1. Ida Pain and Jean Pain, *The Methods of Jean Pain, or Another Kind of Garden*, 7th Edition (Draguigan: Ancienne Imprimerie Negro, 1980).

2. Bruce Fulford, "The Composting Greenhouse at New Alchemy Institute: A Report on Two Years of Operation and Monitoring," *New Alchemy Institute Research Report*, No. 3, November 1986.

3. Agrilab website (http://www.agrilabtech.com).

4. Raymond Henry Walke, "The Preparation, Characterization and Agricultural Use of Bark-Sewage Compost" (PhD diss., University of New Hampshire, 1975).

5. Ehrenfried Pfeiffer, *The Compost Manufacturers Manual: The Practice of Large Scale Composting* (Philadelphia, Pa.: Pfeiffer Foundation, 1956).

6. Ibid.

7. Eliot Epstein, *The Science of Composting* (Lancaster, Pa.: Technomic, 1997).

8. Roger Tim Haug, *The Practical Handbook of Compost Engineering* (Boca Raton, Fla.: Lewis, 1993).

9. Ibid.

10. Ibid.

11. Gaelan Brown, *The Compost-Powered Water Heater: How to Heat Your Greenhouse, Pool, or Building with Only Compost* (Woodstock, Vt.: Countryman Press, 2014).

12. R. A. Aldrich, et al., *Greenhouse Engineering* (Madison: University of Wisconsin NRAES-33 Cooperative Extension, 1994).

13. P. R. Hicklenton, CO_2 *Enrichment in the Greenhouse*, Growers Handbook Series, Vol. 2 (Portland, Ore.: Timber Press, 1988).

14. Aldrich, *Greenhouse Engineering*.

15. Ibid.

16. Ibid.

17. Haug, *Practical Handbook of Compost Engineering*.

18. Ibid.

19. Personal communication, Jason McCune-Sanders, Agrilab Technologies LLC.

Chapter 9: Feed, Fuel, Fertilizer

1. C. Sheppard, "House Fly and Lesser House Fly Control Utilizing the Black Soldier Fly in Manure Management Systems for Caged Laying Hens," *Environmental Entomology* 12 (1983): 1439–442.

2. http://compostmania.com.

3. Larvae are sold under the trade name "Phoenix Worms" and are found at http://www.phoenixworm.com, a site developed for shipping soldier fly larvae as a reptile feed.

4. Sheppard, "House Fly and Lesser House Fly Control."

5. C. Sheppard and L. Newton, "A Value Added Manure Management System Using the Black Soldier Fly," *Bioresource Technology* 50 (1995): 275–79.

6. T. Barry, "Evaluation of the Economic, Social and Biological Feasibility of Bioconverting Food Wastes with the Black Soldier Fly (*Hermetia illucens*)" (PhD diss., University of Texas, 2004).

7. Sheppard and Newton, "A Value Added Manure Management."

8. Ibid.

9. C. Sheppard et al., *Using the Black Soldier Fly,* Hermetia illucens, *as a Value-Added Tool for the Management of Swine Manure* (Report for Mike Williams, Director of the Animal and Poultry Waste Management Center, North Carolina State University, June 6, 2005).

10. M. C. Erickson et al., "Reduction of *Escherichia coli* 0157: H7 and *Salmonella enerica serovar enteritidis* in Chicken Manure by Larvae of the Black Soldier Fly," *Journal of Food Protection* 67 (2004): 685–90.

11. Sheppard, "House Fly and Lesser House Fly Control."

12. J. Tomberline and C. Sheppard, "Lekking Behavior of the Black Soldier Fly," *The Florida Entomologist* 84.4 (2001): 729–30.

13. D. Booth and C. Sheppard, "Oviposition of the Black Soldier Fly, *Hermetia illucens* (Diptera: Stratiomyidae): Eggs, Masses, Timing and Site Characteristics," *Environmental Entomology* 13 (1984): 421–23.

14. J. Tomberlin and C. Sheppard, "Factors Influencing Mating and Oviposition of Black Soldier Flies (Diptera: Stratiomyidae) in a Colony," *Journal of Entomological Science* 37.4 (2002): 345–52.

J. Zhang et al., "An Artificial Light Source Influences Mating and Oviposition of Black

Soldier Flies, *Hermetia illucens*," *Journal of Insect Science* (2010): 1–7. BioOne.

15. Tomberlin and Sheppard used Bioquip Products' Lumite screen outdoor cages for rearing.

16. C. Sheppard et al., "Rearing Methods for the Black Soldier Fly (Diptera: Stratiomyidae)," *Journal of Medical Entomology* 39.4 (2002): 695–98.

17. Ibid.

18. Ibid.

19. Ibid.

20. Ibid.

21. Tomberlin and Sheppard, "Factors Influencing Mating."

22. Zhang, "An Artificial Light Source."

23. Ibid.

Chapter 10: Taking It to the Field

1. J. M. Kemble, et al., *Southeastern U.S. 2014 Vegetable Crop Handbook* (Lincolnshire, Ill.: Vance Publishing Corporation, 2014).

2. Kemble, *Southeastern U.S. 2014 Vegetable Crop Handbook*.

3. Environmental Protection Agency, *National Management Measures to Control Nonpoint Source Pollution from Agriculture* (2003), 157–202.

4. Frederick R. Troeh, J. Arthur Hobbs, and Roy Luther Donahue, *Soil and Water Conservation for Productivity and Environmental Protection* (Englewood Cliffs, N.J.: Prentice-Hall, 1980).

5. Troeh, Hobbs, and Donahue, *Soil and Water Conservation*.

6. Douglas Helms, "Mangum Terrace" (NCpedia, 2006). http://ncpedia.org/mangum-terrace.

7. Ibid.

8. P. A. Yeomans and K. B. Yeomans, *Water for Every Farm: Yeomans Keyline Plan* (Sydney, Australia: K.G. Murray, 1973).

9. P. A. Yeomans, *The Challenge of Landscape: The Development and Practice of Keyline*. (Sydney, Australia: Keyline, 1958).

10. Mollison, *Permaculture*.

11. Adarsh Pal Vig, Geetanjali Rampal, Tarunpreet Singh Thind, and Saroj Arora, "Bio-protective Effects of Glucosinolates: A Review," *LWT — Food Science and Technology* 42.10 (2009): 1561–572.

Chapter 11: The Right Connections

1. A. L. Pulliam, *Farm Layout and Farmstead: Planning for Irrigated Farms in Eastern and Central Oregon* (Oregon State College Extension Bulletin 685, January 1948).

2. Christopher Alexander, Sara Ishikawa, and Murray Silverstein, *A Pattern Language: Towns, Buildings, Construction* (New York: Oxford University Press, 1977).

Chapter 12: The Mini-Innie

1. H. L. Fuller et al., "Comparison of Heat Production of Chickens Measured by Energy Balance and by Gaseous Exchange," *Journal of Nutrition* 113.7 (July 1983): 1403-408.

2. K. Anderson and Carter Thomas, *Hot Weather Management of Poultry* (PS and T Guide #30, North Carolina State University, February 1993).

3. Fuller, "Comparison of Heat Production of Chickens."

4. Aldrich, *Greenhouse Engineering*.

5. L. J. M. Van Der Eerden, "Toxicity of Ammonia to Plants," *Agriculture and Environment* 7.3–4 (November 1982): 223–35.

6. Dev T. Britto and Herbert J. Kronzucker, "NH4+ Toxicity in Higher Plants: A Critical Review," *Journal of Plant Physiology* 159.6 (2002): 567–84.

Chapter 13: Concrete Forest

1. L. Saffarzadeh, L. Vincze, and J. Csapo, "The Effect of Different Levels of Acorn Seeds (*Quercus branti*) on Laying Hens Performance in First Phase of Egg Production," *Acta Agraria Kaposvariensis* 4.2 (2000): 27–35.

2. P. F. Hendrix et al., "Invasion of Exotic Earthworms into Ecosystems Inhabited by Native Earthworms," *Biological Invasions* 8.6 (2006): 1287–1300.

3. Ibid.

INDEX

size of, 104
water level of, 323
wicking hydroponic systems in, 323,
333–34
iron deficiency, 117
irrigation, 264–82
of basin, 55, 57–70
contour bunds in, 275, 278–79
drip. *See* drip irrigation
estimating water usage in, 176–81
furrow flooding in, 264, 271–74
gravity-fed, 161, 163, 166–67, 266
parts and tools for, 312
pond storage of water for, 14, 166
total dynamic head in, 167–68
water pumps in, 167–68
irrigation ditches, 280–82
rotary plow in construction of, 281,
314

Jadrnicek homestead plan, 348
Jang seeders, 312
Jerusalem artichoke, 46, 47, 49
close to chicken coop, 197, 198
in three brothers garden, 48

katniss, 49–50
keyhole design of produce washing area,
306, 307
keyline pattern cultivation, 279, 282–
83, 315
Keyline Scale of Permanence, 293
knives
beet knife, 315
in cultivation, 315
in harvesting, 304, 305
lettuce knife, 304, 305
side knife, 315
koi, 41

labeling of harvest bins, 306
landforms, as permanent part of
landscape, 293
largemouth bass, 136
lawn maintenance
chickens in, 183–84
fuel and time used in, 183
laying boxes, 208–10, 325, 326
nesting material used in, 210

Leaf Eater Advanced downspout filters,
152
leaks in drip irrigation, 268
leaves, 335–43
in compost, 216, 217, 223–24
harvested rainwater system moving,
342–43
manure spreader distributing, 317
in worm basins, 337–42
leeks, ammonia tolerance of, 329
lekking behavior of soldier flies, 255
lemon tree, Meyer, 52
lettuce
chicken dust on, 329
in greenhouse aquaponic system, 103,
112, 113, 117
harvesting of, 304, 305
lettuce knives, 304, 305
lighting, and soldier fly mating
behavior, 256–58
lime
pond water pH adjusted with, 145
in potting soil mix, 244, 245
as soil amendment, 283
Lin, Sandy, 251, 256
live oaks, 337
loamy soil
basin construction in, 62
water availability in, 270
lumber, treated, toxic chemicals in,
114–15

Mangum, Priestly H., 280
Mangum terrace, 280
manifold line
in drip irrigation, 266–68
in hydronic heating system, 95–96, 97
manure
of chickens, 187, 196–201. *See also*
chicken manure
as compost ingredient, 217, 218–19
manure fork, 311
manure spreaders, 317
mixing compost with, 220–21
market building, 300
mating of soldier flies, 255–58
metal drums holding water for thermal
mass, 89
metal tanks for rainwater storage, 155

Mexico, chinampas system in, 9–11
Meyer lemon tree, 52
millet
as chicken forage, 195, 196
as cover crop, 285
as nurse crop for clover, 193
minnows, 14, 39, 41–44
algae bloom for, 41–42
as chicken feed, 14, 39, 41, 42, 43
harvesting of, 42–44
in interior pond, 319
mixing methods
for compost, 220–22, 223
for potting soil, 245
moisture content in compost, 222–24
Mollison, Bill
herb spiral of, 4, 5
on keyline pattern cultivation, 282,
283
permaculture principles of, 3, 4
monthly rainfall, 169–70
mosquitoes
fish eating, 14, 23, 39, 41
in outside ponds, 14, 23, 39
and water storage tanks, 156, 157, 158
in wet conveyance systems, 149, 152
in worm basin, 338
mosquitofish, 39, 44, 329
mowers
flail, 313–14
rotary, 314
sickle bar, 195, 314
mulch, flame weeder use in, 285, 311
multifunctional components, 4–5
municipal roads, placing buildings close
to, 298
municipal water systems, 36, 59, 160
mushrooms grown in basins, 57, 58, 60,
68–70, 340–42
mustard seed, 285–89
walk-behind combine for, 288

Nasturtium officinale, 48, 51, 333
National Centers for Environmental
Information, 169
National Engineering Handbook, 177–78,
263, 277
Natural Resources Conservation
Service, 25, 263, 277, 280

ABOUT THE AUTHORS

Shawn Jadrnicek has nourished his interest in sustainability through work as an organic farmer, nursery grower, Extension agent, arborist, and landscaper, and as the manager of Clemson University's Student Organic Farm. From his earliest permaculture experiments with no-till farming in the Santa Cruz Mountains of California to his highly functional bio-integrated designs in the Southeast, Shawn has learned how to cultivate food crops efficiently in a variety of climates and landscapes. He shares his creative solutions through teaching, consulting, and design work.

Sage Jadrnicek

Stephanie Jadrnicek is an award-winning columnist and journalist. She writes for *The Journal*, a newspaper based in Seneca, South Carolina. Her passion for sustainability stems from her roots in the Shenandoah Valley of Virginia, where her family taught her to grow, can, and preserve the precious things in life. Over the last twenty years she has played the role of assistant, critic, muse, and guinea pig in Shawn's farming experiments. Shawn and Stephanie live in Anderson, South Carolina, with their daughter Sage.

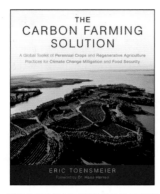

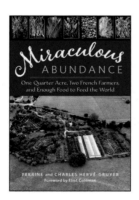
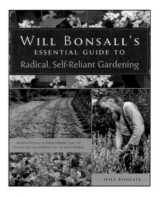